63rd Annual Safety in Ammonia Plants and Related Facilities Symposium 2018

AIChE Technical Manual 2018
Volume 59

Toronto, Canada
16-20 September 2018

ISBN: 978-1-5108-8185-3

Printed from e-media with permission by:

Curran Associates, Inc.
57 Morehouse Lane
Red Hook, NY 12571

Some format issues inherent in the e-media version may also appear in this print version.

Printed by Curran Associates, Inc. (2019)

For permission requests, please contact AIChE
at the address below.

AIChE
120 Wall Street, FL 23
New York, NY 10005-4020

Phone: (800) 242-4363
Fax: (203) 775-5177

www.aiche.org

Additional copies of this publication are available from:

Curran Associates, Inc.
57 Morehouse Lane
Red Hook, NY 12571 USA
Phone: 845-758-0400
Fax: 845-758-2633
Email: curran@proceedings.com
Web: www.proceedings.com

TABLE OF CONTENTS

Successful Startup of the First World Scale Ammonia Plant in Bolivia

The Yacimientos Petroliferos Fiscales Bolivianos (YPFB) fertilizer project is the first fertilizer project executed and first fertilizer plant built in Bolivia. The plant is located in a remote area near Bulo Bulo of Cochabamba Department, Bolivia. In May of 2017, the complex was successfully commissioned and passed the performance test. The 1200 MTPD ammonia plant was constructed by Korea S Company, and is based on the KBR Purifier™ technology with the process air compressor driven by a gas turbine.

This paper shares the commissioning and startup experience, describes the salient process and design features of the ammonia plant, as well as discusses challenges and issues encountered, and key lessons learned during the plant commissioning and startup.

Annie Jing, George Colman
KBR, Houston, USA

Gwanghoon Lee, Mikyung Lee, Youngdae Kim
Korea S Company, Seoul, South Korea

Introduction

Bolivia has been fully reliant on imports for the country's fertilizer demand. To eliminate the dependency on importing fertilizer, monetize the country's abundant natural gas resource, and promote the development of the country's industrialization, the Bolivia government initiated their Bicentenary Strategy. These projects will commemorate the country's independence from Spain as part of the bicentenary celebration in 2025.

The Yacimientos Petroliferos Fiscales Bolivianos (YPFB) fertilizer project was one of the projects in the Strategy and commenced in October, 2012.

The YPFB fertilizer project consists of a 1200 MTPD ammonia plant licensed with KBR technology, a 2100 MTPD urea plant licensed with technology of Toyo Engineering Company (TEC) of Japan, and all offsite and utility facilities, as well as a 5000 metric ton ammonia storage tank, and a 42,000 metric ton urea bulk storage.

In 2017, the first ever fertilizer plant in the country was successfully commissioned and inaugurated into commercial operation, which has introduced Bolivia to the production of fertilizer on a world scale and with state of the art technology.

Location

The YPFB fertilizer plant is located at a greenfield site close to Bulo Bulo, which is a town in central Bolivia in the Cochabamba Department (see Figure 1). It is about 660 kilometer from La Paz (the capital city) and 200 kilometer from Santa Cruz, a major city of Bolivia. The site is shown in Figure 2

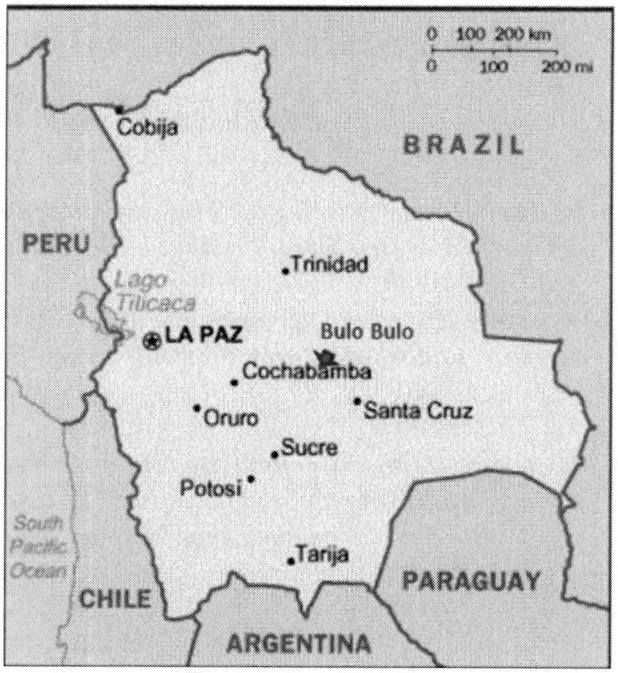

Figure-1. Area Map

Figure-2. Overview of YPFB Fertilizer Complex

Feed Stocks and Transportation

The natural gas for the YPFB fertilizer plant is supplied to the plant through a domestic gas company. The plant design is based on the natural gas composition listed in Table-1.

Composition		
CH_4	mol%	91.046
C_2H_6	mol%	6.480
C_3H_8	mol%	0.029
CO_2	mol%	1.857
N_2	mol%	0.588
Total Sulfur (as H_2S)	mg/Nm3	50
Mercury in Natural Gas	µg/Nm3	0.6

Table-1. Natural Gas Composition

Water for ammonia production and utilities of the plant comes from the Icha River about 5 kilometers away from the plant site.

The plant is designed for balanced ammonia and urea production, i.e. normally all ammonia product converted to urea. The urea product is presently transported from the plant site to the market by container trucks. Railway transportation to and from the plant site is still under construction.

Due to the remote location of the plant site, it was necessary for the plant to be self-sufficient with all the utilities required for the construction, commissioning, startup and normal operation of the facility. Equipment had to be engineered and constructed to meet the size and weight limitations of the transportation infrastructure of the area.

Project Execution

Development of the project was managed by YPFB which was supported by external consulting and project management service providers. A key feature of the project execution strategy was to ensure that a safe, efficient and reliable plant was designed and constructed on the strength of proven technology and experienced contractors, suppliers and vendors.

As the major investor, YPFB assumed full responsibility for the project. Comprehensive and effective management was achieved by adopting international good practices in contracting and project management. The contract for Engineering, Procurement, Construction Management (EPCM), as well as providing a certain period of operation and maintenance services after plant startup was awarded to Korea S Company, under a lump-sum-turnkey, fixed price arrangement. Contracts for the ammonia plant license, basic engineering design packages, and supply of licenser's proprietary equipment were awarded to KBR. KBR also provided an operator training simulator, a steam dynamic study and start-up advisory service to ensure a safe, fast and efficient startup of the ammonia plant.

Process Design Features of YPFB Ammonia Plant

This project had to achieve several objectives including being self-sufficient in steam and power balance, having high efficiency (low energy consumption) and having high reliability. To achieve these objectives, this ammonia plant design was tailored to YPFB requirements and is based on KBR Purifier™ Process with a gas turbine driven air compressor. All the components of the YPFB ammonia plant were designed based on well proven technology features [(1)(2)(3)].

The salient features of YPFB ammonia plant are discussed below:

- ✓ For natural gas pretreatment, in addition to normal sulphur removal, a Mercury Removal Unit is added upstream to prevent mercury from causing damage to the downstream equipment and catalyst.
- ✓ The Process Air Compressor (PAC) is driven by a Gas Turbine (GT), and the hot gas turbine exhaust (GTE) is used as the preheated combustion air for the Primary Reformer as shown in Figure 3. This configuration improved the thermal efficiency of the GT from around thirty percent to over ninety percent, allowed more steam export to the urea plant for its turbine-driven CO_2 compressor and also eliminated the forced draft fan and combustion air preheater (Figures 3 and 4)

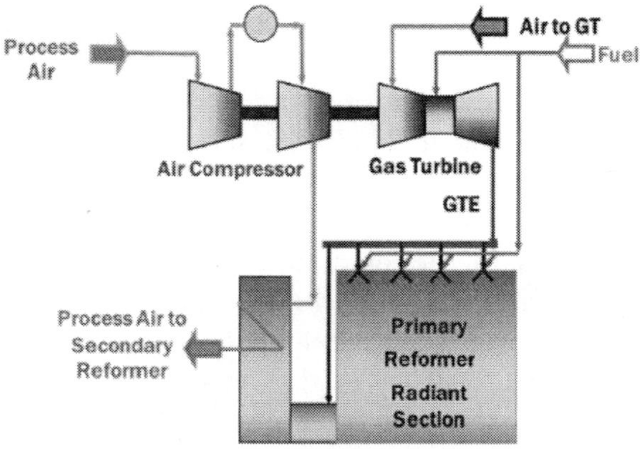

Figure 3. Configuration of GT-PAC and Primary Refromer

Figure 4. YPFB Primary Reformer & GT-PAC

- ✓ Mild operating conditions for the Primary Reformer. The reformer radiant tube outlet temperature is only about 700°C (1292°F), about 100°C (180°F) colder than the conventional plants. This lower temperature increases equipment life, improves operational reliability, and makes the operation with lower steam to carbon ratio (lower energy consumption) viable.
- ✓ 50% excess air used for the Secondary Reformer, shifting more reforming duty from the Primary to the Secondary Reformer. This reduction in primary reforming duty results in a further energy reduction, smaller Primary Reformer and lower equipment cost.
- ✓ Secondary Reformer using the KBR special design of a non-metallic mixing chamber with no metallic burner or mixer leading to increased reliability. (Figure 5)

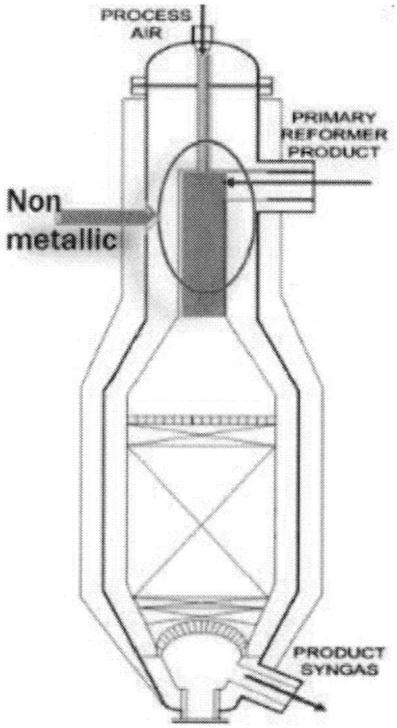

Figure-5. Seconary Reformer

- ✓ Mild operating conditions for the reforming waste heat boiler. The boiler was manufactured based on KBR proprietary natural-circulation, water-tube, and floating-head design with a proven record of high reliability and minimum downtime in cases requiring bundle replacement (Figure 6)

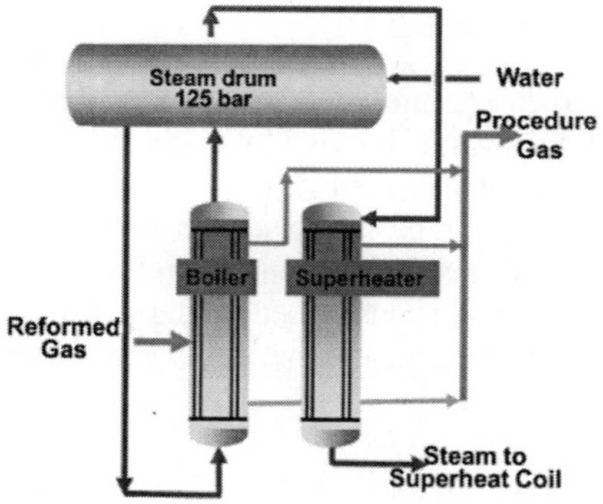

Figure-6. Reforming Waste Heat Boiler and Steam Drum

- ✓ CO_2 removal system used BASF two stage OASE® white design (Figure 7)

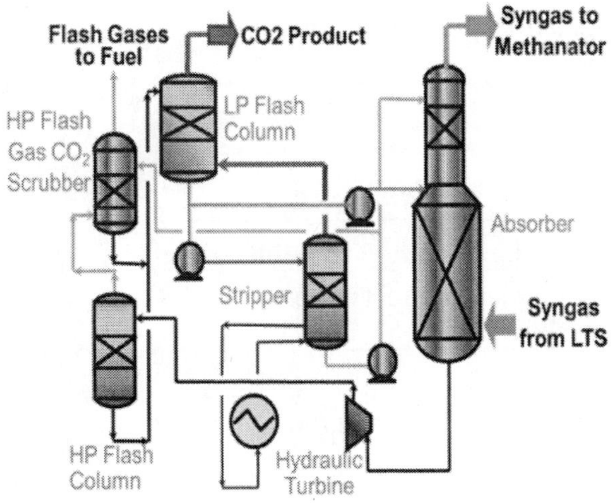

Figure-7. OASE CO_2 Removal System

- ✓ The cryogenic purification unit (the heart of the KBR Purifier™ Technology) located just upstream of the ammonia synthesis loop, serves to remove inerts, excess nitrogen, and generates highly pure and dry syngas with a H_2/N_2 ratio of 3/1 for the synloop. The purifier allows reduction of the energy consumption (both fuel consumption for Primary Reformer and power consumption for syngas and refrigeration compressors are lower), and also stabilizes the whole plant operation (Figure 8).

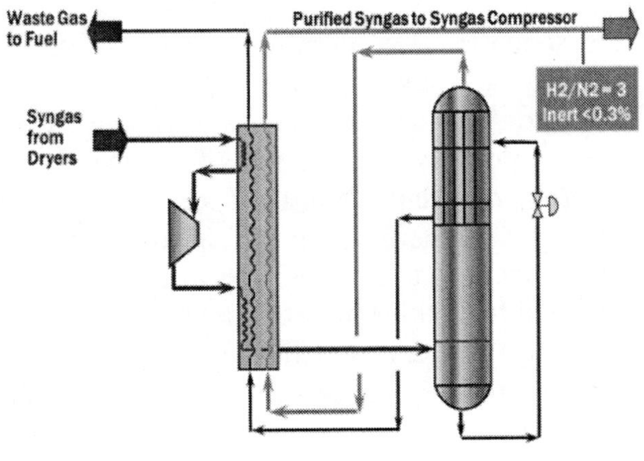

Figure-8. KBR Cryogenic Purification Unit

- ✓ Efficient and compact ammonia synthesis loop configuration (Figure 9) with KBR horizontal converter (Figure 10) and the unitized chiller (Figure 11), with a total of only 7 pieces of equipment in the high pressure loop. The result is lower cost, lower plot area and lower loop pressure drop. KBR proprietary design of the cold wall horizontal converter eliminated the necessity of heavy-duty crane for installation.

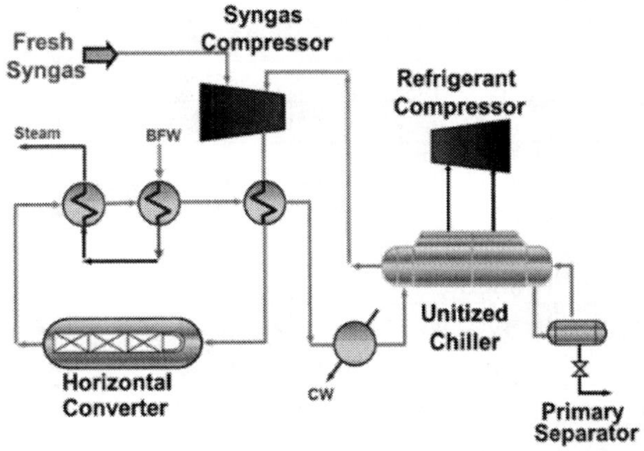

Figure-9. Ammonia Synthesis Loop Configuration

Figure-10. YPFB Ammonia Converter

Figure-11. YPFB Unitized Chiller

Plant Performance

The performance test run for the ammonia plant was conducted in May, 2017. The plant demonstrated that all performance guarantees including capacity, product specification and energy consumption were all met as shown in Table-2.

	Guarantee Value	Test Results
Ammonia Production		
Capacity MTPD	**1200**	**1211**
NH3 wt% *min.*	99.8	99.8
H2O wt% *max.*	0.2	0.2
Oil ppmw *max.*	5	3
Temp °C *max.*	40	24
Press kg/cm2 (g) *min.*	22	27
CO_2 Production		
Capacity MTPD	**1550**	**1591**
CO2 vol% (dry) *min.*	99.5	99.6
Inert vol% (dry) *max.*	0.5	0.4
Temp °C	38	35
Press kg/cm2 (a)	1.8	1.7
Net Specific Energy (Gcal/MT) (feed+fuel-steam credit+power)	**6.52**	**6.51**

Table-2. Comparison of Guarantees and the Average of Measured Values in the Test Run Period.

Project Milestones

The project was kicked off in October, 2012 with the major project milestones as listed in Table-3. First ammonia production was achieved 57 days after initial introduction of the natural gas feed to the primary reformer, and the performance test run was successfully accomplished 38 days after initial ammonia production. The rapid successful startup of the ammonia plant is attributed to the quality of the engineering design, construction, and commissioning by all parties involved.

Engineering	Months (M)
Kick Off Meeting	M+0
Detail Engineering Completion	M+25
Construction	
Mechanical Completion	M+51
Pre-commissioning/Commissioning	
Power Generator STG Commenced	M+49
ISBL Ammonia Plant Startup	
Initial NG Feed-in to Primary Reformer	M+52
Ammonia Produced	M+54
Performance Test Completed	**M+55**

Table-3. Major Milestones of YPFB Project

Safety Performance

The YPFB project team attached great importance to the safety and social impacts of the construction and operation. The plant site HSE orientation and training were required and provided for everyone (employees of Korea S Company and YPFB, subcontractors, vendors, licensers, etc.) working at the site. Safety drills were also conducted periodically for emergency preparedness.

A systematic work permit system was implemented for managing site activities to avoid safety conflicts and accidents which could arise. It was especially challenging with so many people from different countries speaking different languages and performing various kinds of work simultaneously at the site. The work safety system included routine inspections and audits to ensure the workers on the job were trained, equipped with proper protection, and physically fit for their task. Construction procedures also used safety taps, banners, and barricades to restrict access to unnecessary personnel.

Thanks to the project team's commitment to safety, there were no serious accidents that resulted in injury or equipment damage. As a result, the project achieved outstanding safety performance with 18,694,982 work hours without a single fatality and no lost time injury (LTI). This performance is a meritorious achievement for the first project in the country and on a project of this scale.

Challenges and Lessons Learned

The project encountered many challenges. Below are some of the main challenges and lessons learned, as well as good experience gained.

Challenging Climate Conditions

The weather conditions of the area are a tropical climate with a long rainy summer season from September to May. The relentless rain made it impossible to carry out many of the construction and commissioning activities. The high temperature and humidity combination can lead to an intolerable heat index.

These conditions necessitate a very careful and detailed planning of the site activities to minimize the impact on the schedule. A number of measures were implemented on site in order to minimize the impact on construction and commissioning activities. For example, providing temporary rain shelters where feasible, rearranging daytime activities to cooler night hours, enforcing frequent rest periods, providing shaded and cooled areas close to the activity areas and ensuring people were kept hydrated with a proper supply of cool potable water.

As mentioned above, water for the plant was from the nearby river. In the rainy season, days of consecutive downpour lead to the river flooding with thick mud. The thick muddy water led to the blockage of the water intake, overloaded the raw water filtration system and caused demineralized water shortage which in turn limited plant capacity.

Assembling KBR Horizontal Converter at Site

The plant site is remote and inland. Due to limitation on the country's transportation infrastructure, it's impossible to deliver heavy and larger size equipment like the ammonia converter in one piece to the plant site. Careful review and detailed planning among the converter manufacture, ISGEC Hitachi Zosen Limited (IHZL), KBR and Korea S Company overcame the limitation and safely delivered the KBR converter to the site by fabricating the converter shell in two pieces, and shipping the converter in four pieces (two shell pieces, the removable head, and the internal basket). This strategy kept the shipping weight well within the limit.

Once the four converter pieces arrived at site, the converter assembly activity was carried out following the safety and quality procedures developed by IHZL and KBR with close cooperation of all parties involved. After the two shell pieces were welded together, post weld heat treatment (PWHT) and all the required inspection and testing was accomplished, the whole converter was successfully assembled under KBR supervision. Figure 12 shows several photos of the converter welding and assembling activities.

Figure 12. Converter Shell Welding and Assembly at Site

Lost Control of GTE Startup Vent Stack Louver

In our GT-PAC plant, two louvers on the GTE startup vent stack are provided for adjusting GTE to Primary Reformer in a controlled manner. While restarting the GT after a plant trip due to power failure, it was found one louver control was lost, and then after two weeks into operation, another damper's control was also lost. The investigation concluded that the louvers and the control instrumentation of the louvers were not properly constructed and installed to protect the louver control instrumentation from heat impingement of the hot (above 560ºC, 1040ºF) GT exhaust, leading to multiple failures and damage to the system. The louver blade angled down while closing and led to downward flow of the hot GTE gas, causing heat impingement on the louver control instrumentation as well as on the handrail of the platform located close to the louver windows. (Figure 13).

Figure 13. GTE Vent Stack

A temporary repair was done by adding a shield box around the louver window and repairing the damaged instrumentation during one short shutdown period.

A permanent solution was implemented later in the plant's first turnaround with replacement of the louvers with a revised design. The revised design includes the addition of the shield box and relocation of the louver control mechanism. These changes allow the louver to be reinstalled upside down. With this installation, by closing the louver, the blade would guide the hot GTE flow upward instead of downward to avoid the heat impingement on the louver control instrumentation, handrail and platform below louver window. The shield box around the louver window also prevents rain from dropping into the stack. Figure 14 shows the revised GTE Vent Stack Louver Arrangement.

Figure 14 GTE Vent Stack (Revised)

Effective Prevention of Foaming in OASE white CO_2 Removal System

Key to trouble free, quick startup and stable operation of the OASE white CO_2 removal system is the effective prevention of the foaming of the OASE solution. Solution foaming can lead to significant operational upsets, and may cause damage to column internals, solution pumps, and heat exchangers in the system. It may also damage downstream equipment and catalyst due to solution carried over to the CO_2 compressor and/or Methanator. Hence, it is vitally important to ensure removal of the foaming causing agents, such as greases, oils, dusts, and fine particles from the system.

In the pre-commissioning phase, Korea S Company performed very thorough cleaning of the system. Before loading of the internals into the columns, Korea S Company carried out power brushing of the walls of all the columns in the system to remove particles from the walls and degreasing of internal packing materials with installation of temporary dip trays (Figure 15). Thanks to the meticulous planning and preparation, the degreasing (pre-cleaning, water flushing, alkaline degreasing, water rinsing and OASE solution charging) was completed in only 42 days instead of the typical 2 months.

Figure 15. Temporary Dip Trays

In the commissioning and startup phase, two other important steps were carried out before process feed introduction to the system. These are the passivation step for forming a magnetite layer (Fe_3O_4) that protects the carbon steel surfaces from corrosion, and the LTS catalyst de-dusting to prevent fine particles from entering the system.

In operation, periodic dosing of the anti-foam has been executed routinely.

Thanks to all these well carried-out effective measures, the YPFB OASE white CO_2 removal system has been running very smooth and foaming free ever since initial startup.

Lack of Industrial Infrastructure

Due to the lack of industrial infrastructure in the area, there is no local facility near the plant site capable of doing equipment maintenance. Korea S Company maintenance shop at the site could do some repairs, but if repairs of special equipment or valves were required, they would have to be transported to Santa Cruz and/or outside the country. Santa Cruz is 3 to 4 hour's drive one way from the plant site. To get items in and out of the country there was not only long shipping time involved, but also significant time related to customs. YPFB continues to face this challenge today.

Conclusion

Despite the challenging site conditions, Korea S Company and KBR succeeded in constructing, commissioning and starting up of Bolivia's first world-class ammonia plant. The ammonia plant successfully met all its energy consumption, production, and product quality guarantees. The project was a great success with 18.7 million work hours LTI free.

Acknowledgments

Authors acknowledge YPFB, Korea S Company and KBR management for their leadership in bringing this first fertilizer project to successful reality in Bolivia. Further, contributions of the specialists and engineering team members of Korea S Company and KBR as well as the contribution of the operations, maintenance, construction and related suppliers are highly appreciated and acknowledged.

References

1. Yang Yexin, Gosnell J. H.; "CNOOC Chemical Ltd. New Fertilizer Plant"; AICHE Safety in Ammonia Plants & Related Facilities Symposium 2004, Denver, Colorado, USA, September 2004.
2. Jovanovic W., Malhotra A.; "Construction & Commissioning of BFPL's 2200 MTPD – World's Largest Purifier™ Ammonia Plant"; AICHE Safety in Ammonia Plants & Related Facilities Symposium 2006, September 2006.
3. Pattabathula V., Morgan C., Hofman M., Colman G., Malhotra A.; "Commissioning of First World Scale KBR PurifierTM Ammonia Plant in North America"; AICHE Safety in Ammonia Plants & Related Facilities Symposium 2017, New York, USA, September 2017.

The Lessons Learned from the First 100 Safety Hazards In Urea Plants

In 2017, AmmoniaKnowHow.com and UreaKnowHow.com introduced global open source risk registers for ammonia and urea plants. On January 1, 2018, the Risk Register Safety Hazards in Urea Plants contained its first 100 safety hazards. This paper provides the detailed analysis of these hazards: In which plant sections and during which project phase do these safety hazards occur, which safety hazards occur most frequent, and which hazards have the highest risk factor. This paper describes the most critical safety hazards and recommends its prevention and mitigation measures. A surprising conclusion is that most safety hazards may lead to a sudden release of toxic ammonia. The paper provides suggestions on how to safely behave in a urea plant, and how to identify a maximum number of safety hazards.

Prem Baboo
National Fertilizers Limited

Mark Brouwer
UreaKnowHow.com B.V.

Jo Eijkenboom
UreaKnowHow.com B.V.

Majid Mohammadian
OCI N.V.

Introduction

In 2017, AmmoniaKnowHow.com and UreaKnowHow.com introduced to the industry global open source risk registers for ammonia and urea plants. As of January 1, 2018, the "UreaKnowHow.com Risk Register Safety Hazards in Urea Plants" contained its first 100 safety hazards related to more than 100 safety incidents, which led to at least 65 casualties and 217 people injured.

These 100 safety hazards originate from incident databases[1], hazard and operability (HAZOP) study results reported in the public domain and incidents reported by the members of UreaKnowHow.com. We believe these 100 safety hazards represent a significant part of the most critical safety hazards in urea plants as these hazards relate to serious incidents that have the potential to lead to casualties, injuries, loss of containment, risk to the integrity of equipment or cause serious damage to critical and expensive equipment.

The number of safety hazards grows each month, mainly due to its global open source character where all members of UreaKnowHow.com can easily contribute. The more people contribute the more valuable these risk registers become for all of us.

Some safety hazards refer to two or more incidents indicating that its frequency may be higher than other safety hazards. This is valuable input when performing safety assessment exercises.

The risk register provides an overview of the safety hazards involved in a urea plant. Not only are the hazards identified and listed, each hazard is quantified by means of a risk factor, prevention and mitigation measures are suggested, and reference is made to companies who are able to provide support for implementing these measures. Risk registers are a valuable information source for any HAZOP and safety study of a urea plant.

In which plant section most safety hazards occur?

Figure 1 below shows the allocation of the 100 safety hazards per plant section.

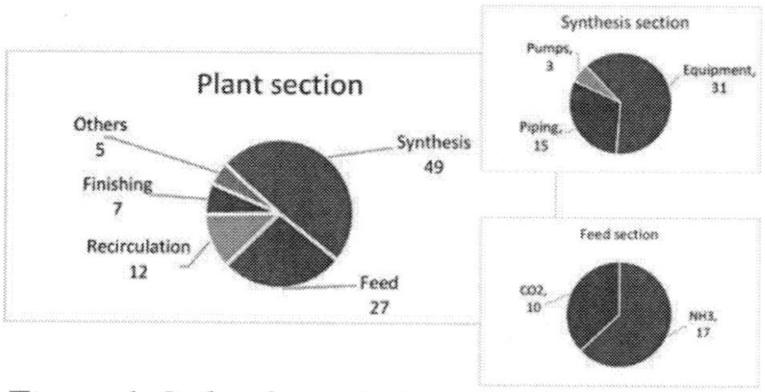

Figure 1. Safety hazards by urea plant section

Safety hazards mostly occur in the high-pressure synthesis section (49), the feed section (27), the recirculation section (12) and finishing section (7).

The safety hazards in the high-pressure synthesis section are 63% related to high-pressure equipment, 31% related to high-pressure piping and 6% related to high pressure pumps. Any leak in this section leads to a critical situation due to the release of toxic ammonia, its high pressure, high temperature and the corrosive ammonium carbamate. High pressure equipment is more vulnerable because of its carbon steel pressure bearing wall; corrosion rates of ammonium carbamate of carbon steel can be 1000 mm per year.

A significant number (27) of safety hazards occur in the feed section out of which 60% are related to ammonia and 40% to the carbon dioxide feed system. A significant part of the feed section is made from carbon steel, and any backflow from the high-pressure synthesis will lead to a situation that corrosive carbamate will come into contact with carbon steel. Note that during a HAZOP it is generally accepted that a backflow scenario is sufficiently protected by applying two non-return valves, each with a different design. This seems in urea plants a questionable assumption/solution as will be detailed out in one of the next chapters in this paper. The carbon steel ammonia lines can also suffer from other failure mechanisms like vibration, weld failure, corrosion from process side, erosion from process side, atmospheric corrosion and corrosion under insulation.

The carbon dioxide feed system has another specific critical safety hazard: a small unnoticed carbon dioxide leak can lead to asphyxiation if it occurs in a built-up unventilated section of plant.

In the recirculation section, various important safety hazards exists such as backflow of carbamate from the high-pressure synthesis to the centrifugal high-pressure carbamate pumps, damage of buffer tanks, and hydrogen explosion risks.

In the finishing section, safety hazards exists related to crystallization risks of the urea melt and cleaning and maintenance activities in the prill tower, granulator and warehouse.

Below are the top 10 equipment items with the most safety hazards:

1. High-pressure heat exchangers
2. High-pressure vessels
3. High-pressure piping

4. Ammonia piping/valves
5. Ammonia pumps
6. Atmospheric tanks
7. Carbon dioxide piping/valves
8. Carbon dioxide compressor
9. Low-pressure piping (melt, flare headers)
10. High-pressure carbamate pumps

When do these safety hazards occur?

In Figure 2 below the split of these 100 safety hazards per project phase is shown.

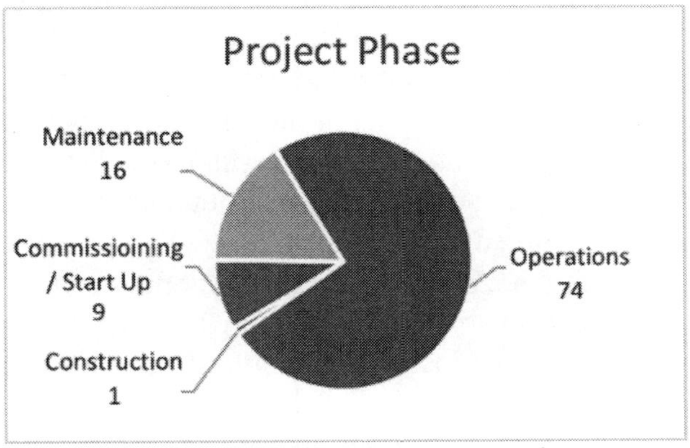

Figure 2. Safety hazards by project phase

Seventy-four of the safety hazards occur during the operations phase. A relatively large number of safety hazards occur during the maintenance (16) and commissioning/start-up (9) phase. Furthermore, a relatively large number of people are affected by these hazards because during these phases typically a larger number of people are present in the plant.

Safety hazards during construction phase

An important safety hazard during the construction phase is related to performing work at elevated heights.

- Although only one construction hazard is identified above related to one accident during the construction of a prill tower, that incident led to at least 12 casualties.
- High elevation work is a major risk during construction and maintenance and needs proper attention.
- Please note that elevated work safety risks also exist when working inside a vertical urea reactor. It is remarkable that no incidents related to working in urea reactors are reported in the public domain (yet) as it is also a confined space.

Safety hazards during commissioning / start-up phase

Nine safety hazards are identified during commissioning and start-up phase, which has led to at least 1 casualty and 20 people injured. Noteworthy are the following:

- Four safety hazards are related to not properly tightening flange connections of high-pressure equipment and piping leading to leakages and even two ruptures. Of these four, one safety hazard is related to a leaking flange connection of a carbon dioxide pipeline causing one person to pass out.
- Two safety hazards are related to the damage of a sight glass leading to one casualty and 18 people injured. Sight glasses are a major safety risk factor in urea plants.
- In one case a high-pressure ammonia pipeline severely vibrated during start up flow conditions: Vibrating pipelines require proper attention.
- Damage to atmospheric storage tanks or loose liners of equipment can occur in case one flushes with steam, which for one or the other reason is able to condense and create vacuum pressure.
- Another safety hazard is related to the high-pressure flush pump.

Safety hazards during maintenance

Sixteen safety hazards are identified during maintenance, which has led to at least 20 casualties and 19 people injured. Similar to construction phase, one sees a relatively high number of casualties and people injured during the maintenance phase, most likely because more people are involved in the job and present in the plant. Additionally:

- Nearly half of the safety hazards during maintenance are related to situations where a part of the plant was in operation or contained some process fluid. These hazards should be mitigated by applying double block and bleed designs and being aware of crystallization properties of urea melt and carbamate (risks of high pressure pockets), the explosion risks of hydrogen and ammonia during hot work and the toxic effects of ammonia.
- Safety hazards can also occur when entering a prill tower or granulator for cleaning activities. Entrants should be aware that urea lumps can fall down, and in a granulator the entrance door can close unexpectedly when the fan is put in operation.
- Further, typical safety hazards associated with maintenance in any chemical process plants are also valid in urea plants such as working at high elevation, asphyxiation in confined areas and accidents with cranes.

Safety hazards during operation

During operation, 74 safety hazards are identified, which has led to at least 28 casualties and 173 people injured.

- A significant part of the 74 safety hazards during operations are related to leaks of process fluids.
- In a urea plant ammonium carbamate present in a crevice leads to crevice corrosion (active corrosion of stainless steel), which will lead to a leak after some time. A crevice can occur due to improper aligning/tightening of flanges or to improper design of valves or accessories.
- Leaking ammonium carbamate is very corrosive and ongoing corrosion of stainless and carbon steel parts and piping must be avoided. Proper design, operation and maintenance of leak detection systems for high pressure equipment, high pressure flanges, and high-pressure valves are very important.
- Leaking ammonium carbamate from higher pressure forms very hard crystals, which erode the sealing areas. Once a leak occurs it is nearly impossible to stop the leak by tightening the bolts. In a urea plant proper tightening of flanges and other connections is challenging because of the limited choice of suitable materials of construction. Proper attention by experienced fitters is important. Hot bolting/torqueing and installing clamps during operation to stop leaks has led to five casualties and four people injured and should be avoided at any time.
- Safety hazards like backflow, contamination, vibration, atmospheric corrosion, or corrosion under insulation should receive proper attention.
- Seven safety hazards are related to hydrogen, which is present in the feed streams to the urea plant. Hydrogen can also enter the urea plant when one integrates the ammonia plant with the urea plant. For example, via process water and/or drain systems. Also note that hydrogen dissolves to a certain extent in the liquid streams leading to the presence of hydrogen in unexpected sections of the urea plant.

Which safety hazards have the highest risk factor?

Table 1 below shows the risk matrix applied on each safety hazard and shows in each cell the number of safety hazards with a certain risk factor.

Risk factors before Prevention and Mitigation Measures			Likelihood				
			Rare	Unlikely	Moderate	Likely	Almost Certain
			<1% Chance	1-20% Chance	21-50% Chance	51-80% Chance	>80% Chance
			1 per 100 years	1 per 10 years	1 per 5 years	1 per year	1 per 3 months
Consequence	Category	Safety	Never head of in the industry	Head of in the industry	Happened in the organisation or more than once in the industry	Happened at the facility or more than once in the organisation	Happened more than once at the facility
	Catastrophic	≥1 fatalities / 6 serious injuries requiring hospitalization for		16	37	7	3
	Major	<6 serious injuries requiring hospitalization for ≥24hr		2	10	4	2
	Moderate	Single serious injury requiring hospitalization for ≥24hr		1	7	3	3
	Minor	Injury requiring hospitalization for <24hr			2		1
	Insignificant	First Aid / minor injury		2			

Table 1. Risk matrix of 100 safety hazards before implementing any mitigation measures.

In total, 56 safety hazards have in our view intolerable risk factor (red color in above graph). Find in Appendix 1 a list of these most critical safety hazards.

Even if one would implement all recommended prevention and mitigation measures, a significant number (64) of the above safety hazards would still remain a high-risk factor (catastrophic consequences and rare likelihood). It is not feasible to mitigate these consequences further against reasonable efforts. Refer to Table 2 below.

Risk factors after Prevention and Mitigation Measures			Likelihood				
			Rare	Unlikely	Moderate	Likely	Almost Certain
			<1% Chance	1-20% Chance	21-50% Chance	51-80% Chance	>80% Chance
			1 per 100 years	1 per 10 years	1 per 5 years	1 per year	1 per 3 months
Consequence	Category	Safety	Never head of in the industry	Heard of in the industry	Happened in the organisation or more than once in the industry	Happened at the facility or more than once in the organisation	Happened more than once at the facility
	Catastrophic	≥1 fatalities / 6 serious injuries requiring hospitalization for	64				
	Major	<6 serious injuries requiring hospitalization for ≥24hr	14				
	Moderate	Single serious injury requiring hospitalization for ≥24hr	14				
	Minor	Injury requiring hospitalization for <24hr	3				
	Insignificant	First Aid / minor injury	5				

Table 2. Risk Matrix after implementing all prevention and mitigation measures

This means continuous attention and awareness of these 64 safety hazards is very important.

Leaks with toxic ammonia release

In 76 of the 100 safety hazards a leak could occur whereby toxic ammonia will be released. This has already led to at least 36 casualties and 198 people injured. It is remarkable to notice that in nearly 90% of these incidents (during which a toxic ammonia leak occurred), the leak occurred suddenly, without any pre-warning. Here one should think about failure of high pressure flange connections, rupture of high-pressure equipment and piping, failure of an isolation block valve, atmospheric corrosion phenomena, backflow of corrosive ammonium carbamate into the carbon steel feed lines, failures of high-pressure equipment and piping due to carbamate corrosion phenomena, tube ruptures and explosions caused by hydrogen.

In 5 incidents, there was a pre-warning in the form of heavy vibrations or a small leak, but the incident still happened because one tried to stop the vibrations or leak instead of shutting down the plant. This led to 9 casualties and 4 people injured. In case of a leak, the only advice is to stop the plant, trace and repair the leak. Do not try hot bolting/torqueing or to install a clamp. Pay attention to heavy vibrating pipelines by performing a pulsation and mechanical analysis.

Main safety hazards and its prevention and mitigation measures

Find below a list of preventive and mitigation measures to avoid leaks with toxic ammonia release:

- Proper design related to backflow, crystallization and explosion risks, dead-ends in pipelines, crevices, double block and bleed and blind flanges for isolation purposes, carbamate safety valves, avoid cooling jackets of carbon steel vessels;
- Proper design and fabrication quality related to high-pressure equipment, piping, accessories, valves and pumps;

- Proper quality of welds in ammonia pipelines;
- Proper operational procedures related to heating up with steam, chloride and nickel analysis, flushing equipment for maintenance entry, blocking in, sampling, carbon dioxide contamination of ammonia and operation of flares, ammonia detection sensors;
- Continuous leak detection systems for loose liners, for heat exchanger tube-tubesheet connections and for rupture discs and safety valves;
- Operator plant tour inspections for leakages from flange connections, overlay welding, valves, pump seals severe vibrations and proper housekeeping;
- Risk based corrosion inspections for high-pressure equipment, piping, accessories, valves and corrosion under insulation and atmospheric corrosion;
- Risk based preventive maintenance related to sight glasses, flange connections (proper installation of gaskets and tightening procedures) and seals/stuffing boxes of pumps;

Also, we believe it is important to pay more attention to pre-warnings or precursor failures so that more incidents can be avoided.[2]

Once an ammonium carbamate leak occurs it will not stop anymore; one may be able to stop the leak path but a crevice will remain. In this crevice ongoing corrosion will take place leading to another leak or possible rupture in the future. It is risky to stop flange leaks by hot torqueing. Installing a clamp to stop the leak is not done in the urea industry as one cannot assure no crevice will remain. The best and only advice is: In case of a leak one should stop the plant and repair the leak.

To realize a safe and reliable urea plant, it is therefore important to pay proper attention to the design of flange connections and tighten flange connections in the right manner and with the right torque and right sequence. Best practice is to use (hydraulic) bolt tensioners for the high-pressure equipment and to use torque wrenches for flanges in piping in the high-pressure and other sections of a urea plant. The optimum choice of materials of construction and a good understanding of the sealing mechanism and failure modes also have a large impact on the reliability of flange connections. It is best practice to perform a tightness test before introducing ammonia and carbon dioxide feed. Large flanges can be taped off and via a small hole a flange leak can be identified by means of a soap solution. Refer to the Figure 3 below.

Figure 3. Tightness test for large size flanges

Crystallization risks

Ammonium carbamate and urea easily crystallize at lower temperatures. During shut down conditions it is possible that pockets with ammonium carbamate and urea remain in the equipment or piping even when flushing has been applied. Incidents occur because one unexpectedly encounters these pockets during maintenance activities. Also, incidents occur when trying to open up clogged urea melt pipelines. This is a difficult and risky exercise requiring proper personal protection as the main mitigation measure, and proper attention to tracing and insulating cold parts of the pipelines is the main preventative measure.

Also pay attention to a proper design, flush system and quality of safety valves handling an ammonium carbamate stream assuring no passing and clogging of upstream and downstream piping.

Vibration risks

Certain pipelines are prone to vibrations like two-phase flow pipelines (for example the pipeline from high-pressure stripper to the recirculation section, minimum flow pipelines of high-pressure pumps) and discharge lines of high-pressure reciprocating pumps. A proper pulsation and mechanical analysis during the design phase of these critical lines is an important prevention measure. These pipelines should be included in a risk based inspection program (fatigue cracks) and in case of abnormal vibrations one should stop the plant and rectify the situation.

Backflow risks

The high-pressure synthesis section is fed with carbon dioxide and ammonia. Typical material of construction in the feed section is carbon steel while the high-pressure synthesis contains the very corrosive ammonium carbamate. Backflow of this carbamate is therefore a critical safety hazard and several incidents did occur. A further complicating factor is that crystallization can easily occur, which can lead to non-return valves not functioning. Backflow can more easily happen when the feed line enters at the bottom of a vertical reactor which is completely filled with liquid and continuously a liquid head pushes back the feed flow.

Best practice for the ammonia feed system are two non-return valves in series and two fast closing motor-operated-valves. Preferably the non-return valves should be of different designs and high quality, but do not apply a non-return valve design with vanes as the weld of the vanes with the body is prone to corrosion failure. Non-return valves should be installed in a vertical line and a flushing connection should be installed upstream the non-return valve to be able to flush away any solids. Install fast closing motor-operated-valves controllable from DCS and apply the material specification break upstream these motor-operated-valves. Pay attention also to bypass/start-up lines. For carbon dioxide, additional measures like a third non-return valve and a buffer vessel can be installed to protect the carbon dioxide compressor against the corrosive carbamate.

Hydrogen explosion risks

Hydrogen is present in both the carbon dioxide as well as the ammonia feed. Further, oxygen is added in the plant to assure a passive chromium-oxide layer on the applied protective layers. In several sections of the urea plant the concentrations of hydrogen and oxygen increase and the composition may enter the explosive area, so the risks for hydrogen explosions become realistic. Recommended preventive measures are adding a hydrogen removal converter in the carbon dioxide feed, an atmospheric flash step in the ammonia feed and adding methane to reduce the explosive area. Recommended mitigation measures are an expansion volume or a design pressure seven times higher than the operating pressure. Note that part of the hydrogen and oxygen will dissolve in the liquid phase and can reach and accumulate in less expected sections like the ammonia water tank and other low-pressure equipment.[3] Also, when one connects the ammonia plant with the urea plant, hydrogen containing streams can enter the urea plant creating explosion risks.

Carbamate corrosion risks

Ammonium carbamate is the intermediate product when converting ammonia and carbon dioxide into urea and is the most common and most corrosive component in a urea plant. Corrosion rates can be controlled by assuring sufficient oxygen being present, but corrosion rates can become very high in cases where oxygen gets depleted. There are several reasons why this can occur (for example design with dead-ends or crevices, loss of oxygen supply, condensation of carbamate vapors, etc.) leading to several corrosion failure mechanisms. In some cases, carbamate corrosion leads to a "controlled" leak first via a leak detection system and proper action can be taken. In other cases, a sudden leak or failure occurs and one has to rely on preventive maintenance procedures, risk based inspection techniques, regular visual inspections and ammonia leak warning systems.

For example, a leak in a loose liner of a high-pressure vessel will lead to a situation that a large area of the carbon steel is exposed to corrosive carbamate and typically a break-before-leak scenario will be valid. The integrity of overlay welding in high pressure equipment cannot be monitored continuously and leaks should be avoided by preventive maintenance (corrosion inspections during a turnaround). A leak in the overlay welding will lead to a situation that only a limited area of the carbon steel will be exposed to corrosive carbamate and typically a leak-before-break scenario will be valid. Regular inspection rounds by operators to identify such a leak are recommended.

Atmospheric corrosion / corrosion under insulation risks

Atmospheric corrosion or corrosion under insulation is an often underestimated form of failure. In urea plants, atmospheric corrosion risks do occur. Special attention should be paid to the carbon steel high pressure carbon dioxide and ammonia feed lines and to stress corrosion cracking risks of nitrates and (bi-) carbonates with carbon steel and chlorides with stainless steel. Neighboring plants and prevailing wind directions are important and sufficient attention to proper insulation is vital especially after a turnaround. A protective shield against rain above high pressure equipment is recommended.

How to safely behave in a urea plant?

The fact that many leaks and failures can occur suddenly without any pre-warning has in our view consequences on how to safely behave in a urea plant. One should avoid any leak and adhere to a zero-tolerance philosophy regarding leaks. This is vital for ammonia and carbamate leaks but also good practice for steam, carbon dioxide and other leaks. In a sudden large ammonia cloud one will only remain conscious a few seconds. Know the escape routes blindly, escape lateral/upwind direction, have readily available a wet cloth or better a full face mask and move away quickly. Assure that all safety devices are ready to work such as safety showers and eye washes. Further, in case of a leak, stop the plant, do not apply hot bolting/hot torqueing, do not apply clamping and limit the number of spectators.

How to identify a maximum number of safety hazards?

It is commonly accepted that by performing a proper HAZOP study with experienced participants at regular time intervals one is able to identify most safety hazards. It is good practice to invite an experienced urea expert from outside into the HAZOP. We see that for urea plants, HAZOP studies do not always identify commonly accepted assumptions. For example the assumption that backflow can be avoided by installing two non-return valves with a different design as described in one of the previous chapters.

Also, corrosion is typically excluded from a HAZOP study and should be covered by maintenance and inspection programs. In a urea plant however, corrosion from the process side and also atmospheric corrosion or corrosion under insulation are major causes for safety hazards. For example the Risk Register Urea Reactor identifies 50 safety hazards of which most are related to corrosion.[4]

Finally, one is able to identify more safety hazards by performing a HAZOP study also for transition phases like commissioning, start-up and shut-down situations and by using transition guide words like too fast, too slow, wrong sequence and control changes.[2] For example, in urea plants the blocking-in of the high pressure synthesis is a specific situation, which can occur quite often and needs proper attention in a HAZOP study.

Furthermore, one is able to improve the quality of the HAZOP by making use of available incident databases and risk registers, by keeping track of incidents and near misses within the company. We invite all to share incidents and near misses within the industry to support each other improving safety studies. We believe global open source risk registers[5] and specific industry operational risk databases[6] are best practice to keep track and share safety hazards in our industry leading to a more safe workplace for all of us.

Conclusions

Most safety hazards in a urea plant involve a release of toxic ammonia of which 90% can occur suddenly, without any pre-warning. Other main safety hazards are crystallization risks, vibration risks, backflow risks, hydrogen explosion risks, ammonium carbamate corrosion risks, atmospheric corrosion and corrosion under insulation risks. The fact that in a urea plant many safety hazards can suddenly occur releasing toxic ammonia has in our view consequences how to safely behave in a urea plant. To identify the maximum number of safety hazards, involve an experienced urea expert in the HAZOP, include transition phases in the HAZOP and make use of available incident databases, risk registers and specific industry operational risk databases.

References

1) 2015 AICHE Ammonia Safety Conference, Brouwer a.o., UreaKnowHow.com Urea Incident Database: Share for Safety.
2) 2017 AICHE Ammonia Safety Conference, J. Redman a.o., Koch, Achieving equipment reliability through understanding precursor failures
3) 2017 AICHE Ammonia Safety Conference, Cousins a.o., Catalyzing a path to a safer place
4) http://www.ureaknowhow.com/ukh2/images/rr/rr_urea_reactor_v7.pdf
5) http://www.ureaknowhow.com/ukh2/risk-registers.html
6) http://fiorda.eu//

Appendix 1: Most critical safety hazards.

Catastrophic consequences and almost certain likelihood:

- Rupture of carbon steel ammonia pipeline due to backflow from corrosive carbamate from synthesis
- Rupture of the high-pressure scrubber due to hydrogen explosion
- Rupture of high-pressure equipment due to high process pressure

Catastrophic consequences and likely likelihood:

- Large ammonia leak due to blowing out of a plunger from the stuffing box of a high-pressure reciprocating pump due to broken bolts of plunger yoke
- Leak/rupture of high-pressure flange connection in synthesis section
- Leak in loose liner of urea reactor due to no proper leak detection system*
- Failure of urea reactor due to no proper leak detection system*
- Failure of high-pressure flange connection due to hot bolting
- Asphyxiation/unconscious while working on tank
- Asphyxiation/unconscious when entering a pipeline/vessel (confined space) while welding was performed and welding gas was applied

* In our view a proper leak detection system means an active vacuum based leak detection system with a continuous, accurate and reliable ammonia detector.

Catastrophic consequences and moderate likelihood:

- Failure of a valve in an ammonia line due to human error to solve a vibration problem
- Ammonia leak due to a sight glass failure during start up
- Isolation valve fails during maintenance of high-pressure ammonia pump
- Significant wall thickness reduction of high-pressure ammonia line due to atmospheric corrosion
- Ammonia leak due to loosening packing ring of stuffing box on a high-pressure reciprocating pump
- Explosion due to atmospheric corrosion of ammonia pipeline and/or erosion from process side
- Large uncontrolled ammonia emission when suction line of high-pressure ammonia pump fails due to unwanted backflow scenario
- Rupture of pipeline due to clogging of pipeline upstream or downstream safety valve due to crystallization of carbamate
- Rupture of 316L Urea Grade (UG) high-pressure pipeline at inlet of reactor (conventional plant) due to erosion corrosion close to thermowell
- High-pressure butterfly valve twists leading to leakage
- High-pressure carbamate gas line ruptures due to strain induced intergranular cracking
- 316L UG liquid outlet line of high-pressure ammonia stripper leaks due to relatively high temperature and as a consequence higher corrosion rates
- Rupture of high-pressure drain valve due to corrosion of threads during leakage
- Crack in stripper liquid outlet pipeline due to heavy vibrations
- Leak of high-pressure flange connection with titanium gasket manway cover of urea reactor
- High vibrations of discharge line of high-pressure reciprocating carbamate pump
- Major leak at main flange of high pressure carbamate condenser during start up
- Failure of flanged head connection of a urea reactor
- Reactor ruptures due to bad maintenance practice of closing the manway cover
- Vent line of ammonia water tank explodes during grinding
- Rupture of valve due to high pressure pocket during maintenance
- Failure of sight glass
- High-pressure equipment ruptures due to not commissioning all leak detection holes
- Severe damage of loose liner in high-pressure equipment (cause not clear)
- Liner bulges and ruptures due to too high pressure behind the liner or due to draining the synthesis
- Damage of carbon steel pressure bearing wall due to leakage of protecting sleeve in nozzle
- Rupture of high-pressure stripper due to fabrication defect
- Rupture of reactor due to active corrosion at a weld defect in weld clip to liner
- Rupture of reactor due to stress corrosion cracking of carbon steel pressure bearing wall due to flushing of leak detection system with water and bad design of leak detection holes in multi-layer vessel
- Rupture of urea reactor due to cracks in carbon steel behind the liner
- Rupture of urea reactor due to defect in titanium liner
- Tubesheet of pool condenser damaged due to erosion corrosion steam side
- Tube rupture in high-pressure heat exchanger leads to high pressure on steam or cooling water side
- Failure of the tube to tubesheet weld, potential damage of carbon steel tubesheet
- Failure of tube in tubesheet due to chloride stress corrosion cracking
- Chloride from atmosphere (cooling water effluent, marine atmosphere) entering the steam condensate system via atmospheric steam condensate tank

- Rupture or crack in pressure bearing wall due to atmospheric corrosion (nitrate stress corrosion cracking)

Major consequences and almost certain likelihood:
- Rupture of carbon steel carbon dioxide feed line due to backflow from corrosive carbamate from synthesis
- Carbon dioxide breakthrough with ammonia recycle leading to failure of pipeline

Major consequences and likely likelihood:
- Excessive ammonia emission due to mechanical seal failure of high-pressure centrifugal ammonia pump due to dirt, oil in flush lines
- Significant ammonia and carbamate leak when plunger flows out from stuffing box of high-pressure reciprocating pump due to broken bolts of plunger yoke
- Rupture of tank due to vacuum pressure created
- Crane collapses during maintenance job

Moderate consequences and almost certain likelihood:
- During cleaning or maintenance activities in prill tower or granulator heavy lumps fall down causing casualties
- While trying to open clogged melt line, melt splashes out of 3-way valve or flange
- Granulator door jammed close unexpectedly

Achieving Excellence in Process Safety Management (PSM)

In line with Fatima Group's vision and HSE policy, Fatima Fertilizers Company Limited (FFCL) stipulated a Process Safety Management program (PSM) as the major corner-stone. It was purely a self-initiative to implement this model rather following any regulatory requirement. FFCL believes that robust PSM implementation equates to enhanced safety, reliability and productivity, and this is evidenced by no Fatality/Lost Time Injury after implementation, lowest Total Recordable Injury Rate, and above 42 million safe man-hours without lost time accident. A PSM program based on Dupont's 22 PSM elements was implemented (in house) within 5 years. DuPont, being the external consultant, audited the site twice and provided a 4.0 rating (OSHA Excellence equivalent) making FFCL the first in Pakistan to achieve this milestone in the shortest time.

Noman Irshad CSP, GradIOSH, NEBOSH IDIP
Senior Process Safety Engineer
Fatima Fertilizers Company Limited (FFCL)

Ahsan Sarfaraz
Unit Manager HSE
Fatima Fertilizers Company Limited (FFCL)

Introduction

Process Safety is about establishing a safety culture, providing management leadership and commitment and implementing a comprehensive Process Safety Management (PSM) program. The Process Safety Management Program implemented at Fatima Fertilizers Company (FFCL) is based on the OSHA Standard (Process Safety Management of Highly Hazardous Chemicals), and it is directed towards preventing serious process related chemical incidents that might affect:

- Plant personnel
- Off-site communities
- Environment
- Result in significant loss of property or business

The program also supports the application of management controls that apply to the operations involving hazardous materials. These management controls ensure that process hazards are identified, understood, and controlled by adequate measures so that process-related injuries and incidents are eliminated.

This paper describes the steps that were taken in the successful implementation of Process Safety Management at one of the largest fertilizer complexes in Pakistan. Included are strategies that were implemented along with the monitoring tools utilized to achieve the desired goals. This paper will also describe future strategies and how to sustain the excellence. Finally, this paper discusses how implementing the PSM system at the FFCL site resulted in a safer work environment as well as increased productivity.

Description of Facility

Fatima Fertilizer Company Limited (FFCL) is one of Pakistan's largest fertilizers manufacturing complexes that processes natural resources like natural gas, rock phosphate, limestone and water into vital products for farmers and industrial customers. FFCL produces Sarsabz Urea, Sarsabz Nitro phosphate (NP) and Sarsabz Calcium-Ammonium-Nitrate (CAN). A total of 1733 employees work at FFCL, including 250 management, 468 staff and 1015 contract employees.

A 1500 MTPD Ammonia Plant is currently operational using natural gas as the basic raw material. This ammonia and its by-product carbon dioxide are then used for the production of 1500 MTPD prilled urea. The nitric acid production facility has a capacity of 1500 MTPD (based on 100% conc.), which along with ammonia and limestone, is utilized for the production of 1200 MTPD prilled nitro phosphate and 1400 MTPD granulated calcium ammonium nitrate. In addition to the main process units, independent small units are installed to aid their smooth operation which include a nitrogen plant, instrument air and plant air plant, reverse osmosis (RO) plant, demineralized water treatment and steam generation unit. The complex is self- sufficient in power generation and also supplies a residential colony. Figure 1 illustrates the main process at FFCL.

Due to the hazardous operations of all these units, the probability and consequences of a serious process safety incident are high. Operating this type of facility comes with a direct responsibility to identify hazards associated with the process and ensure the facilities are designed, operated, and maintained to minimize hazards. Process safety management (PSM) is the systematic tool used to prevent the unexpected release of toxic, reactive, or flammable liquids and gases in various industrial processes worldwide.

The FFCL Process Safety Management System is aligned with and certified by DuPont to ensure compliance and improvement in our product manufacturing and raw material processing. Fatima Fertilizer elected to implement the DuPont PSM 22 elements model at its site.

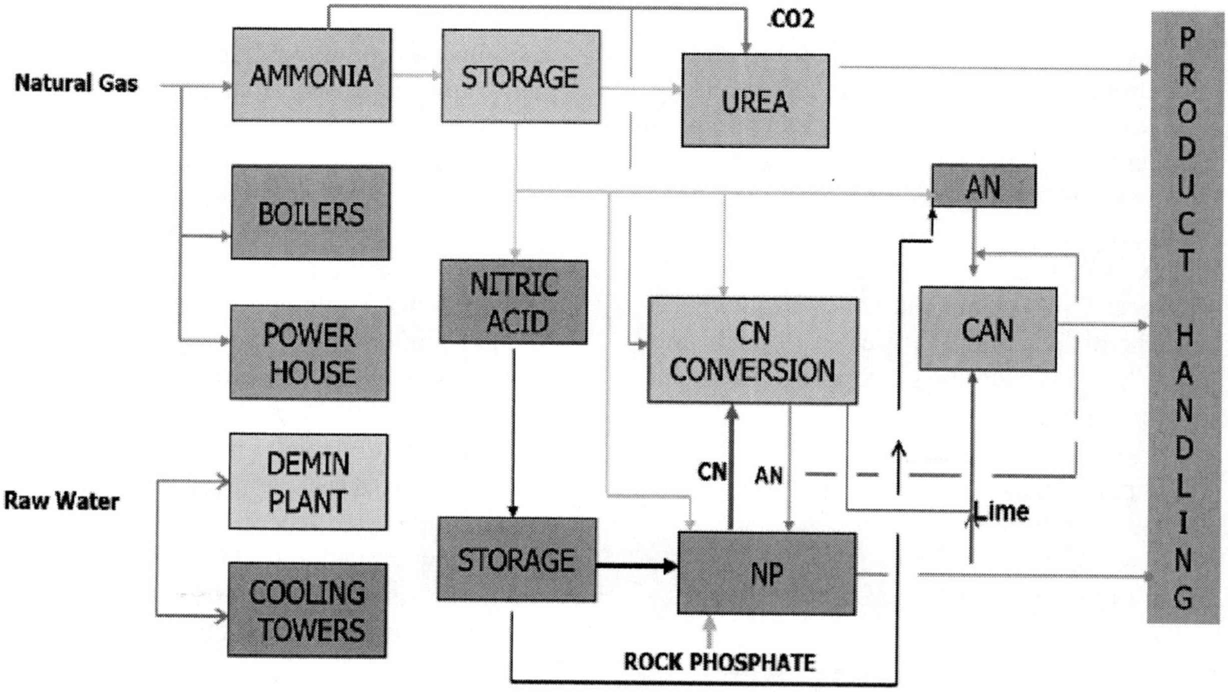

Figure 1. FFCL Process Block Diagram

Getting Management Commitment

Commitment is the basic component of a successful safety program. For a program to be most effective, this commitment must exist at the top and works through the line organization. For this commitment to be demonstrated throughout the entire organization, top management must believe that employee safety is as important as:

- Asset availability
- Reliability
- Cost
- Quality
- Production targets

The Corporate Health, Safety and Environment Manager of Fatima Group was the pioneer to conceive the implementation of the PSM idea at Fatima Group in 2012. In this regard, a detailed presentation was delivered to the CEO of Fatima group covering the following topics:

- Why PSM/World class safety standards, and why DuPont Process Safety Standards?
- Fundamentals of process safety management with a brief background on basic principles and components, provided to ensure consistent understanding of what is being proposed.
- Statement of PSM goals for Fatima and how these basic principles apply to Fatima's business operations and structure, and support its commitment to continuous improvement.
- Benefits to Fatima of PSM systematic approach including discussion of efficiencies to be achieved through interdivisional teamwork, framed in terms of savings and/or avoided losses.
- Support in terms of resources and budget required to implement the PSM program at FFCL.
- The role of each individual in achieving excellence in Process Safety Management.
- Commitment to promptly develop a detailed action plan for management review and approval, including clear deliverables and realistic timeline.

The CEO showed keen interest in the PSM drive, approved the proposal and ensured his full commitment and endorsement from the Board of Directors.

Figure 2. PSM 22 elements model

PSM 22 Elements Model

The DuPont 22 Elements Model is represented in Figure 2 and includes the following:

Audits and Observations: This element covers the importance of effective auditing in site safety management and provides guidelines for conducting and evaluating safety audits

Contractor Safety Management: The intent of this element is to make contractors responsible for effectively meeting the safety, health and environmental requirements. It covers safety expectations of contractors with safety performance of the contractor as the top most priority.

Effective Communication: The purpose of this element is to emphasize and elaborate the importance of effective two-way communication in prevention of occupational accidents/illnesses and achieving safety goals and objectives.

Emergency Preparedness and Contingency Planning: The intent of this element is in-depth planning for potential emergencies ensuring effective response by site personnel. The outcome of these efforts is mitigation of the impact of incidents on personnel, environment / facilities and prompt control of emergency.

Goals, Objectives and Plans: The purpose of this element is to provide guidelines for establishing realistic, achievable and quantifiable safety goals and objectives. Managing safety, like managing other aspects of a business, includes setting of performance goals and objectives which should be Specific, Measurable, Attainable, Result Oriented, Time Bound (SMART) and within the sphere of influence of the person and group who is to be held accountable for achievement.

Incident Investigation and Communication: The purpose of this element is to document the process for investigating incidents that occur on-site or off-site in a way that promotes thorough and efficient investigation in a timely manner; uniform, accurate, clear, and concise documentation and reporting; identifies and implements recommendations to prevent incident recurrence; involves the right people to get the information; ensures a clear understanding of key factors and key learnings; participating personnel obtain a positive learning experience.

Line Management Accountability and Responsibility: The purpose of this element is to delegate HSE responsibility and accountability to each level of the organization. Line supervisors and managers are totally responsible and accountable for safety as well as cost, quality and productivity.

Management Commitment: Management is responsible for safety of personnel and protection of company property. Management will direct the establishment and implementation of safety programs through participation in various safety committees and conducting plant safety audits. Management should have specific, quantifiable, personal safety goals/targets and implementation plans, which must be stewarded regularly

Mechanical Integrity: This element addresses equipment tests and inspections including predictive and preventive maintenance, reliability engineering, maintenance procedures, quality control procedures, training and performance of maintenance personnel. All of these mechanical integrity efforts ensure an incident free and reliable operation, and they help to pin point root causes and avoid incident recurrence and pre-mature failures.

Management of Change - Facility and Technology: Manufacturing plants are designed according to standard engineering practices. The changes to the documented process safety information (e.g. hazard of materials, equipment design basis and process design basis), even if subtle or temporary, can lead to catastrophic events. Therefore, these changes must be managed in such a manner that safety, the integrity of the plant and the environment are not compromised. All changes must receive appropriate review and authorization before being implemented.

Management of Change - Personnel: Safe operations of facilities require an effective personnel change management system as people are the essential ingredient in "Process Safety Management" and play the most important role in its implementation and day to day compliance. It is essential that personnel changes at all levels are controlled according to a pre-established criteria so that minimum levels of experience and knowledge are maintained at the site.

Motivation and Awareness: The purpose of this element is to discuss and provide guidelines on different concepts and recommended practices on progressive motivation. Internal motivation is necessary to sustain high-level safety performance once that level of performance has been reached. External motivation is necessary to make the initial transition to high level safety performance because of established behavior patterns in the individual.

Policies and Principles: To improve safety, a deliberate safety policy must be established and applied daily by each member of the work force, whether manager, supervisor, or contractor employee. Top management must establish the policy that will spell out the principles that are to govern all decisions regarding safety. Without such a policy, safety tends to be pushed aside when other concerns become pressing.

Procedure and Performance Standards: This element provides standards of performance including such items as rules, procedures, and design criteria that specify how activities are to be done. They should be written, practical, and available at the point of action, reviewed regularly, followed, and enforced. Adherence to standards must be enforced, even to the point where adherence becomes a condition of employment.

Process Safety Information: It provides a foundation for identifying and understanding the hazards involved in the process. It ensures that PSM goals of HSE are achieved by providing process safety documentation. A PSI package shall be prepared for each process unit. Documents of the PSI package should be maintained up to date for the life of each process unit.

Risk Assessment and Process Hazard Analysis: A systematic and comprehensive study to identify and evaluate the significant hazards of the process and the safeguards associated with Highly Hazardous Processes (HHP) and Lower Hazard Operations (LHO). Process Hazard Analysis systematically identifies the safety hazards such as potential for fires, explosions and / or release of toxic materials, and is a well-defined program to remove or lower these hazards.

Pre Startup Safety Review(PSSR): PSSR provides a final checkpoint for new and modified equipment and facilities to confirm that all appropriate elements of Process Safety Management have been addressed satisfactorily and the equipment / facility is safe to start-up. It is mainly intended to make sure that alterations / additions to the process or system do not create hazards to personnel at the site, surrounding facilities, community and environment by inadequate, incomplete, or unauthorized design or installation.

Quality Assurance (QA): QA is important for new facilities and revisions or repairs to existing facilities to ensure that safety critical equipment which handles hazardous material (as it is fabricated) is suitable for the process application. It also ensures that safety critical equipment installed is consistent with design specifications and manufacturer's recommendations.

Safety Personnel: This element highlights the main responsibilities of safety personnel so that all members of line organization understand the role and place of safety personnel in the organization.

Training and Development: This element signifies that all personnel whose work could affect the safety of the site must have, and maintain, the necessary knowledge and skills to execute their job functions in a manner consistent with the safe operation of the site.

Integrated Organization for Safety: The purpose of the overall safety organization is to mobilize all available talent in the interest of safety, health, and environmental protection. It does not, in any way, relieve individual members of the line organization of their safety responsibilities. Various committees are staffed principally by members of the line organization supplemented by safety staff members and other specialists.

Initial Gap Analysis

After getting the commitment from top management, the initial gap analysis was conducted by Corporate HSE led by the Corporate HSE Manager. The gap analysis revealed the following major deficiencies in the exiting safety system:

- The formal HSE Goals and Objectives along with reward / reprimand system were found to be inadequate.
- The increased number of injuries: 22 recordable injuries were reported in 2011.
- Elevated Total Recordable Injury Rate of 0.51 (per 200,000 hours).
- High number of fires with 29 process and non-process fires occurring in 2012.
- High rate of loss of containment incidents and process releases.
- Low significance and importance of safe practices.
- Inadequate Contractor Safety Management System.
- Weak emergency response and lack of trust between the site and the local community.
- Low employee morale.

Implementation Strategy

In a developing country like Pakistan, implementing a new process safety system which is not obligated by legislation, compulsory for the production and doesn't earn extra profit is not welcome. So, implementing PSM, which is not directly involved with generating extra profit for the company, could be subjected to the unwillingness at all levels.

Since the site had inadequate PSM knowledge and awareness, the development of a comprehensive implementation strategy was needed to enhance the PSM awareness. The initial step of the strategy was to impart PSM awareness down the line to all levels of management and staff employees.

The PSM Task Force

A multi-disciplinary PSM task force led by the Corporate HSE Manager was established with the charter to initiate the PSM awareness campaigns. The task force converted the initial gap analysis into a smart and tangible action plan. The element wide action plan consisted of 332 action items, segregated into long term, short term, procedural change and capital expense (CAPEX) categories. The evergreen action item list was updated with the addition of new recommendations after every 2nd party and 3rd party audit. A meeting of the task force was conducted on a monthly basis to monitor the completion status of the action plan.

PSM Champions

Every department nominated two individuals as "PSM Champions." Three full days of interactive training sessions were conducted by the Corporate HSE manger. The training session, which was based on "Train the Trainer Concept," was very successful, and the PSM Champions were validated at the end of these sessions by a written test. The champions were assigned to complete not only the PSM related tasks, but also to increase the knowledge level awareness by conducting independent training sessions with in their respective departments. Later, these champions were involved in the development of PSM procedures.

Conversion into Local Language

The Task Force converted the PSM Procedures into one page crisp leaflets. These leaflets were translated into local languages and distributed among the unskilled and lower level of staff/labor for ease of their understanding and to increase awareness of the PSM system. Communicating the PSM awareness into local languages down the line to unskilled labor gained their interest and boosted the moral among the staff employees.

Weekly safety talk

A weekly safety talk is considered to be one of the most effective modes of communication for the lowest level of hierarchy. The PSM elements leaflets were shared in the weekly safety talks to increase the awareness and knowledge level of the site. The feedback from the safety talk was shared with management in order to close the communication loop.

PSM Quizzes

Quizzes are the appropriate way to test the knowledge and recognize the top performing individuals. PSM quizzes were arranged so that departments competed with each other in a very positive and encouraging atmosphere. This healthy competition proved to be extremely beneficial in engagement of staff employees and gaining their buy-in for the commitment towards PSM Implementation.

PSM Validation

Each employee at FFCL has gone through online testing and validation for all 22 PSM Elements. The top scores from each department were recognized and awarded.

Spot Awards

The Spot Award Program applauded the employees complying with safe working practices and fulfilling the PSM Element Procedure requirements every month. An employee identified as complying with these requirements would be recognized on the spot in the field, awarded with a special safety award, and acknowledged by sharing the observation with all employees at site. This program increased the motivational level of all staff employees.

Awareness Campaigns

PSM awareness campaigns were run within departments and also collectively for the whole site by respective department champions and the PSM Task Force. The campaigns interacted with all levels of staff and management employees and encouraged the openness and knowledge sharing among them.

Contractor's Engagement

Contractors play a vital role in the development of safety culture in any organization. To increase PSM awareness, contractors were involved in all the PSM campaigns, Spot awards, and safety talks. They were encouraged to share their knowledge and implications (if any) of the implementation of PSM protocols during their routine activities. The contractors who demonstrated excellent PSM knowledge and practices were recognized by special awards of appreciation in HSE annual Awards Ceremony by FFCL's CEO.

Measuring and Monitoring

The action plan monitoring and percentage completion was stewarded at the Executive Health, Safety and Environment (EHSE) Forum, the most prestigious meeting chaired by the Director of Operations that includes the whole top management of FFCL. The guidance and directions from the EHSE forum kept the whole PSM implementation drive on the accurate track.

Audits

Internal Audits, 2nd party audits and 3rd party audits proved to be instrumental in fine tuning the PSM implementation drive. The recommendations from these audits defined the direction towards achieving PSM excellence. A total of four internal audits were conducted by the PSM Task Force at the site, and six 2nd party audits were carried out by Corporate HSE and other Fatima Group Affiliates. The audits were conducted in a manner which enlightened the way towards excellence and were always welcomed. Gaps in the system were identified and recommendations were punched in the detailed action plan for proper stewardship.

Outcome and Achievement

FFCL believes that robust PSM implementation equates to enhanced process safety, reliability and productivity, and the same is evident as follows:

Zero Fatality

Since the effective implementation of the PSM System, there has not been a fatality at site.

Safety Performance

Further, the following improvements have been achieved:

- Lost Time Incident Rate (LTIR) also remained 'ZERO' in the last 5.5 years including FFCL and contract employees (see figure 3). It's only possible due to the higher level of management commitment, safe systems of work and employees' efforts at all levels.

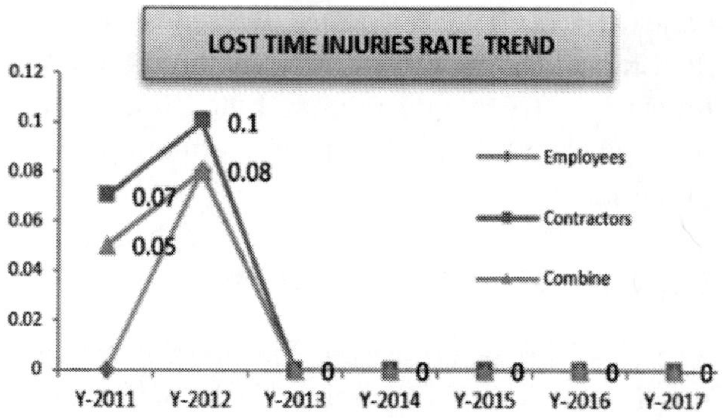

Figure 3. LTIR 6 year trend

- The 42 million safe man hours and counting since 2012 shows the implementation of best safety standards and practices being executed at site. See figure 4.

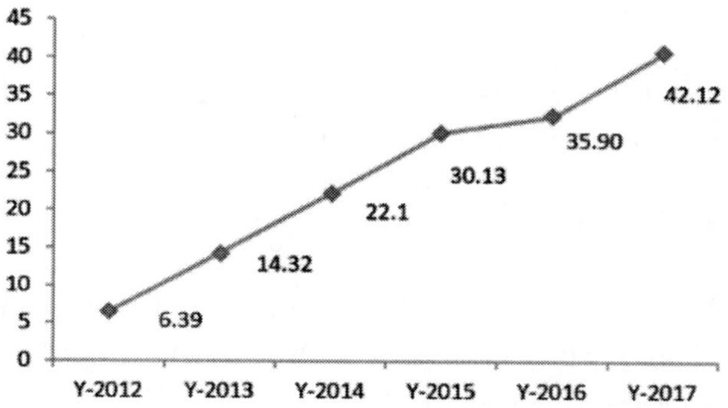

Figure 4. Safe million man-hours 5 year trend

- Improved Emergency Response.
- The decreasing trend of Total Recordable Injury Rate (TRIR) with 2017 closing on 0.096 is the clear depiction of high safety performance of FFCL (Figure 5).

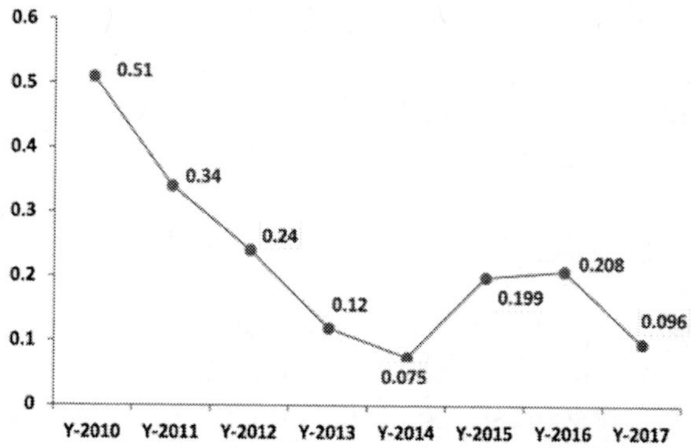

Figure 5. TRIR 7 year trend

- Improved service and capacity factors of FFCL. The on-stream factor improved after implementation of PSM, thus resulting in higher fertilizer production figures. See figure 6.

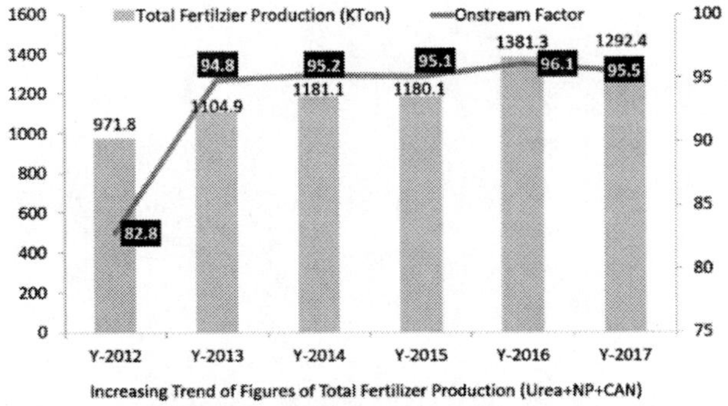

Figure 6. Total fertilizer production and on-stream factor

- The decreasing trend of Fires, Process Releases in past 5 Years.
- Reduced risk to nearby communities, improved reputation and community engagement.

DuPont (3rd part) Assessment

In May 2015, DuPont as a 3rd party assessor carried out the first assessment of FFCL. The first assessment rated FFCL at **3.6** which shows that the PSM System is very well implemented. The DuPont final report was received and its recommendations were inserted into the master PSM Action Plan. In November 2017, DuPont conducted its 2nd assessment of FFCL Site, and the final assessment report rated FFCL at Level **4.0** "Excellence" Level. The DuPont Pakistan contribution and guidance throughout the implementation of the PSM system are much appreciated. DuPont Pakistan expressed their interest in jointly presenting the FFCL achievement as a showcase study in their upcoming seminar. This was very much acknowledged by the CEO and all the top management.

Challenges Ahead

Achieving excellence is not the end of the story, rather the beginning of a new era to explore further advancement and to excel towards a world class industry. FFCL management is keen to uplift the process safety standards to their maximum. In this regard, the following are the key challenges to be addressed in future.

- Sustainability of excellence level 4.0 and strive for world class.
- Inter-dependent culture development at the site.
- Increasing Safe Million Man-Hours and preventing complacency.
- Contractor Safety Excellence and future improvements.
- Off the job safety improvement.
- Occupational Health and Industrial Hygiene Program compliance.
- Environmental Management System benchmarking

Figure 7 represents the roadmap of FFCL towards world class organization is safety standards.

Conclusion

Effectively applying PSM at FFCL has clearly demonstrated an uplifting of the safety standards and morale of employees with a reduced number of injuries and serious process safety incidents. Strong management commitment served as the backbone of whole PSM drive, and, combined with employee participation, helped in securing the objective "To be Excellence in Process Safety" in the remarkable minimum time span of 5.5 years.

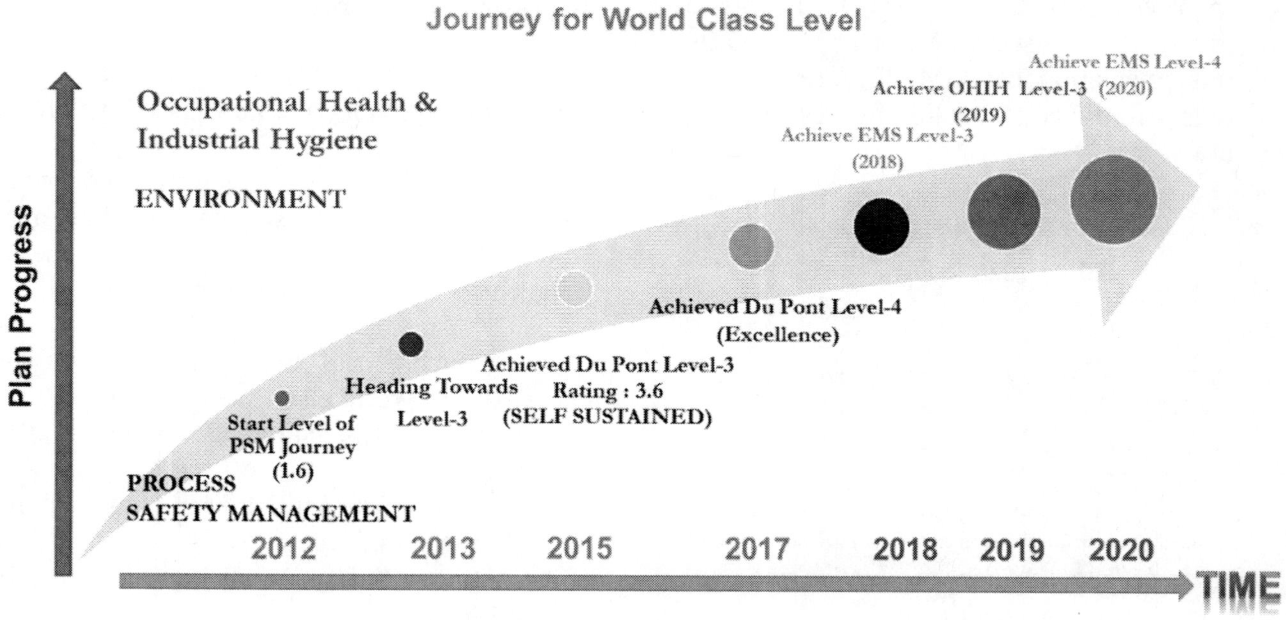

Figure 7. FFCL Roadmap Towards World Class Level

Iowa Fertilizer Company (IFCo) Ammonia Plant Uses First Application of the KBR Shift Effluent Scrubber for Reduced Methanol Emissions

Methanol is a byproduct produced from side reactions in the front-end of ammonia plants, mainly across the Low Temperature Shift (LTS) catalyst. Some of this methanol passes through to the CO2 Removal System where it eventually comes out with the CO2 product and atmospheric vent.

Methanol is classified as a Hazardous Air Pollutant (HAP) in the US. New plants emitting more than 10 ST/year of methanol qualify as a major HAP source, generally requiring the use of Maximum Achievable Control Technology (MACT) to limit the emission. For ammonia plants with significant process CO2 venting, limiting emissions below the trigger for MACT regulations is challenging.

The IFCo 2200 MTPD Ammonia Plant represents the first application of the KBR Shift Effluent Scrubber system for controlling methanol emissions. This paper presents the system in detail and reviews performance of the unit as measured during the plant's first year of operation.

Harrie Duisters
OCI N.V.

Stephen Noe and Stephen Lang
Kellogg Brown and Root (KBR)

Introduction

The Iowa Fertilizer Company (IFCo) complex in Wever, Iowa (Figure 1) was commissioned in 2017 and is the newest site owned by OCI N.V., a global producer of natural gas-based fertilizers and industrial chemicals with headquarters in the Netherlands. Orascom Construction Limited was the overall EPC company and responsible for construction of the plant site. IFCo's diversified nitrogen fertilizer product portfolio includes design annual capacities of 1.5 million metric tons of UAN, 875 thousand metric tons of ammonia, and 420 thousand metric tons of granular urea, in addition to 315 thousand metric tons of diesel exhaust fluid (DEF). Ammonia production is provided by the KBR Purifier™ Process with a design capacity of 2200 MTPD (2420 STPD).

While securing permits for the new complex, the Iowa Department of Natural Resources (IDNR) placed strict environmental requirements on the plant's emissions. These requirements necessitated the installation of several state-of-the-art emission reduction technologies in the plant:

- Special scrubber for the granulator exhaust to reduce the urea dust emission (particulate matter <0.1 kg/t or 0.2 lb/ST urea)

- A catalytic De-N_2O reactor in the nitric acid plant tail gas to reduce greenhouse gas emissions (98% N_2O reduction)
- A Selective Catalytic Reduction (SCR) unit in the primary reformer convection section to reduce flue gas NO_x emissions (< 9 ppmv)

An additional challenge was presented by the process CO_2 vent. Methanol is a byproduct produced from side reactions in the front-end of ammonia plants, mainly across the LTS catalyst. While most of the methanol condenses with the process condensate, some of the methanol passes through to the CO_2 Removal System where it eventually comes out with the CO_2 product and atmospheric vent.

Methanol has been classified as a Hazardous Air Pollutant (HAP) in the US since the 1990 Clean Air Act Amendments and is controlled under both federal and state regulations. When OCI began the permitting steps for the new complex in Iowa, KBR was consulted for the best approach to achieve the permissible level of methanol emission from the CO_2 vent. This was a challenge that KBR had long been prepared for.

End-of-pipe technologies such as catalytic oxidation are available but involve additional equipment, expensive catalyst and energy penalties for heating and pressure drop. Alternative source control measures such as chilling the process stream also require additional equipment and extra power for refrigeration.

The IFCo ammonia plant represents the first application of the KBR Shift Effluent Scrubber system for controlling methanol emissions. Let's take a closer look at why such an economical system is needed and how it works.

Figure 1. IFCo Fertilizer Complex in Wever, Iowa

USA Regulation of Ammonia Plant Methanol Emissions

Before 1990, methanol was typically regulated in the USA as a non-specific volatile organic compound (VOC). Its source, presence or concentration levels in ammonia plant CO_2 vents were generally not well known by regulators unless a state had a regulatory framework or ambient air quality standards for hazardous or toxic chemicals. After 1990 this all changed with federal HAP regulations specifying that either 10 ST/year (9.07 MT/year) of a single HAP or 25 ST/year (22.7 MT/year) of all HAPs combined at a facility triggered designation as a major source, potentially requiring the use of Maximum Achievable Control Technology (MACT) to limit HAP emissions. Such a major source must apply MACT as defined per its source categories or subcategory developed by the US Environmental Protection Agency (EPA).

While the EPA has developed dozens of MACT standards and limits across many industrial sectors, they have not directly addressed CO_2 vents from steam-methane reforming (SMR) systems. Thus regulatory authorities are faced with the need to make case-by-case MACT determinations unless a state's regulatory framework otherwise covers the methanol emissions, e.g., Louisiana's Toxic Air Pollutant (TAP) program.

Since there had been no recent new USA ammonia plants and only a few hydrogen plants built for refineries prior to the IFCo project, these determinations had not yet been developed for permitting SMR systems as a category. However, the recent surge in domestic SMR-based projects will likely trigger the EPA's attention to formalize limits and requirements under future HAP regulations.

It should be noted that MACT for HAPs is a far stricter standard than the Best Achievable Control Technology (BACT) standard for VOCs. MACT shall be at least as stringent as the emission control that is achieved in practice by the best controlled similar source and which reflects the maximum degree of reduction in emissions, regardless of cost. In other words, MACT is concerned with what is "achieved in practice" by a benchmark source in that source category, whether through end-of-the-pipe technological controls or by other means, such as source control. MACT for a given source may be stricter than what has been achieved in practice by the best controlled similar source, provided that increase in stringency is economically feasible and technologically achievable for that source.

If the major threshold triggers cannot be avoided for a HAP, then the permit applicants' MACT determination and its proposed limits are incorporated into a Federal or State Title V Operating Permit where conditions will be federally enforceable. There are a variety of options for permit conditions, but they must provide a means of measurable assurance for record-keeping and reporting.

For large ammonia plants with significant process CO_2 venting, even the use of newer LTS catalyst to minimize byproducts formation cannot limit methanol emissions below the trigger for MACT regulations. Additional steps to either control the emission at the source or limit the emission with end-of-pipe treatment are therefore required.

For the IFCo ammonia unit, the challenge was to streamline the permitting process by proactively reducing methanol emissions from the CO_2 vent below the threshold for triggering MACT regulations. KBR's recommendation for doing this in the most cost-effective manner possible was to apply untested technology developed by KBR 15 years earlier. The successful application of this technology would allow the plant to meet the following emission requirement specified in the IDNR permit:

- CO_2 vent, methanol < 0.019 lb/ST (<0.0095 kg/MT) ammonia product

The Shift Effluent Scrubber

A new source control technology was developed and patented by KBR in 2000 for removing methanol upstream the CO2 absorber [1]. The technology required no new equipment but simply increased the size of existing equipment along with some modification of the equipment operating conditions for minimal cost impact. Energy penalty was also small compared to other methods. A bit of history provides context to its development.

In 1997 KBR had anticipated future permitting difficulties with methanol emissions from the CO2 vent and studied available technologies for reducing emissions below the 10 ST/year limit required to avoid classification as a major source falling under MACT regulations. Source control scenarios using cooling and chilling of the CO2 Absorber feed to drive more methanol into the process condensate were examined as well as end-of-pipe catalytic oxidation upstream the CO2 vent. These methods were effective but generally included a considerable cost penalty for the new equipment as well as an energy penalty for additional pressure drop, refrigeration power or lost heat.

In a conventional flowsheet, process condensate is separated from the raw synthesis gas in the Raw Gas Separator and treated with steam stripping in the High Pressure Condensate Stripper (HPCS) before sending the stripped condensate to the off-sites water treatment facilities (Figure 2). The raw synthesis gas entering the CO2 Absorber contains methanol at equilibrium with the process condensate, and the condensate may have up to 1000 ppmw methanol. Process steam is used for stripping in the HPCS, so methanol stripped out of the condensate is recycled and disassociated in the Primary Reformer. Methanol in the stripped condensate is typically reduced to around 30 ppmw.

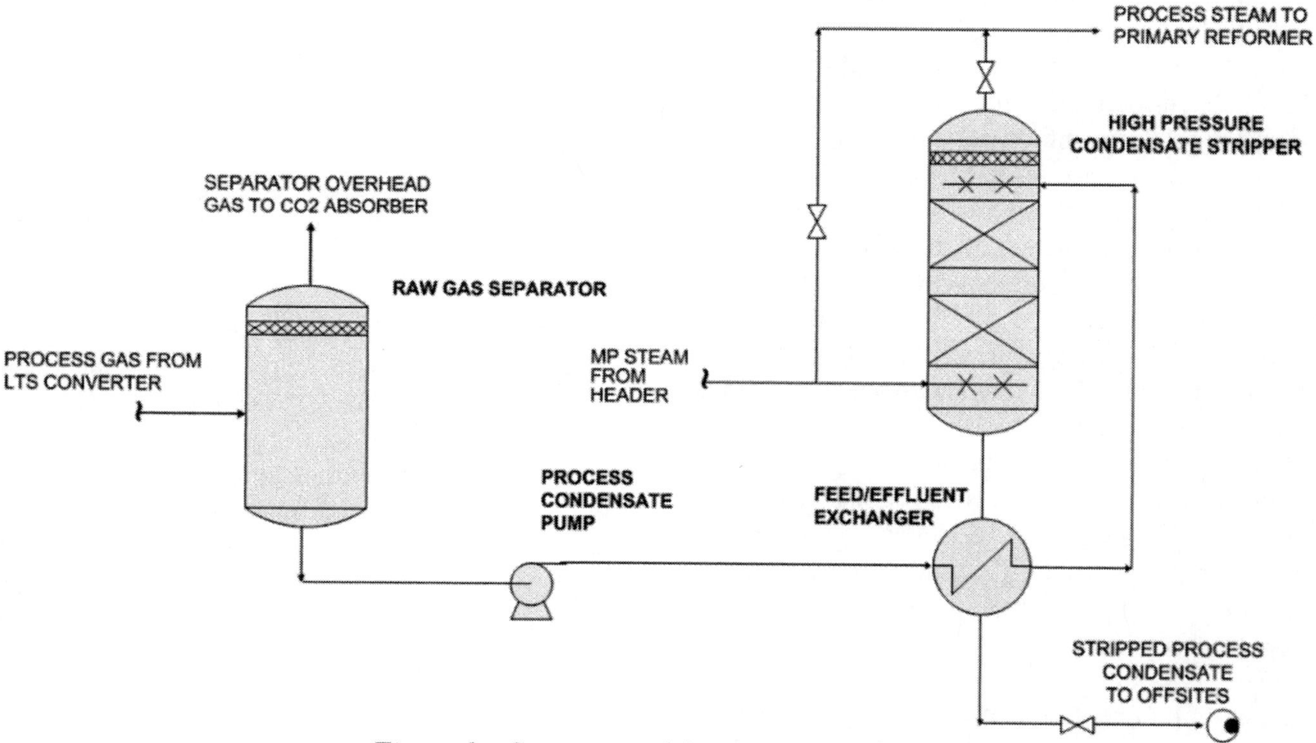

Figure 2. Conventional Separator Flowsheet

A new approach for reducing methanol emissions in the CO2 vent was developed by KBR using water scrubbing of the CO2 Absorber feed to capture most of the methanol into the process condensate. In this system, the Raw Gas Separator drum immediately upstream the CO2 Absorber is "stretched" to include a packed bed and associated internals (Figure 3). A portion of the stripped process condensate from the HPCS is recycled to the converted scrubbing column and used to wash methanol out of the raw synthesis gas. Exiting the packed bed, the used scrubbing water with absorbed methanol combines with the process condensate and is sent back to the HPCS, completing the scrubbing/stripping circuit. Methanol and other VOCs are then stripped out and disassociated in the Primary Reformer.

If the packed bed provides a sufficient number of stages, gas exiting the scrubbing column will be near equilibrium with the stripped condensate. Simulations indicated that with a reasonable column design and stripped condensate purity, methanol carried to the CO2 Absorber could be reduced by over 90%, reducing the emission in the CO2 vent below the MACT trigger. The "Shift Effluent Scrubber" was born.

The energy penalty is very low since the additional packed bed has a low pressure drop. Flow for the condensate pump feeding the HPCS increases, but this is a low-head pump so the power penalty is also small. The cost penalty is also much lower than with other scenarios examined since no new equipment is required. The separator grows in height when converted to a scrubbing column and the Process Condensate Pump, HPCS Feed-Effluent exchanger and HPCS all increase in size due to the condensate flow increase of about 40%. Increasing the size of existing equipment is generally cost effective compared to adding new equipment.

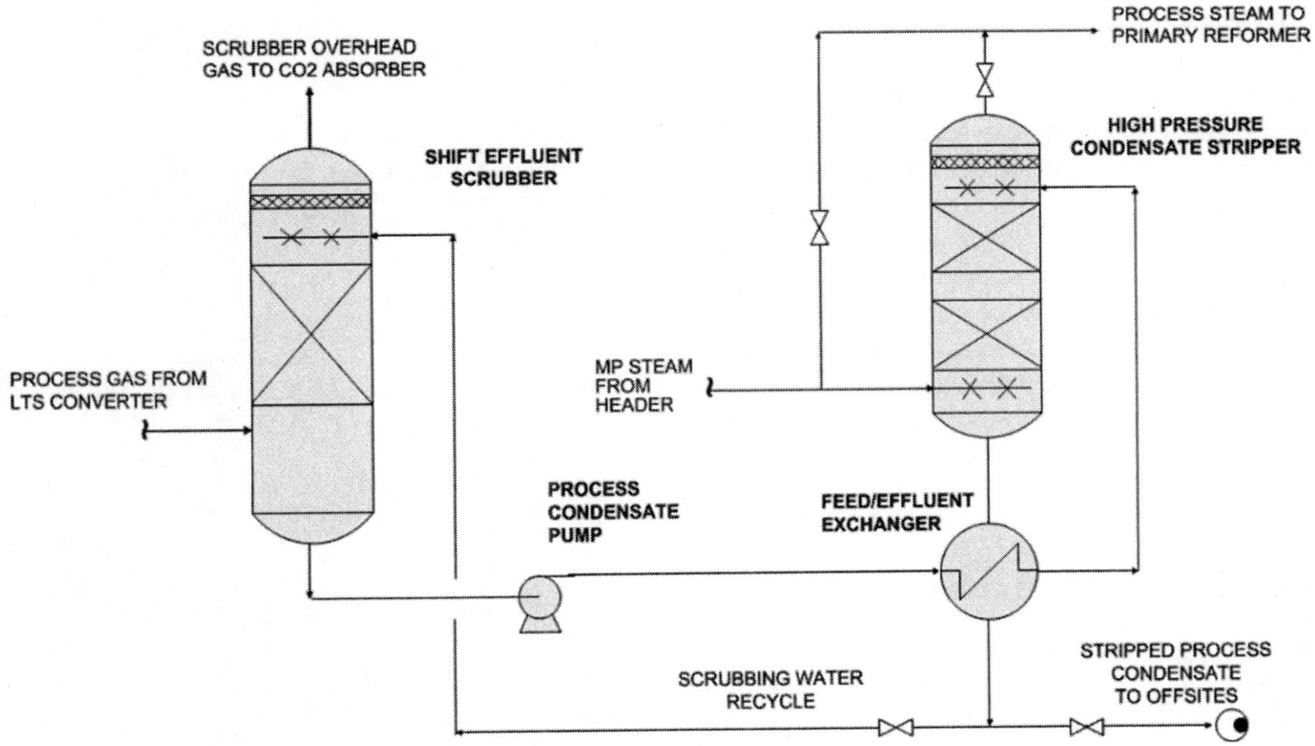

Figure 3. Shift Effluent Scrubber Flowsheet

The new scrubber system is relatively benign for operators and maintenance personnel. There are no new catalyst systems to monitor. There are no additional pollutants to process in the plant off-sites. Additional captured methanol is decomposed in the primary reformer. Compared to operation with a conventional separator drum, the only added task is setting the scrubbing water flow to the new column.

The Shift Effluent Scrubber system is KBR proprietary technology detailed under US Patent 6,015,450A. However, prior to the recent surge in domestic projects, there had been little interest in such emission control systems for ammonia plants as most new facilities and revamp projects since the late 1990's had been installed overseas. The system had therefore not yet been demonstrated commercially when offered for the IFCo ammonia plant.

Simulations of the scrubber column were key to OCI and KBR's confidence in the new system. The KBR proprietary property package used to simulate the system had been developed and tuned over decades using vapor-liquid equilibria (VLE) data from raw gas separators and HPCS systems operating in scores of ammonia units provided by KBR. Trace components such as methanol and ammonia being modeled in the new water scrubber were the same as those in the HPCS, so the new aqueous system could draw on KBR's wealth of experience in modeling HPCS systems. With such a solid technical foundation, the risk from being untested was considered low and OCI decided to incorporate the unit in the IFCo plant (Figure 4).

Scrubber Performance

In order to comply with the emissions permit, the plant would need to demonstrate during initial operation that the methanol concentration in the CO2 vent stack did not exceed the limit specified in the IDNR permit. An independent lab qualified for such testing would be required. IFCo employed ARI Environmental Inc. (ARI) to perform the first test which was conducted on Sept. 14, 2017. Plant rate was about 99% of design. The test result for the CO2 vent was as follows:

- Methanol = 2.97 x 10^{-6} lb/scf (dry) (50.3 mg/Nm3 dry) vent flow
- Methanol = 0.033 lb/ST (0.0165 kg/MT) ammonia product

The test result was at first disappointing until it was discovered the flow of stripped scrubbing water had not been tuned for operation at full capacity. A sample of the stripped process condensate (scrubbing water) taken during the test also showed the concentration of methanol was slightly high, indicating the flow of stripping steam to the HPCS was slightly low during the test.

Figure 4. Item 120-D Shift Effluent Scrubber, dwarfed by the CO2 Removal System columns

Note that other tests were being performed on the vent stack at the same time as the test for methanol, so the new scrubbing column was not yet the primary focus.

A second test was quickly scheduled and performed by ARI on Nov. 29, 2017 at a plant rate of about 106% of design. Prior to the second test, both the scrubbing water flow to the scrubbing column and the stripping steam flow to the stripping column (HPCS) were slightly increased The second test result was as follows:

- Methanol = 0.046 x 10^{-6} lb/scf (dry) (0.78 mg/Nm3 dry) vent flow
- Methanol = 0.0005 lb/ST (0.00025 kg/MT) ammonia product

During both vent stack tests, the urea unit was online and about half of the process CO2 was actually being vented. The reported values per ton ammonia product include this benefit in lowering the emission. Therefore, if all of the CO2 was being vented, the reported emissions would have been about double the above values. However, for the second test, this would still only be about 5% of the limit in the IDNR permit.

The second test result clearly demonstrated the new scrubber system would allow the plant to meet its emission target of 0.019 lb/ST (0.0095 kg/MT) ammonia product and in fact showed that during the test the column was being operated more conservatively than required. Additional testing has since been performed to better define optimum scrubber operation.

Operating stability of the Shift Effluent Scrubber has been very good during the first year of operation. There have been no instances of column flooding, foaming or changes in pressure drop. Should the column ever experience flooding and carry-over, the main impact would be a slight dilution of the circulating solvent in the CO2 removal system. In such a case, make-up water to the system could be reduced for a short time to restore the original concentration.

Conclusions

Incorporating first-of-a-kind technology into either a grassroots or revamp project is a difficult decision which must weigh the potential benefits against a long history in the industry of unwelcome surprises with new concepts. OCI and KBR candidly discussed these issues prior to incorporating the new design into the IFCo ammonia unit. OCI's confidence in the technology and decision to use it has been rewarded with a new emission control system with minimal energy penalty that comfortably meets the IDNR permit requirement.

The Shift Effluent Scrubber system has performed very well during the first year of operation, allowing IFCo to easily pass the emission test for methanol in the CO2 vent stack. To date there have been no operational problems or upsets associated with the new system. OCI and KBR are both pleased with the unit.

With successful operation now demonstrated, the KBR Shift Effluent Scrubber system can be presented in permitting applications as a proven MACT for limiting methanol emissions from the CO2 vent stack in new and retrofit plants. The system may also allow a plant to stay under the EPA trigger for a MACT determination, simplifying the permitting process. The system has sufficient design flexibility to accommodate even larger capacities being explored for the next generation of ammonia plants. The KBR Shift Effluent Scrubber is the model MACT technology for Ammonia Plant CO2 Vent methanol HAP control.

Acknowledgment

OCI and KBR are grateful to the personnel at the IFCo Fertilizer Complex for their assistance in collecting and analyzing plant data and sample sets for evaluating the performance of this new technology.

References

1. Joshi, G.C. and Noe, S.A., "Reducing Methanol Emissions from a Syngas Unit", U.S. Patent 6015450A, issued 18 January 2000

Failure of the HP Boiler Feed Water Pump on an Ammonia Plant

A catastrophic failure occurred on the main High-Pressure Boiler Feed Water Pump Train on an Ammonia Plant. The direct cause of the failure was flow reversal from the Boiler Feed Water system and/or the Steam Drum, to the turbine driven High Pressure Boiler Feed Water Pump. While the failure of the pump is the focus of the paper, the identified causal factors involve failures of several auxiliary operating systems, which are also discussed. The paper presents the sequence of events and associated root causes that resulted in the failure of the pump as well as recommendations to prevent a re-occurrence.

Godfrey Harrylal, Shiva Maharaj
Industrial Plant Services Limited, Trinidad and Tobago, W.I.

Introduction

The high-pressure boiler feed water pump train of an ammonia plant experienced a failure on 25th September 2015. The failure occurred in a 1,850 MTPD (2,040 STPD) ammonia plant. This paper describes the failure that occurred, the investigations and the lessons learnt.

System Configuration

As shown in Figure 1, boiler feed water (BFW), leaving the deaerator, is directed to the steam drum for the purpose of high pressure (HP) steam generation using two (2) identical boiler feed water pumps. The main HP BFW pump is driven by an API 611 condensing turbine which utilizes medium pressure (MP) steam, and the standby HP BFW pump is driven by an electric motor, which operates at the 4,160V level.

The HP BFW pumps are API 610 centrifugal multi-stage, opposed impeller, horizontally split volute pumps. Automatic recirculation (ARC) valves in the discharge piping of the pumps are designed to prevent liquid vaporization in the casing and pump damage, by ensuring positive flow. The ARC valves on each pump also act as check valves to prevent reverse flow. The pumps are also furnished with minimum flow lines from the ARC valves that converge to a single line back to the deaerator.

The turbine utilizes a mechanical hydraulic speed governor for speed control. The governor controls turbine speed by adjusting the quantity of MP steam supplied to the turbine. The governor has an output between 0 – 100% representing minimum (3,026 rpm) to maximum (3,728 rpm) governor speed settings. The DCS boiler feed water flow controller, acts in two ways. It controls the speed set point of the turbine and also the output to the control valve on the discharge of the standby pump.

Deaerated boiler feed water from the BFW pumps is fed to the steam drum via the three paths each with a control valve and a tilting disc check valve. Two of the paths include BFW pre-heating and steam generating shell and tube heat exchangers.

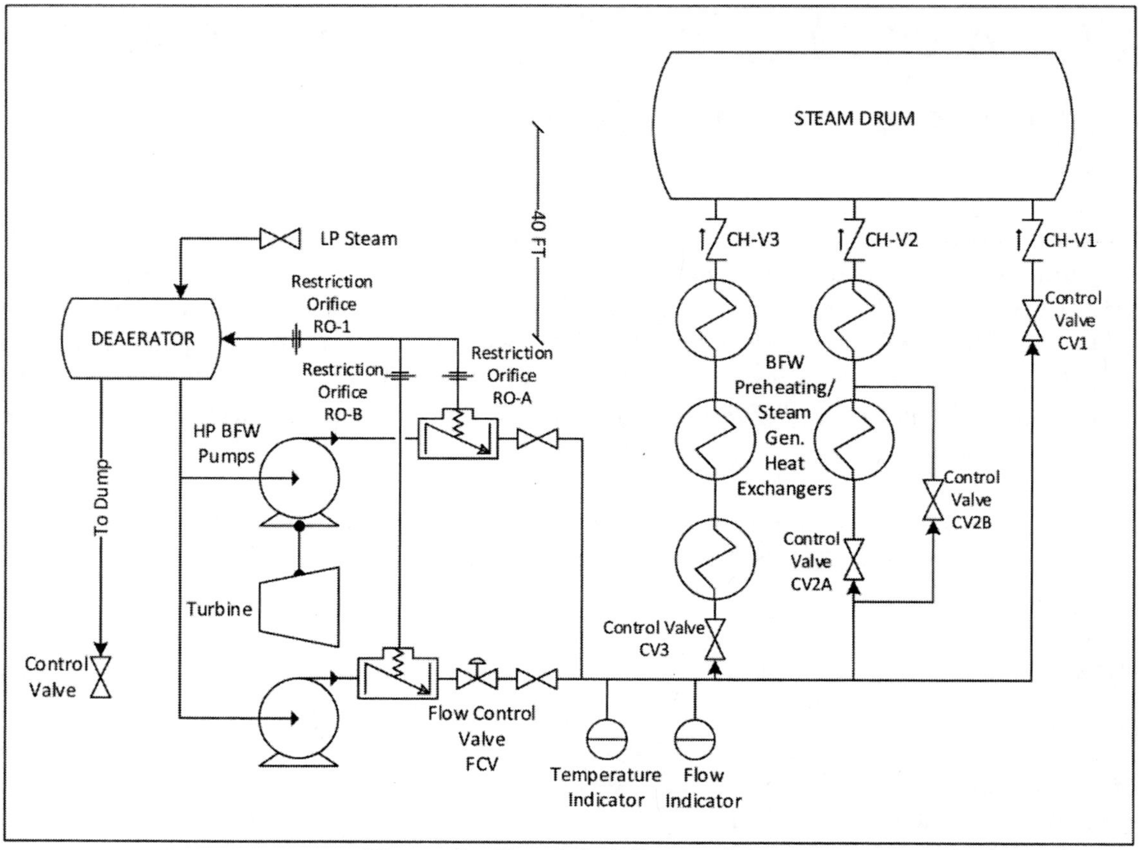

Figure 1. Simplified System Diagram

Incident Description

On 25th September 2015, the plant experienced a spurious activation of the process air compressor shutdown signal resulting in the synthesis gas compressor and downstream steam generating equipment to shut down as per the plant's design interlock logic. Despite the shutdown signal that initiated these events, the process air compressor remained online which resulted in a deficit MP steam mass balance since the process air compressor is one of the largest consumers of MP steam.

The ensuing steam system upset resulted in the following sequence of events over a 40 minute period: a high steam drum level, low BFW flow to the steam drum, the automatic start of the standby HP BFW pump on low flow, no BFW flow to the steam drum and eventually low steam drum level. These events culminated in the activation of the steam drum low level interlock which initiated a shutdown of the primary reformer and downstream systems (total plant trip) as per design.

With the main process systems offline a plant operator was dispatched to initiate shutdown of the main and standby BFW pumps which were still online. The operator utilized the mechanical trip lever on the turbine driver for the main HP BFW pump and initially observed a speed reduction. Within seconds, the operator heard an accelerating sound and observed the turbine speed to be increasing past 5200 rpm (well over the maximum of 3728 rpm). An instruction was given by the shift supervisor to evacuate the area immediately and within seconds the catastrophic failure of the main BFW pump and turbine unit occurred. Thankfully, only minor injury was sustained by an operator which was treated by on site first aid. Figure 2 shows the resulting damage.

Figure 2. A – BFW pump & turbine view from pump discharge side; B – Dispersion of components; C – BFW pump turbine damage; D – BFW pump shaft damage

Actions Before the BFW Pump Failure

During the steam upset with the trip of the synthesis gas compressor train, the steam drum level was rising. The plant operators attempted to control the steam drum level by reducing the BFW flow from the main pump. To reduce the BFW flow, the speed of the BFW turbine was reduced by using the governor control to bring the turbine speed to minimum governor. With the level still rising, the flow through the steam inlet valve for the turbine driver of the main BFW pump was reduced by a plant operator partially closing the valve. This action only momentarily dropped the main pump's speed from 3500rpm to 1900rpm. The pump's speed returned to 3500rpm while the closing of the valve was still in progress. It is believed that the turbine's mechanical hydraulic governor compensated for the reduction in steam flow by opening to allow more flow as evidenced by a return to 3500rpm. The corresponding drop in discharge flow also had the unintended side effect of auto-starting the standby pump against its closed discharge control valve. The steam inlet valve was then reopened from the partially closed position.

As the steam system was now being managed, the steam drum level began to fall. The plant operators at this time did not observe any flow being indicated on the outlet of the pumps. Without isolating the main pump, the discharge control valve on the standby pump was opened fully, but still no flow was observed. The steam drum level continued to fall until the eventual plant trip due to low-low steam drum level.

Findings

Inspection Findings

The steam turbine was a total wreck with the turbine casing disintegrated and the rotor components dispersed across the site. The damage noted on the pump skid and foundation indicated clockwise rotation from the drive end of the train, indicating that the machine was experiencing reverse rotation as it usually operates in a counter-clockwise direction from this frame of reference. This damage, combined with the dispersion of parts shown in Figure 2, indicated reverse rotation of the unit. Additionally, more evidence of reverse rotation was observed upon inspection of the pump's internal components. The shaft was bent toward the non-drive end as shown in Figure 2D with significant deterioration to the casing wear rings, inter stage rings, impeller rings, impeller and bushings at the discharge end of the pump, some of which is shown in Figure 3A.

Inspection of the ARC valve on the main pump revealed the presence of debris lodged between and around the main check valve and lower body seating surfaces, likely preventing the check valve from closing fully, shown in Figures 3B & 3C. The vortex plug on the recirculating/bypass end of the ARC was also found to be stuck in the valve body, possibly restricting bypass flow. The debris was confirmed to be a section of the pump's casing inter-stage ring. The pump suction strainer, shown in Figure 3D, was found to be distorted in a collapsed inverted condition which suggests flow reversal and was found to be fouled on both the upstream and downstream sides with the pump rotor's wear ring material also found in the ARC valve.

Of the three check valves on the lines to the Steam Drum, two (CH-V1 & CH-V3) were found stuck in the open position with debris lodged in the seating area – shown in Figures 3E & 3F. The debris that was found in one of the valves was identified to be SMAW welding electrodes. On the other valve, twisted tape turbulators from within the tubes of a BFW pre-heater/steam generator were noted between the disc and seating area.

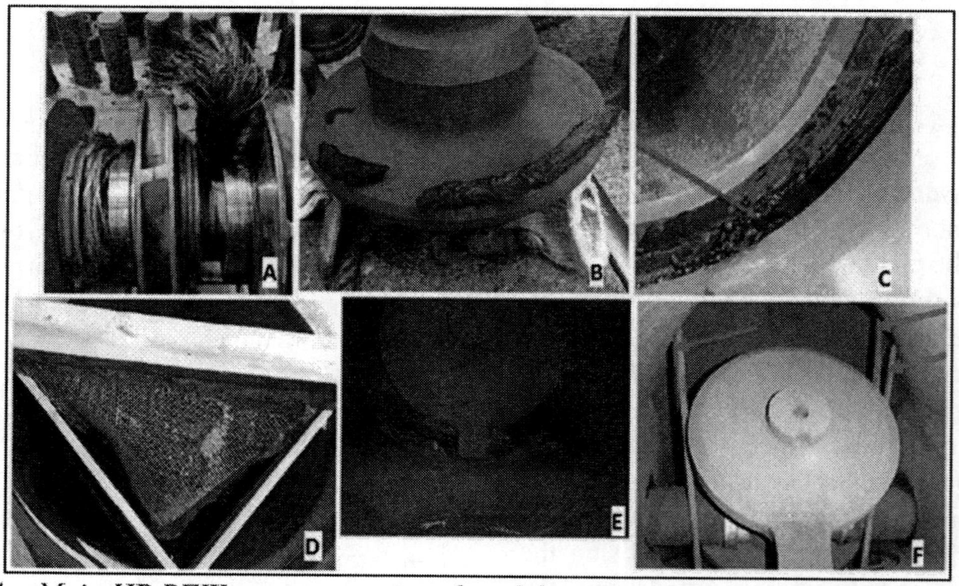

Figure 3. A – Main HP BFW pump wear rings found damaged; B & C – Wear ring debris lodged in main BFW pump ARC check valve seat; D – Main HP BFW pump strainer found fouled on both sides; E & F Check valves found open

Discussion

Analysis of the Sequence of Events

The trip signal generated by the air compressor and resulting steam upset led to the shutdown of the synthesis gas compressor. The output of the HP BFW pump began falling automatically to match the steam generation by the plant's logic, causing the steam drum level to begin falling. In an effort to manage the steam drum level, the combined HP BFW flow controller was placed in manual, and the control valves to the steam drum were operated in an effort to maintain a suitable level in the drum. This had the desired effect of causing the steam drum level to begin rising.

The steam drum eventually reached 98%, at which point the combined HP BFW flow controller was reduced to 0%, causing the turbine driven pump to reduce to minimum governor speed. In an effort to further reduce the output of the pump, an attempt was made to partially close the HP BFW pump turbine's steam inlet valve, causing only a momentary drop in speed and flow. As a result, the standby HP BFW pump automatically started. It is suspected at this instant, damage was done to the main pump's wear rings, causing fragments to lodge in the seat of the ARC valve.

The standby pump's flow controller was in manual at 0%. This gives the operator finite control of the BFW output such that when this pump starts automatically, there are no unwieldy swings in the BFW output flow. Based on this, there were three pathways available to the fluid being pumped – the common recirculation line, the common discharge line, and most importantly, because of its reduced output, back to the main pump. With both pumps running on the common recirculation line, which was designed for the flow of one pump (via RO-1), both pumps began to gradually deteriorate as the minimum flow requirements were not being met. Both pumps were also flowing to the common discharge, and with a compromised ARC valve on the main pump, the output of the standby pump is expected to have caused continuing damage to the main pump.

The HP BFW flow dropped slowly to 0 klb/h, and in an attempt to restore flow, the standby pump's discharge controller was set to 100%. This caused a momentary increase in flow, then a similar gradual decrease as this pump was also damaged. The steam drum level fell to the point of low-low alarm, tripping the plant. The turbine was then manually tripped, resulting in the reverse flow of a high energy BFW and steam two phase medium that caused the pump and turbine to rapidly increase in rotational velocity in the reverse direction. This reverse rotation likely caused the turbine discs to loosen, vibrate, rub and separate from the shaft and ultimately causing the disintegration of the turbine.

Figure 4 shows the eventual flow path from the steam drum through the pump and to the deaerator.

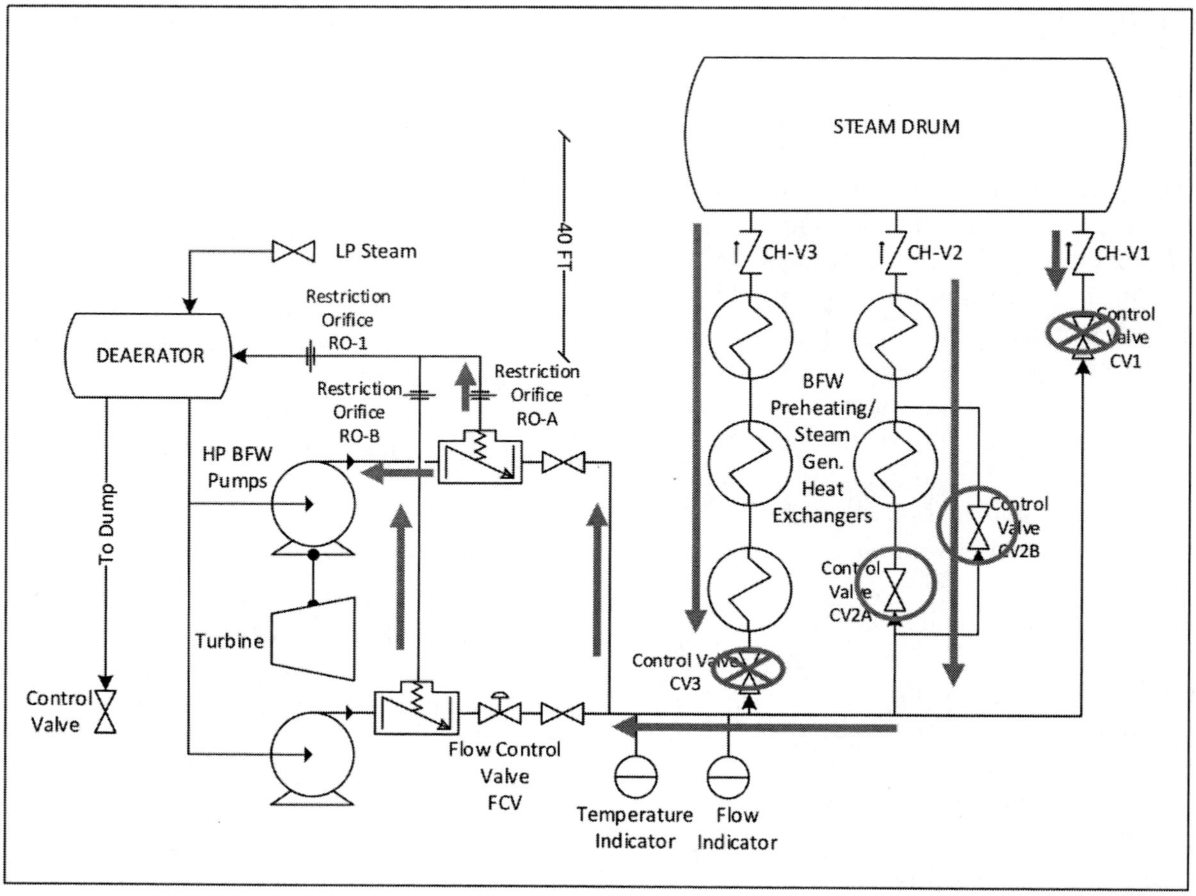

Figure 4. Reverse Flow

Failure of the ARC

Inspection of the ARC valve revealed debris trapped between the main check valve and lower body seating faces. Results from positive material identification checks of this debris were similar to that of the pump wear rings. The gap created was 7mm of the 48mm gap for the full valve travel. The equivalent cross sectional flow area of this 7 mm gap is estimated to be that of a 3" orifice which with the expected pressure drop in the reverse condition would allow for more than 400 klb/h flow.

Loss of Forward Flow

The throttling of the main BFW pump turbine's steam inlet valve resulted in reduced capacity of the pump, to the point where it could not deliver a forward flow into the header. From this point, the pump's output was directed to the deaerator via the recirculation line. In parallel, when the standby BFW pump automatically started with the discharge control valve (FCV) manually closed, the flow from this pump would also have been directed to the deaerator through the same recirculation line. The recirculation lines for both the main and standby pumps join and pass the flow through a restriction orifice, which was found to be sized for only a single pump's recirculation flow as confirmed by the plant designers after the incident. As a direct result of this, it is suspected that for a period of time, both pumps operated below the minimum required flow. In the post incident analysis, it was found that the current practice of the plant designers dictates separate recirculation lines for each pump.

It is known that subjecting a pump to low flow conditions will result in increased temperatures in the pump due to recirculation, opening of the clearances due to rubbing of the wear rings and therefore, degraded performance. With the installed thermoplastic casing wear rings of this pump, rubbing and damage is expected to occur more easily than for pumps with metal wear rings due to the reduced clearance.

Severe damage to the thermoplastic wear rings was noted in the inspection findings of the rotor. This type of wear ring allows for smaller pump clearances and greater efficiency in operation. However, notably, API 610 refers primarily to metal wear rings but does not explicitly prohibit the use of non-metallic wear components.

It is also known that, prior to the temporary partial closing of the turbine's steam inlet valve, the pump was online with no observed performance issues based on the boiler feed water flow rate, but, subsequent to this, the pump was unable to deliver any forward flow. Therefore, it is certain that a change in pump performance occurred over that period and it is likely due to the low flow conditions created by the temporary partial closing of the steam inlet valve and the restrictions created by only one recirculation line, when the bypass pump started.

It is notable that after this time, the discharge temperature indication for the BFW pumps did in fact reach its maximum saturation point. However, plant design does not call for an alarm for this condition, and it remained unrecognized until post incident analysis.

Reverse Rotation and Flow

Inspection findings and the observation of the turbine deceleration followed by reacceleration after being tripped suggest that reverse rotation of the BFW pump and turbine occurred.

The maximum speed that can be attained by the BFW pump with reverse flow of service fluid was calculated. The suction specific speed for the BFW pump was calculated from the normal operating point on the pump datasheet and was applied to the Hydraulic Institute's curve[1] for reverse runaway speed vs specific speed – shown in Figure 5.

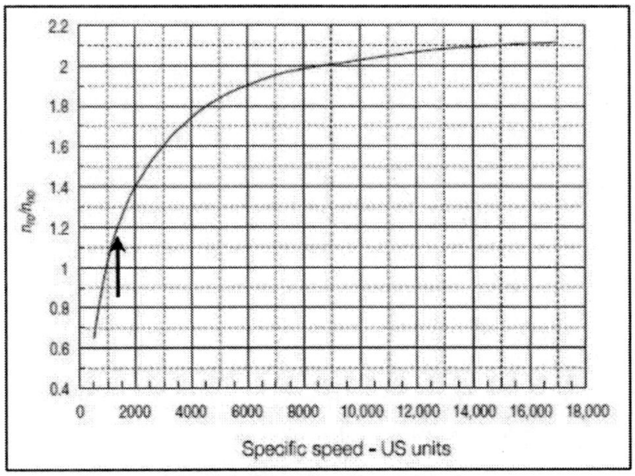

Figure 5. Hydraulic Institute's curve

A reverse runaway speed of 1.18 times normal operating speed or 4,095 rpm was estimated. A literature review for reverse runaway speeds in centrifugal pumps suggest that 1.25 times normal operating speed is a typical finding[2,3,4]. For the HP BFW pump, the maximum reverse runaway speed with service fluid (BFW) was therefore taken as 4,450 rpm.

The observation of the plant operator was that a speed of at least 5,200 rpm was achieved prior to the failure. This observation when compared to the maximum expected runaway speed of 4,450 rpm indicates that the medium involved in the reverse flow must have been not only BFW but a combination of steam and BFW which are available in the downstream steam generating heat exchangers and steam drum.

Additional evidence supporting reverse flow of BFW and steam from the downstream equipment include:

a. A mass balance of the deaerator in which a 107 klb/h additional input was noted. A rise in deaerator pressure was also observed to the point of transmitter saturation.

b. It was found that -4.2 klb/h BFW flow was recorded for the no flow conditions observed. Post incident field checks of the flow indicator reveal that the actual value transmitted was reproduced by simulating reverse flow and this value is in fact the lower saturation limit of the instrument.

c. Even though it is known that low flow conditions will cause increased pump temperatures, reverse flow was the more likely cause of the temperature excursion observed on the BFW transmitter. This is supported by the rate of the temperature increase as well as the location of the damage on the pump shaft.

It should be noted that the control valves on the lines with the two compromised check valves (CV1 & CV3) were closed at the time. However, the control valves have shut off classifications of ANSI Class II and III, which allow some leakage across the valves. As a result of this, the quantity of BFW/steam in reverse flow from the steam drum is not certain. Notwithstanding this fact, the volume of BFW/steam available in the steam generating heat exchangers is significant enough to reverse rotate the pump train to failure.[5]

Pump and Turbine Failure

The tripping of the turbine removed the last of the resistance to reverse rotation and allowed full reverse flow of the steam and BFW media from the downstream equipment and piping, back through the pump. This directly resulted in the rapid acceleration and unstoppable gross overspeed of the turbine in the reverse direction and consequent catastrophic failure.

Based on observations, both the pump and turbine appear to have failed completely at around the same time. However, the damage to the turbine was of greater magnitude. After considering several possible scenarios, the likely explanation is that the machine speed in reverse was high enough for the turbine discs to become loose on the shaft.

Calculations show that this is possible at as low as 130% of the normal forward operating speed, which was certainly exceeded, as the failure occurred at approximately 146% of the operating speed. Operating with loose discs would lead to significant and unpredictable vibration due to rubbing between the rotor and the stators, completely freeing the discs, leading to the turbine failure. Once the turbine failed, the pump shaft would be left running in reverse at this speed with an overhung coupling that would likely quickly bend the shaft and cause the pump to seize.

Examination of the pump design documentation shows that design for reverse rotation was considered. However, the recommended factor did not consider the specific pump configuration of a double impeller suction as well as actual suction and discharge pressures. Despite this, reverse flow design does not consider the conditions or the two phase medium that this pump was exposed to. It is also notable that the option on the turbine datasheet to indicate that the machine was designed for reverse rotation was not selected. It is however not a normal practice to select a turbine designed for reverse rotation. Additionally, the coupling was not designed to fail in case of reverse rotation.

Analysis and Cause

Loss of forward flow and failure of the ARC valve allowed a reversal of a high energy steam/BFW flow and the subsequent failure of the pump and turbine. The loss of forward flow was caused by the temporary closure of the turbine's steam inlet valve and it was exacerbated by the failed steam drum inlet check valves. Testing of the ARC was not historically done unless issues were identified. The root causes of this incident were determined to be:

1. Inadequate understanding of the potential consequences of reverse flow in the Process Hazard Analysis.
2. Inadequate safeguards to prevent reverse flow in the HP BFW system.
3. Inadequate management of the Steam Drum level.

1. Inadequate understanding of the potential consequences of reverse flow in the Process Hazard Analysis.

The design of the HP boiler feed water system relies on a single engineering control, the ARC valve, to safeguard against reverse flow through the pump. The check valves at the inlets to the steam drum cannot be considered as a second dissimilar check valve in series due to the inventory of BFW and steam available between the ARC and these valves.

The PHA performed at the detailed engineering phase for this plant identified check valve failure as a cause of reverse flow but failed to recognize the potential consequence and its severity; as the magnitude of the resulting reverse flow was not evaluated. As a result, additional safeguards were not identified.

The 2015 revalidation PHA addressed the scenario of a failed check valve and reverse pump rotation, but only for the case where the pump is started while rotating in the reverse direction. It did not recognize that reverse rotation on its own could have resulted in the failure of the turbine and was not identified as a consequence. Hence, additional safeguards were again not identified.

2. Inadequate safeguards to prevent reverse flow in the HP BFW system.

Adjusting of the steam inlet block valve should not have been performed as it is against operating procedure to safeguard the boiler feed water pump. Review of the SOP shows that the procedure was also reliant on the functionality of the single safeguard, the ARC valve as a non-return valve.

The minimum governor speed is set at the lower safe operating limit for the pump based on a minimum flow. As such, the only method for reducing the speed lower than this is by partially closing the turbine's steam inlet valve. However, closing the inlet steam valve decreases the pumps ability to achieve minimum flow and can result in pump failure. The use of the turbine's steam inlet valve to control the output of the main BFW pump below the minimum governor speed was not warned against in the SOP. This oversight was also subsequently clarified.

3. Inadequate management of the Steam Drum level.

The SOP guiding actions following a trip of the air compressor identifies specific actions for control of steam drum level. Further procedure review identified additional actions that would have allowed effective management of the steam drum level. However, none of these were documented in the SOP. It was concluded that the SOP could have provided enhanced guidance for this emergency situation. The SOP was subsequently amended to include the additional steps.

Recommendations

To prevent a reoccurrence, the following recommendations were made:

1) Review and amend the PHA of the HP BFW System:

   Although the PHA did identify the possibility of reverse flow, the consequence and severity were underestimated. Because of the inventory between the discharge of the pumps and the steam drum inlet check valves, the severity is elevated, and the consequence of turbine failure was not considered.

   The PHA was reviewed, and it was recommended that separate recirculation lines as well as automatic shut-off valves be installed.

2) Amend SOP's:

   Closing the pump discharge prior to shutting down the turbine or motor is now included in the SOP. This will prevent flow from one pump to the other through the common discharge line. The closing of these valves will be done manually until the automatic shut-off control valves are installed.

   Clarification on the actions the operator is allowed to take with the steam turbine inlet valve was included in the SOP. This information will prevent attempts to use the turbine's steam inlet valve to lower the machine's speed below the speed required for the pump's minimum flow specification.

   Emergency procedures have been amended to include all identified methods of managing the steam drum level. This information will allow informed, safe actions that can be executed based on current plant conditions.

3) Implementation of automatic shut-off valves:

   Automatic shut-off valves are to be installed on the pump discharge. These valves, when activated, will immediately close on low pump discharge pressure. After pump start-up, these valves will positively isolate the discharge of the pump should the discharge pressure fall below a predetermined value.

4) Installation of separate recirculation lines for each pump:

   A separate recirculation line is to be installed for each pump. This will allow both pumps to establish proper minimum flows when both are online at the same time. This can occur under normal circumstances such as plant startup (when the motor driven standby pump is used before steam can run the turbine), and when switching pumps to conduct routine PM's and other maintenance activities.

5) The ARC's and several other critical check valves have now been identified for periodic testing.

Conclusion

Consequence of failure of a high-pressure boiler feed water pump train can be significant in certain situations. Safeguards for existing systems and procedures governing pump operations should be evaluated in light of this incident.

References

1. Hydraulic Institute Standards, "ANSI/H1 2.4 - Installation, Operation and Maintenance-Vertical Pumps", 2000
2. Igor J. Karassik, "Centrifugal Pump Clinic 2nd Edition", 1989
3. John Davidson, Otto von Bertele, "Process Pump Selection- A Systems Approach", 2005
4. Robert Talbot, "Hazards of Reverse Pump Rotation", 27th Texas A&M Pump Symposium, 2011
5. D. P. Wallace, P. J. Nightingale and A. P. Walker "Failure of Boiler Feedwater Pump Turbine following Site Power Failure" Boston, MA, - AIChE Ammonia Safety Symposium, 1996.

Assessing Safety Culture in PT Pupuk Indonesia Group

In the last five years, accidents resulting in injuries to personnel and/or damage to properties or the environment have been occurring in PT Pupuk Indonesia's (PI) subsidiaries. The root cause of the accidents was mostly lack of awareness that led to unsafe action.

To minimize accidents, we focus on changing people's behavior. PI developed methods to assess safety culture within its subsidiaries based on DuPont's Safety Training Observation Program (STOP) and the Royal Dutch Shell's Hearts and Minds. In addition, PI adapted Parker and Hudson's model of the HSE Culture Ladder as a reference for safety culture level. The effort was led by a Safety Champion team that developed a safety culture questionnaire to measure safety perception and suggest an action plan to improve safety culture. The result found that only one Pupuk Indonesia's subsidiary, Petrokimia Gresik, had over 90% safety culture positive perception. The others have below 75% safety culture positive perception. The survey will be conducted annually and serves as a basis of safety continual improvement programs in Pupuk Indonesia Group.

Nendra Gunawan
PT Pupuk Indonesia (Persero)

Wildan Johardi
PT Pupuk Indonesia (Persero)

Edi Wibawa
PT Pupuk Indonesia (Persero)

Introduction

PT Pupuk Indonesia (Persero) (PI) is a government owned Indonesian fertilizer holding company which consists of 10 subsidiaries, namely; PT Pupuk Sriwidjaja Palembang, PT Pupuk Kalimantan Timur, PT Pupuk Kujang, PT Petrokimia Gresik and PT Pupuk Iskandar Muda which are fertilizer companies, PT Rekayasa Industry, an Engineering, Procurement, Construction and Commissioning company, PT Mega Eltra, a construction and trading company, PT Pupuk Indonesia Energy, a power generation company mainly focusing in power plants, PT Pupuk Indonesia Logistik, a shipping and logistic company, and PT Pupuk Indonesia Pangan which focuses mainly on distribution of food and agricultural product. Aligning with its vision to be the leader in the agro and petrochemical industry, PI places its employee's safety as the utmost priority in doing business. To ensure the Health, Safety and Environmental (HSE) management system is implemented properly in its subsidiaries, several efforts have been done including the following:

- Routine plant visits from PI board of directors in which one of the agenda items is to enforce the implementation of HSE management systems,

- Safety Champion teams, consisting of safety representatives and HSE Managers from PI subsidiaries, make regular visits to each company within PI group to assist with conducting Process Safety Management training.

However, during 2014 to 2017, PI had 12 major incidents causing lost time injuries (see Figure 1).

Figure 1. Major Incidents Occurred in PT Pupuk Indonesia

Our incident root cause analyses showed that most of the incidents were caused by lack of safety awareness. This fact raised a concern to management in how well personnel are aware for their own safety. In an effort to ensure personnel have the right attitude towards safety, management is looking to determine the current safety culture level in PI, especially in the fertilizer companies within PI holding.

Safety Culture

Safety culture is the attitude, beliefs, perceptions and values that employees share in relation to safety in the workplace. Safety culture needs to be a part of organizational culture and described as "The way we do things around here." Studies show that workplace related accidents are the result of a breakdown in organizational policies and procedures, and the breakdown occurs from negligence to safety issues. A good safety culture stems from senior managements' commitment to safety, realistic practices for handling hazards, continuous organizational learning, and care and concern for hazards shared across the workforce.

PT Pupuk Indonesia (PI) through its subsidiaries strives to ensure the safety of its employees through various efforts. PI encourages its subsidiaries to comply with government regulations and implement HSE management systems properly. Implementation may vary between subsidiaries, but each subsidiary must have a HSE Policy in place and it must be implemented. Each subsidiary shall also have the following: a hazard identification and risk assessment program, a set of HSE procedures implemented to ensure hazards and risk in the workplace are controlled and mitigated, and audits and inspections in place to ensure the effectiveness of the HSE management system implementation. Training is also required to be given to personnel and contractors to ensure competence and awareness with HSE management systems. Despite these efforts, PT Pupuk Indonesia still has accidents throughout its subsidiaries.

PT Pupuk Indonesia's safety culture maturity level is still in need of improvement to achieve a proactive level. The workforce needs to be encouraged to have the initiative in implementing HSE management systems. Monitoring and evaluation of implementation of HSE management systems needs to be used as a tool of continuous improvement for the organization.

Determine Safety Culture Level

In 2017, PI recorded a total of 1655 unsafe conditions throughout the company. PI also recorded 987 Health, Safety and Environmental (HSE) improvement recommendations. 595 (60%) of these recommendations are now closed.

HSE improvement recommendations are part of PI's effort to involve employees and contractors in improving HSE management systems implementation throughout its operations. Improvement recommendations is an ongoing program in increasing Safety Culture perception. It's a method of hazard reporting and leadership

In addition to the efforts described above, PI decided to develop a program to increase the level of safety awareness throughout the company and its subsidiaries. The programs that were implemented included the Safety Culture program and Process Safety Management (PSM). The new programs focus on the importance of behavioral aspects in safety, which has been an issue for quite a while. Figure 2 below shows that accidents are the result of unsafe behaviors, and the behavioral and cultural approach will effectively reduce the number of incidents and accidents that occur within an organization. Figure 2 also shows that the at-risk behavior is a recipe for incidents and accidents that could prove fatal to the employee.

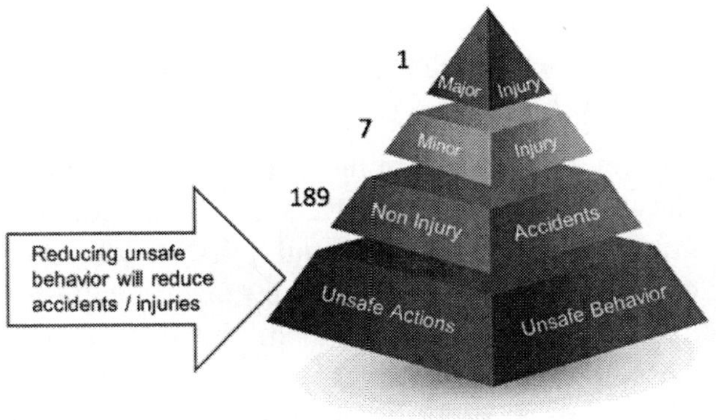

Figure 2. Accident Pyramid in PT Pupuk Indonesia (Persero)

Research suggests that we can reduce the number of accidents and incidents by reducing the number of unsafe actions. PI realized that a change was needed to have a positive safety culture throughout the company.

To effect change, the following steps were taken to understand the culture, identify the problem, and then select the right tool to fix the problem:

1. Assess the company's culture and determine the company's safety culture level.
2. Determine who needs to lead the safety culture change.
3. Identify specific problems in relation to the safety culture within the company.
4. Focus on leadership.
5. Focus on supervision of workers and contractors.

Safety Culture Maturity Model

Safety Culture Maturity Model (SCMM) is a tool to help an organization achieve steps in increasing their safety culture level. SCMM was developed to assist companies in the following:

1. Determine the safety culture level
2. Identify efforts needed to increase the safety culture level.

SCMM is a model used to determine safety culture level, but it is not used a tool to audit or assess the HSE Management System. The SCMM concept focuses on behavioral and cultural factors, which can only be done when the engineering aspect of the HSE Management System has been properly implemented. SCMM can also be utilized to measure safety culture improvement in organizations. The measurement results may help organizations to determine a basis for developing a suitable safety culture improvement program.

To effectively implement SCMM, an organization needs to comply with the following criteria:

1. HSE Management System has been effectively implemented.

2. Technical failure is not the root cause of the majority of accidents/injuries that occur.
3. Organization complies with HSE legal requirements.
4. HSE aspects are not just implemented to comply with government requirements.

Should an organization fail to comply with the above criteria, the organization should focus on improving its HSE Management System implementation effectiveness prior to improving its safety culture. A positive safety culture should have three key elements as follows:

1. Clear and accessible procedures, work instructions, and applicable rules to control and mitigate hazards in the workplace.
2. Positive attitude towards risk management and compliance to risk mitigation and control at the workplace.
3. Ability to learn from failures, accidents, near misses and safety performance indicators to ensure continuous improvement.

SCMM serves to measure perception towards the 3 elements above. As described by Fleming and Lardner, 2002[1] for the Keil Center, Safety Culture Maturity Model has the following five stages (shown in Figure 3):

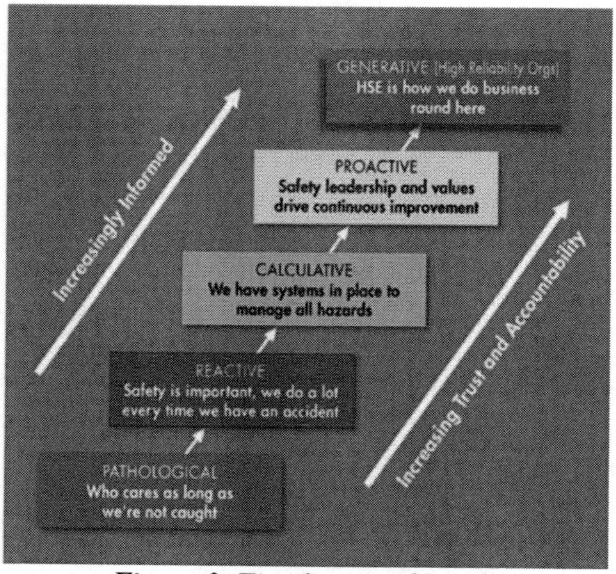

Figure 3. Five Stages of SCCM

1. **Pathological** - a condition where people don't care for their own safety and the safety management system is there only to comply with government regulations.
2. **Reactive** - a condition where safety is implemented only when an accident or injury occurs.
3. **Calculative** - a condition where HSE management system implementation only focuses to the system and numbers. Safety only exists as a key performance indicator. However, at this stage, personnel and management are starting to show concern to safety implementation.
4. **Proactive** - At this stage, steps to prevent accident and injuries are starting to show, and involvement of personnel and management is apparent in safety implementation.
5. **Generative** - At this stage, the organization has a high standard for safety implementation. Personnel within the organization know their role to ensure high safety standards are met. Personnel communicate safety concerns to management, and the organization acts on problems and gaps to continuously improve the safety implementation.

Figure 4 from Fleming and Lardner, 2002[1], shows that to implement a safety behavior program, an organization should first assess their cultural maturity level. Once the safety culture maturity level is determined, the organization needs to draw commitment from its management. Management commitment is needed to ensure resources are readily available for safety culture improvement implementation. Program implementation and development then can continue to the training stage to improve awareness and competencies of the workforce towards safety. In determining the baseline data for improving its safety culture maturity level, PI conducted a safety culture level measurement.

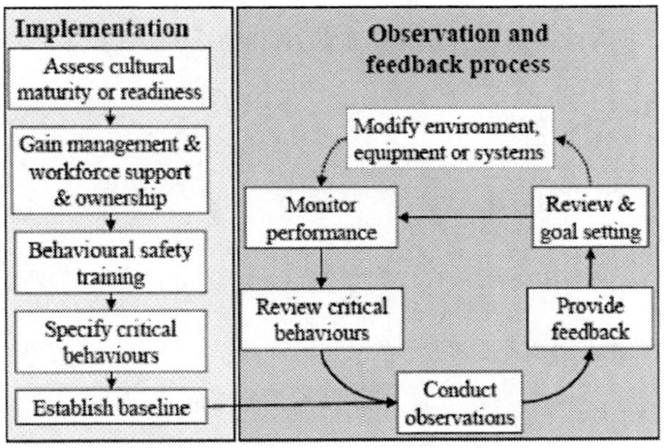

Figure 4. Implementation of Safety Behaivor Program

Safety Culture Maturity Level Measurement

A Safety Culture Level measurement is conducted with the following method.

1. Questionnaire
2. Focus group discussion with personnel of management level, supervisory level and operator level.
3. Behavior observation.

Data collected is analyzed using both a quantitative and a qualitative approach. The quantitative approach consists of primary data from the questionnaire filled out by participants. The qualitative approach uses secondary data from interviews, focus group discussion results, and analysis from behavior observations conducted.

Benefits and Limitations to Surveys

The survey questionnaire is designed to elicit workforce attitudes on key aspects of safety culture. Generally, the whole workforce is surveyed or sampled to represent the demography of the organization, thus allowing all the workforce views to be considered. The survey also reveals current attitudes and perceptions towards safety and safety management. A successful survey should clearly point to issues that need to be addressed.

There are limitations to the survey, as the survey may only identify some inadequacies of safety management systems. It also requires a high response rate to ensure results fully represent an organization's culture. Responses also represent workforce expectations to the organization, thus may lead to management losing credibility if responses are perceived as being inadequate. Further investigation may be required to interpret results.

Pupuk Indonesia Safety Culture Maturity

Demography

Safety culture measurement within PT Pupuk Indonesia group subsidiaries was conducted in 2014 and again in 2016 to 2017. Both measurements were intended to measure safety culture level within each subsidiary. The first measurement in 2014 was conducted by a third-party consultant to ensure the measurement had minimum bias and could capture the baseline of PI's safety culture. The second measurement was conducted internally to measure changes from the first measurement.

The sample for the questionnaire was random but with specific criteria based on the position of the sample within the organization. The measurement involved a sample from management, operation level employees, and contractors. The sample population was at least 10% of the total employees of each subsidiary, which consisted of top management, middle management, supervisory level and operator level. Criteria for the participants were as follow:

1. Role within the company
2. Department or Unit with higher working risk
3. Employment status

The purpose of the sample criteria was to map the level of safety culture within each level of the organization to enable PI to ensure more effective programs are implemented. The survey was based on the role of the personnel in the organization. Rather than having a program based on e.g. gender, age or education level, the role of the personnel in the organization reflects the organization's perceptions towards safety. The total sample was 1791 personnel from 5 companies within PI Holding. The demographics are shown in Figure 5.

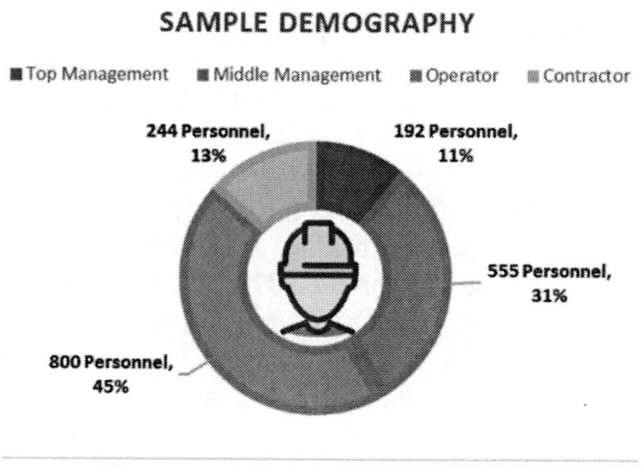

Figure 5. PI Sample Demography

PI's Safety Culture Perspective

Survey results taken in early 2017, shown in Figure 6, show that most of the PI personnel have reached a proactive stage. At this stage, steps to prevent accidents and injuries are starting to show and involvement of personnel and management is apparent in safety implementation.

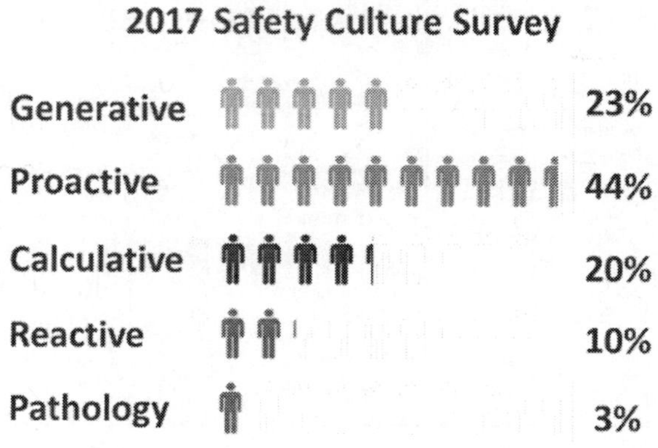

Figure 6. 2017 Survey Results

These results also show that participants have better safety perception than during previous survey in 2014, shown in Figure 7.

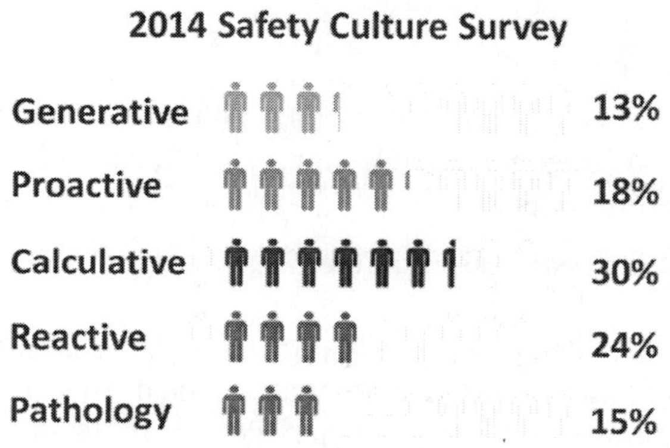

Figure 7. 2014 Survey Results

A slight shift in safety culture was observed, and perception also shifted from mostly calculative to mostly proactive shown in Figure 8.

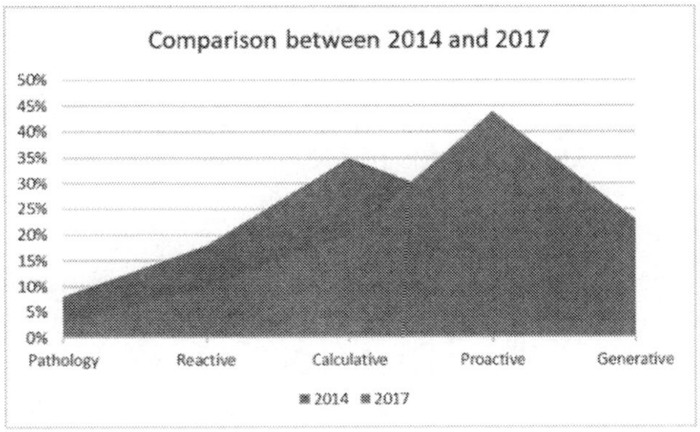

Figure 8. Comparison of 2014 to 2017 surveys

Even though PI has shown a positive change towards proactive safety culture, there are still improvements to be made to get to the next level. The survey results in Figure 9 show that the organization has implemented a proper HSE management system, the organization technically complies with HSE requirements, and it has a HSE management system in place. PI has shown a maturity in the calculative level. Variables show that continuous improvement is happening.

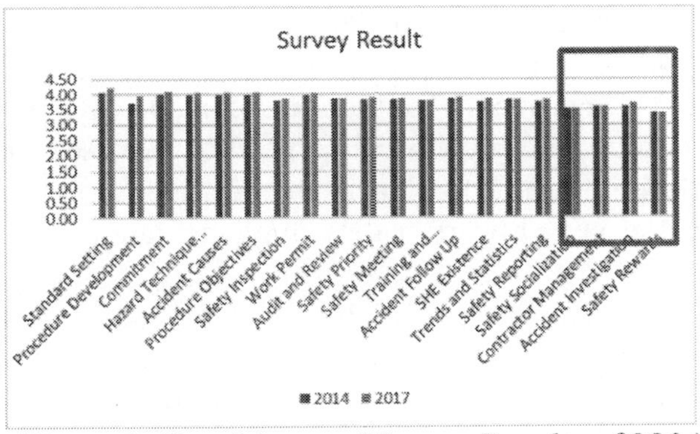

Figure 9. Breakdown of Survey Results of 2014 and 2017

Highlighted in Figure 9 are areas of improvement that still exist such as accident investigation, safety socialization/campaigns, safety rewards and contractor management. These variables suggest that HSE implementation is still based on compliance with the standard and government regulation.

As shown in the survey results, contractor management in PI's subsidiaries needed immediate attention. Contractor's personnel are exposed to more hazards compared to company employees. Contractors need to be treated as part of the company and need to be informed of the site HSE management system and hazard information as applicable to the respective contractor's scope.

The survey also showed that training and HSE existence needs immediate attention. HSE existence is a variable describing how far the organization incorporates HSE aspects to daily activities. A low HSE existence shows that HSE management implementation is dependent on HSE personnel or HSE department.

Safety Culture Perception in Top Management Level

The top management survey results show that, on average, top management has a good perception of safety. Compared to middle management and operators, top management has the highest score in the survey results, demonstrating that top management has a good knowledge of safety. PI has attempted to internalize safety as an organization's culture by having documented safety policies and procedures, by raising safety commitment and communicating safety to the workforce. Top management has a mature safety culture perception in the proactive level. This result is normal and suitable to their role as top management - HSE needs to start from the top management, cascade down to middle management, the operator level, and then to contractors. However, as described above, contractor management is an issue which top management needs to resolve - Contractor Safety Management System (CSMS) is still not consistently implemented throughout the organization.

Safety Culture Perception in Middle Management Level

Results of both surveys in 2014 and 2017 show middle management is at the calculative level, with only slight improvement. Both surveys show that middle management still focuses on compliance to HSE rules and regulations and compliance to procedures, which may be a result because it is their Key Performance Indicator. Middle management also shows good commitment to safety and shows a good attention to incident and injury prevention. However, improvements are still needed for training and competencies, accident follow up, reporting, safety socialization and contractor management.

Safety Culture Perception at Operator Level

Operator safety culture perception, on average, shows that safety culture is at the calculative stage. Changes in culture from 2014 to 2017 show that there is an increase in safety culture level, but the changes are not significant. Operator level shows a good attitude towards aspects in preventing accidents and incidents such as programs in engineering practices and safety management. The variables reflect attention towards technical and management aspects of safety management system implementation such as standard setting, work permit implementation, hazard technique management, commitment, procedure development, accident causes, safety priority, and audit and reviews. However, to ensure safety culture continues developing at operator level, there are still improvements to be made on variables such as safety training, safety socialization, contractor management and safety reward.

Safety Culture Survey Result

Assessment results show that in general, PI group is still in the calculative level. PI has a HSE management system in place and implemented properly. However, the HSE management system is still implemented as the organization's requirements, it is a part of personnel's job description, or it is being done to comply with procedures. HSE management system implementation is only being done as accident or injury prevention and has not been implemented as a means to continuous improvement.

Safety culture maturity level survey result reflects that the workforce has not incorporated safety as a value, and it is not reflected in daily activities. Safety is still enforced as part of complying with rules and regulations using discipline. Improving the value leadership can have a great influence on HSE implementation and HSE performance. However, leadership needs to be reflected throughout all levels of the organization.

Training and competency is an issue that needs immediate action as the training and competency perception among top management is not shared to middle management and to the operator. Top management has a perspective that the training and competencies have been correctly implemented throughout the entire organization. Yet, middle management and operator level perception is that training and competencies, especially with the soft skills in safety culture (e.g. safety leadership and safety communication) still need to be improved. Training and competencies regarding safety culture also need to be cascaded down to contractors to ensure they have the same perspective towards safety implementation.

The safety reward variable is the lowest scoring variable during both surveys at all levels, and the reward variable shows that the organization has not shown attention to safety culture motivation. Safety reward reflects satisfaction of the workforce with the level of HSE performance. Safety reward has not been regarded as a motivation for the workforce. An effective safety reward program may serve to maintain and even to increase workforce motivation towards HSE implementation.

The Contractor Management score shows signs of improvement is needed. The effort to improve contractor management serves to improve the mutual relationship between organization and its contractors to ensure a more effective safety management and safety communication. Contractors need to be more involved in the implementation of the entire HSE management system.

The quality of the accident investigation process, as well as communication of lessons learned, need to be improved to ensure effective incident/injury prevention.

Incident Rate and Relation to Safety Culture

Incident rates showed a significant decrease between 2014 and 2016. However, there was a significant increase in lost time incident frequency rate in 2017, as shown in Figure 10.

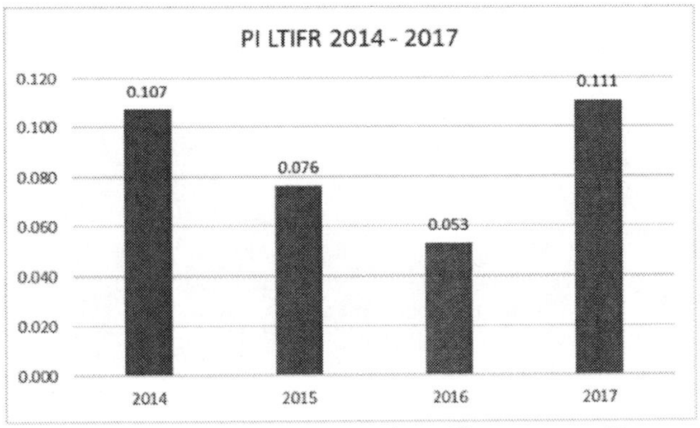

Figure 10. Lost Time Incident Frequency Rate (LTIFR)

There were 3 loss time incidents recorded during 2017, with 2 out of 3 incidents resulting in injury to PI contractor personnel.

Thirty accidents resulted in recorded injury to personnel, of which 60% or 18 injuries were PI contractor personnel. 80% of the injuries were caused by unsafe acts or unsafe behavior. The incidents and injuries reflect a safety culture where contractor management has not been effective at preventing incidents and injury. Results of the incidents highlight the fact that the contractor management variable needs immediate improvement. Additional focus is needed on improving the competency of contractor personnel at implementing safety practices. Short term workforce such as contractors are not as familiar to safety implementation in the organization as PI's workforce, yet contractors are more exposed to hazard and risk in the workplace.

Conclusion and Recommendation

During the period of 2014 – 2017, major injuries were caused by unsafe acts. These unsafe acts were a result of the safety culture variables that are strongly influenced by safety competence and safety motivation of the personnel.

Several programs have been initiated since the survey conducted in 2017. The programs are:

1. Safety Leadership Training aiming to raise awareness and improve leadership towards safety was conducted for the top management level in November 2017. This training was attended by managers, general managers and board of directors, especially operational/production directors.
2. CSMS in house training for Procurement, HSE and Pengembangan Sumber Daya Manusia (or Human Resources Development) personnel. This training also covers effective communications to contractors to ensure safety is the integral part of the work given to contractors.
3. Contractors have been included in the inspection and safety walkthrough programs as well as routine HSE meetings.
4. Implementation of "We Care", a behavior based monitoring program. We Care is a customized behavior based monitoring program aimed to ensure personnel are caring to each other by correcting and reporting unsafe actions and unsafe conditions found during work. This monitoring program serves as an intervention, ensuring that unsafe actions and unsafe conditions don't escalate to near miss and/or incident/injury.

The survey of safety culture maturity showed that the following variables are in need of improvement.

1. Safety campaigns have been conducted, however an evaluation needs to be done to ensure its effectiveness in relation to its goal to convey safety related information to the workforce.
2. Safety meetings have been held on a regular basis, however an evaluation needs to be done to ensure safety meetings effectively address concerns of safety implementation within the organization. The surveys show that safety meetings have not effectively addressed concerns regarding safety management system implementation, hazard control and mitigation.
3. Top management has a good perception of safety training and competencies, however the view is not fully shared within middle management and operators. An evaluation to ensure the effectiveness of training conducted and its ability to convey the safety message and information to the workforce needs to be done.
4. CSMS has been implemented, however the effectiveness of the CSMS process still needs to be evaluated and improved. The CSMS process needs to address hazard and risk specific to the workplace, as well as identify, assess, mitigate and communicate specific hazard in the work place related to contractors

References

1. Fleming and Lardner, "Strategies to promote safe behaviour as part of a health and safety management system", Prepared by The Keil Center for the Health and Safety Executive, Contract Research Report 430/2002, 2002

Introduction to a Predictive Analytics System in Ammonia Plants

Complex, capital-intensive industries face a pressing demand to minimize unplanned maintenance, achieve higher asset uptime and meet regulatory compliance constraints.
The use of basic monitoring systems are limited in scope and provide isolated views. The processes are driven mostly by domain knowledge and biased judgment, and are resource-intensive, time consuming and create many false alerts. This is truly reactive maintenance, which means fixing a problem as quickly as possible after it occurred.
Some production facilities – such as the Yara Ferrara ammonia plant highlighted in this paper – are realizing the tangible benefits of predictive maintenance using systems that provide automated monitoring and alerting, predictive modeling, advanced analytics and reporting. They are reducing maintenance costs and minimizing maintenance-related disruptions to their operations.

Andrea Padovani, Maxime Loviat
Yara Ferrara

E. Vanhaecke, Michel Warzee
Yara International

Introduction

This paper discusses Yara's journey to implement a predictive maintenance early monitoring system – SmartSignal from BHGE (Baker Hughes General Electric) and GE Digital – at the Yara Ferrara Ammonia plant. It will describe some of the technical outcomes, the organizational findings, and the value of early warnings of impending problems.

Pilot phase

Production losses reduce the efficiency of Yara operations, and two significant causes of these losses are unexpected failure of rotating and static equipment. These type of failures result in disruption of production, and can potentially affecting worker's safety and the environment.

In 2014, Yara started to review the possibilities of predictive monitoring systems and invited companies to present their solutions. Two case studies from past process incidents were selected in order to challenge the proposed solutions

The first case evaluated was the failure of a gas turbine inlet nozzle which occurred in a Yara ammonia plant in March 2011. The wheelspace temperature drifted in May - July 2010. This early warning could have been an opportunity to inspect the gas turbine earlier if an effective predictive monitoring flagged the temperature deviation. A trip in December 2010 worsened the damages and led to a shutdown for inspection in March 2011 (Figure 1).

Figure 1. - Gas Turbine inspection in December 2010

The predictive monitoring system showed the possibility to detect the start of the deviation seven months before the December trip and could have led to at least a better anticipation of the repair or extension of the operation to the next scheduled maintenance stop (Figure 2)

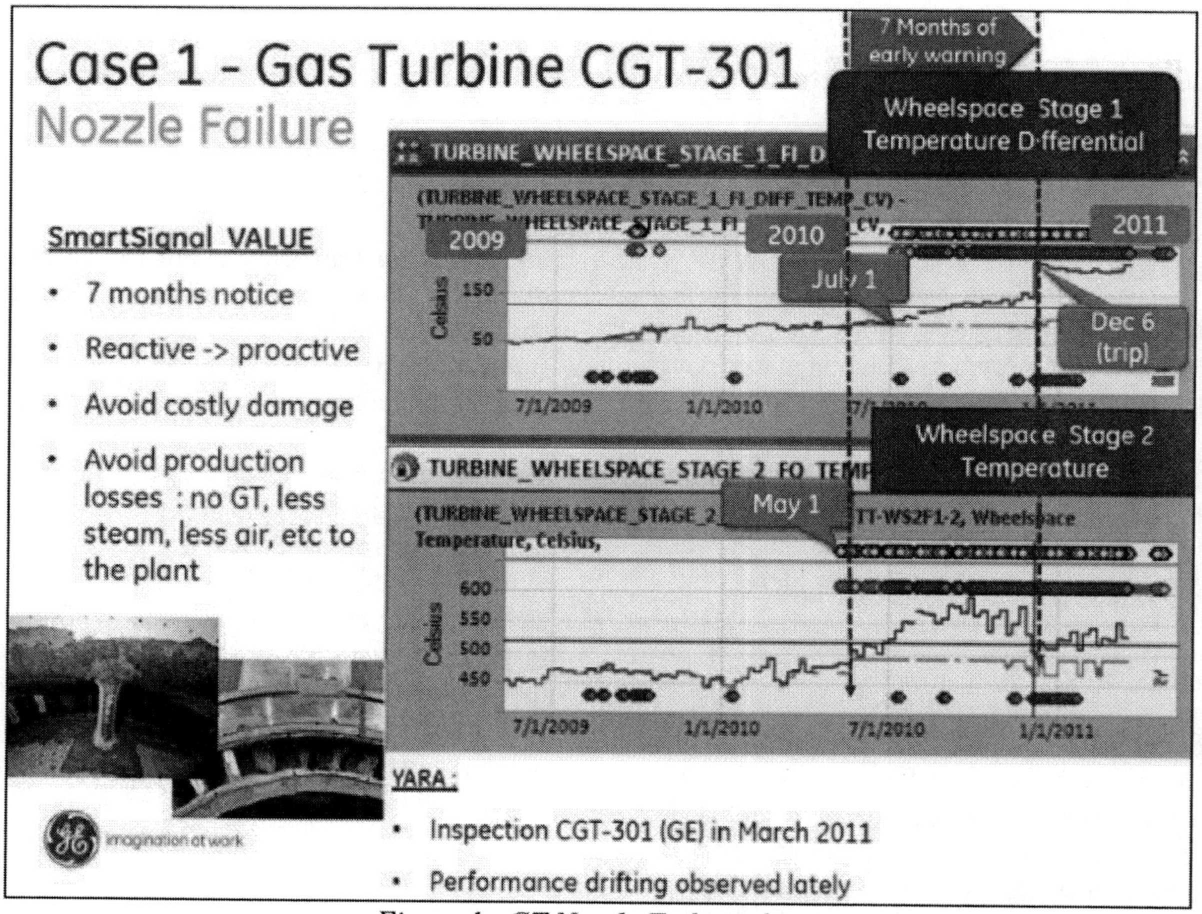

Figure 1. GT Nozzle Failure charts

A second evaluated case was a leak of the waste heat boiler (WHB) downstream of the secondary reformer in October 2014 which was discovered during the turnaround by equipment inspection. Since the leak was first detected during the turnaround inspection, the decision to exchange this major WHB with its available spare could only be made in the middle of the turnaround (Figure 3).

Figure 3. Waste Heat Boiler

The turnaround scope and spare part strategy could have been anticipated with early detection using the predictive monitoring system.

Detection of the WHB leak was much more challenging than the one indicated on the rotating machines mainly due to the high process integration and the limited instrumentation installed. Nevertheless, it was possible to notice a small deviation several months before the scheduled maintenance turnaround of October 2014, when the leak was discovered (Figure 4).

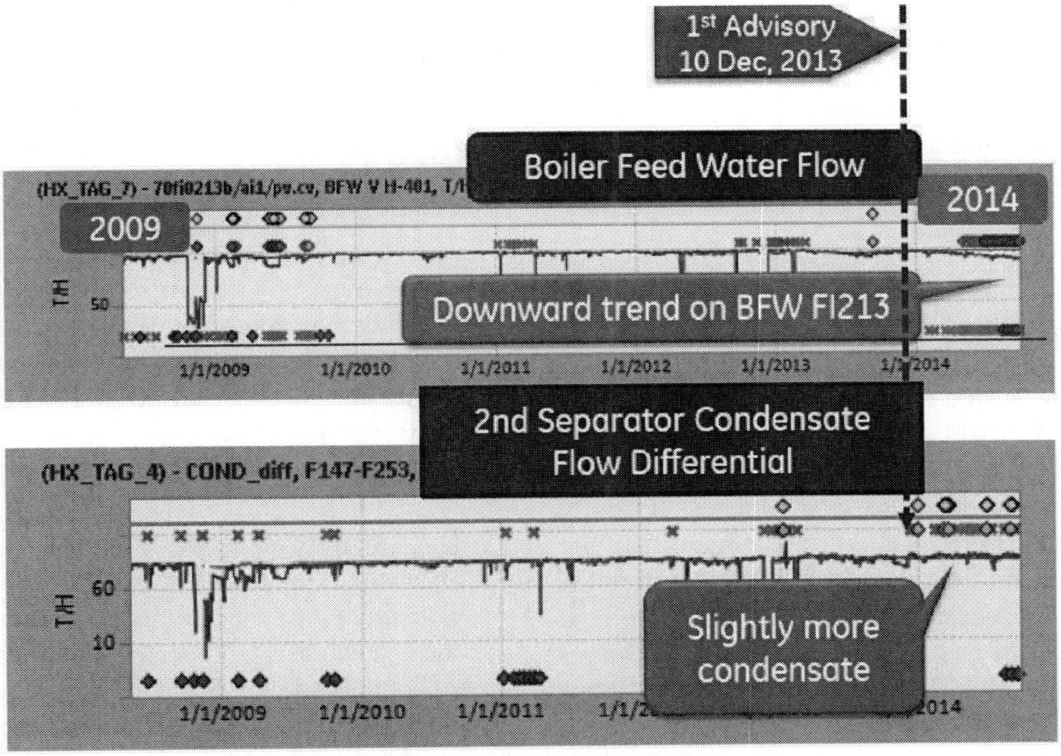

Figure 4. Waste Heat Boiler Charts

"Early monitoring" would have allowed anticipation of the change-out for this major equipment, and optimized preparation and associated costs.

Based on these positive outcomes of the two case studies, Yara decided to perform a pilot project in cooperation with BHGE and GE Digital in order to evaluate the SmartSignal Predictive Analytics Solution for the ammonia plant at Yara Ferrara. The proposed project included both software running in GE's servers in Chicago, and consulting services supporting the system usage.

The pilot project allowed YARA to determine how GE and SmartSignal can help achieve some key targets, namely:

- Anticipate potential failures
- Decreased production disruptions
- Increased assets availability
- Improved personnel safety due to reduction of disruptions
- Decreased maintenance costs.

Ferrara Plant description

The Yara Ferrara site is located in Emilia-Romagna region, northern Italy and includes an Ammonia and Urea production plant started in the late 1970's.

The Ammonia plant is a natural gas based Haldor Topsoe plant with an initial name plate capacity of 1500 MTPD (1652 STPD). The front end is driven by a side-fired Topsoe design furnace, with 672 burners and 416 Catalyst tubes. The production capacity has been increased through several major revamps and reaches currently 1730 MTPD (1905 STPD). The reliability of the plant, especially the front end, has been improved by upgrading the design of the waste heat boilers. The upgraded design drastically reduced the related downtime. The specific energy consumption has been significantly reduced over the years, with the most recent modification being in 2015 as part of a major energy revamp for both the Ammonia and Urea units. The 2015 project brought the Yara Ferrara site to a very good energy level for plants of this age.

The carbon dioxide produced by the ammonia plant is mainly consumed by the urea plant and a liquid carbon dioxide unit. A part of the urea solution is sent to an AdBlue unit started in 2008. The Ferrara site also includes an Argon Recovery Unit and hydrogen is exported to third parties. Figure 5 shows the process flow of the Yara Ferrara site.

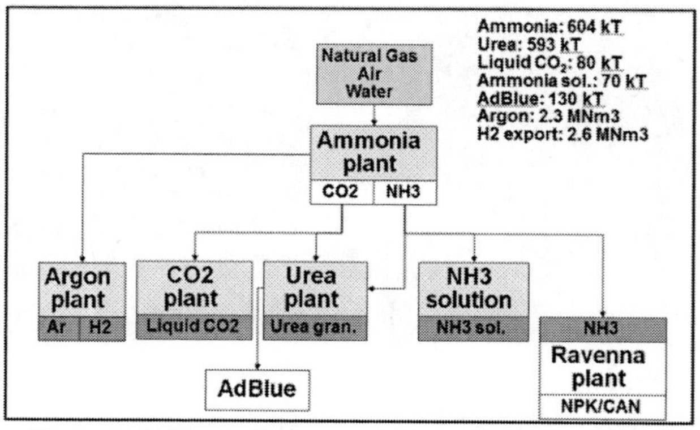

Figure 5. Yara Ferrara Block Flow Diagram

Predictive maintenance and early monitoring

SmartSignal from BHGE and GE Digital is a predictive maintenance system to protect critical assets, providing problem localization, diagnostic guidance, enhanced notification, prioritization and escalation, transient suppression and auto-adaptation. The system works as a tool that gives early warnings about the degradation of plant performance.

The way the system works is that it generates an estimate for each sensor or variable used (shown in green in Figure 6 and Figure 7) considering the current values of other sensors involved in the model. The actual behavior of that sensor is expected to lay inside a tolerated band centered around the estimated value. If the actual value from the field (depicted in blue in Figure 6 and

Figure 7) deviates outside the tolerated dynamic band, an advisory alert is generated, and if it persists long enough, an advisory notification – much more reactive compared to traditional fixed alarm threshold – will be posted in the system's incident view. The band width is tuned based on experience.

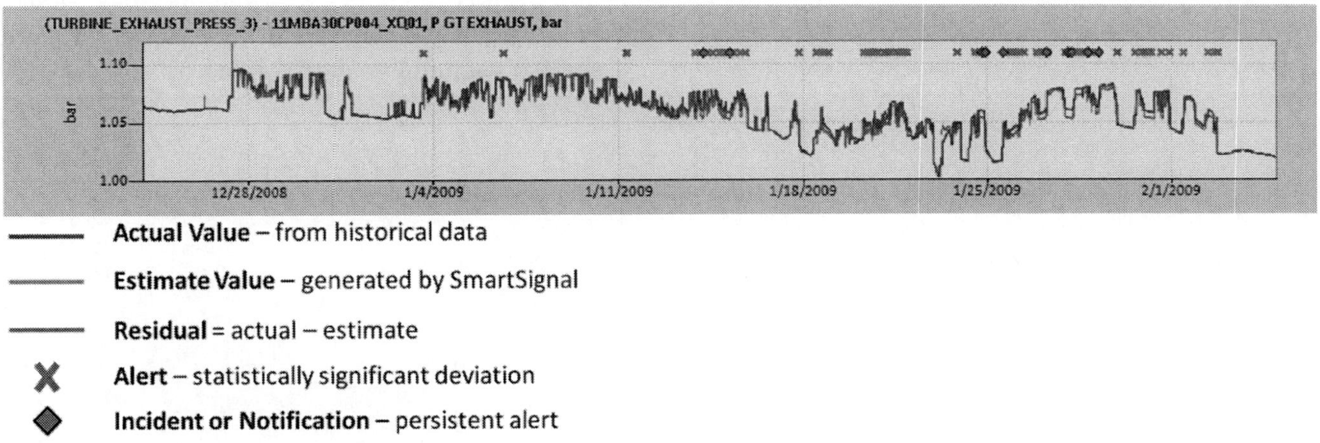

Figure 6. SmartSignal Color Codes

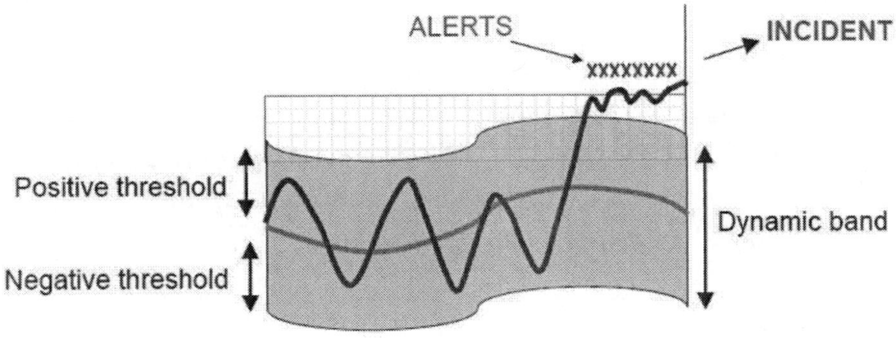

Figure 7. SmartSignal Dynamic Band tuning

Similarity Based Modeling (SBM)

SmartSignal model capability is based on empirical techniques using only historical operating data of existing instrumentation rather than detailed process or equipment engineering data.

The software uses patented advanced, mathematically compact modeling techniques known as Similarity Based Modeling (SMB) to analyze data in real time with tolerance to faulty data. Many modeling techniques require excessive computing power; the SmartSignal solution performs advanced modeling on thousands of data points every 5 or 10 minutes using commercial, off the shelf standard hardware and software.

The asset is divided into several logical models which contain measurements that are correlated, and for which it's desired to generate their predictions in context of each other. Within these models, SmartSignal predictively identifies apparent cause, and localizes these causes to specific components.

SmartSignal develops models using reference data (State Matrix) across the normal operating envelope. This empirical data regression technique is applicable across a wide range of equipment types, and it is scalable across a large fleet of equipment with minimal need for resources.

SmartSignal Similarity Based Modeling (SBM) produces a mathematically-efficient representation of the original system that provides enough precision to provide fast estimates for comparison with actual live data (Figure 8).

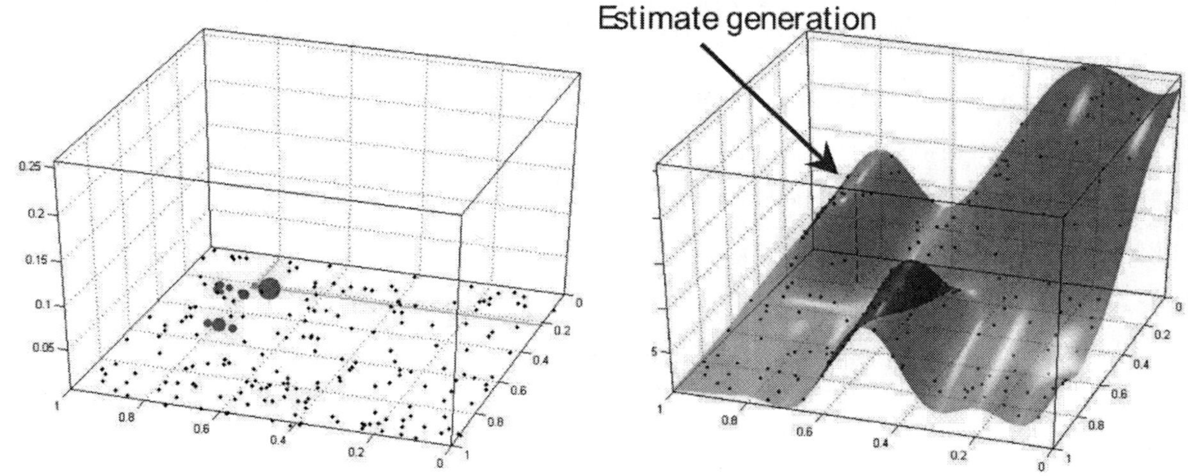

SmartSignal compares the known current conditions to the historical State Matrix values.

Modeling generates an estimate based on the current value and the State Matrix values.

Figure 8. Similarity Based Modeling basic principle

Implementation Phase

The actual implementation of the system tailored to the Ferrara plant was mainly about the definition of the model to be used. The plant was split into *assets* – small well-defined sections of the unit – and for each of these a set of instruments (also called tags) was identified which could collect all the relevant information available. These sets of tags are called *BluePrints* and are the basis of the model. The adequate identification of the assets is a critical task and is the base of a successful system.

The involvement of local experts in this stage is necessary to effectively segregate the plant sections into a homogeneous set of information with the capacity to be descriptive of the good operational mode of the section itself. The historical experience of the plant behaviour and its problems should help identify both the dimension of the assets and the tags needed to monitor specific issues: measurements that are affected by defective behaviours need to be included in the "BluePrint" under the penalty of nullifying the predictive capacity of the system.

The work was split in two phases: one covering rotating equipment, and the second one covering static equipment of the front-end section of the plant.

Rotating machines are easily identified, highly instrumented and the boundary process conditions are completely collected with few measurements (inlet and/or outlet flow, pressure and temperature). The assets identification is immediate and linked to mechanical realization of the equipment itself. Defective behaviours are easily detected by several instruments installed on the machine itself.

During the implementation there were available existing *BluePrints*, already tested and with a really flexible definition covering the most common situations; the tag identification was straight forward.

Static equipment assets present a greater challenge in their identification. A single piece of equipment was not a good candidate, either for the small number of available measurements, or for several interactions that are present between equipment. Defective behaviours can hardly be detected by the instrumentation on the equipment itself, while defects in equipment can affect measurements physically far away from the originating point. A leakage can affect even different circuits. A mass unbalance can be measurable with instruments as far away as the front end of the plant to the final products.

The definition of the assets is not trivial, and local expertise is necessary to balance the asset dimension and its capability to include all measurements that contains leading information. Assets too big can bring problems to identify the root causes for a single abnormal measurement detected by the system (too many possible causes). On the other side, reducing the dimension (and the sensors number) can cut out important early warning sources.

An effective working methodology is to focus on the main failures that are required to be identified in advance and looking at which measurements can lead to its detection. Each asset should include all the equipment that are possible sources of the problem with their sensors as well as all the instruments whose measurement is potentially affected, even if physically or logically far away from the concerned equipment.

Existing *BluePrints* are less reusable due to several differences in the plants layout, but they are a great time saving because they avoid the need to start from scratch. It is far more common that they need some adaptation compared to rotating equipment.

Once the *BluePrints* are defined, historical data must be collected to be used in the SBM calculation. The data needs to be descriptive of the typical situation of the plant, and should take into consideration variations due to ambient constraints (weather, temperatures, etc.) and to different load conditions.

At least one year of data should be included to cover the different seasons, but still some uncommon, although not abnormal, process behaviours could be missing. If few uncommon situations are covered by data there will be a higher probability that the system will identify these situations as abnormal, leading to additional tuning during the commissioning of the system.

Go-live Phase

The system went live with the support of BHGE and GE Digital using a "weekly meeting service" model of work.

A dedicated team of GE experts has access to the system and can analyze it, exclude the less realistic findings, and prepare a report for the local team of plant experts. The local team will review the report, and, finally, during the weekly meeting with the supplier team, they will discuss it trying to identify root causes.

This working model provides several advantages. First of all there is no need to create an in-house knowledge of the system, neither from a software point of view nor from a system management point of view. When it is found that the model needs a review due to poor data coverage of a specific situation or due to minor process changes, the situation can be handled by expert technicians keeping the plant local force focused on the plant itself. The additional number of experts that can be involved in this way is an added value when trying to identify the root causes of problems and abnormal behaviors.

In addition, the weekly meeting enforces the use of the system that due to the lack of an objective dead-line, can easily be relegated to a lower priority compared to the everyday maintenance and workload. This scenario is quite realistic and is valid potentially for all predictive maintenance tools. Such tools can give a real advantage only if actively used, and if time is dedicated to them for field analysis and review of the findings.

The local experts will treat only the cases that pass a first filtering given by the supplier experts. The first filter will lead to a time saving for the local team. The local involvement is not negligible due to the fact that most of the investigation can only be done physically at the plant level. The checks of instrumentation involved in all raised cases, the collection of background information and comparison with well-known historical events can only be done locally. In several cases, the data evaluation results in no findings. This is not avoidable when evaluating predictive maintenance.

The choice of a different work model with no support of the supplier requires an in-house team experienced on the system. The size of the team must be according to the size of the monitored assets. This model could be more economically effective, but it is completely dependent on the availability of local forces both for creating the knowhow and for handling the first filter of identified cases. Both work models require the use of local plant experts, whose practice and historical memory is fundamental in issue detection and in cause identification.

Outcome from the System

In the early weeks after implementation, the model correctly identified an issue in the control oil circuit of the air compressor. The system found an unexpected behavior on the pressures measured compared to the predicted ones, revealing in particular a step change in the differential pressure measured across the downstream oil pump filter (Figure 9). The change was as small as 0.03 bar (0.5 psi) and was well below the existing alarm threshold. It obviously went unnoticed by the operators. The system also detected anomalies in another pressure measurement in the circuit.

The first field analysis focused on the filter, resulting in no findings. Several preventive actions were taken, such as cleaning the filter and checking oil quality, with no results. Additional analysis found a small bypass in the holder of a second micro filter protecting the oil pneumatic electrical converter (whose differential pressure is not measured). This damage altered the amount of oil flow affecting the differential pressure across the main filter.

The situation had only a minor impact, and was correctly identified far before it could affect the control capability, although it required quite an effort to correctly identify the disguised root cause.

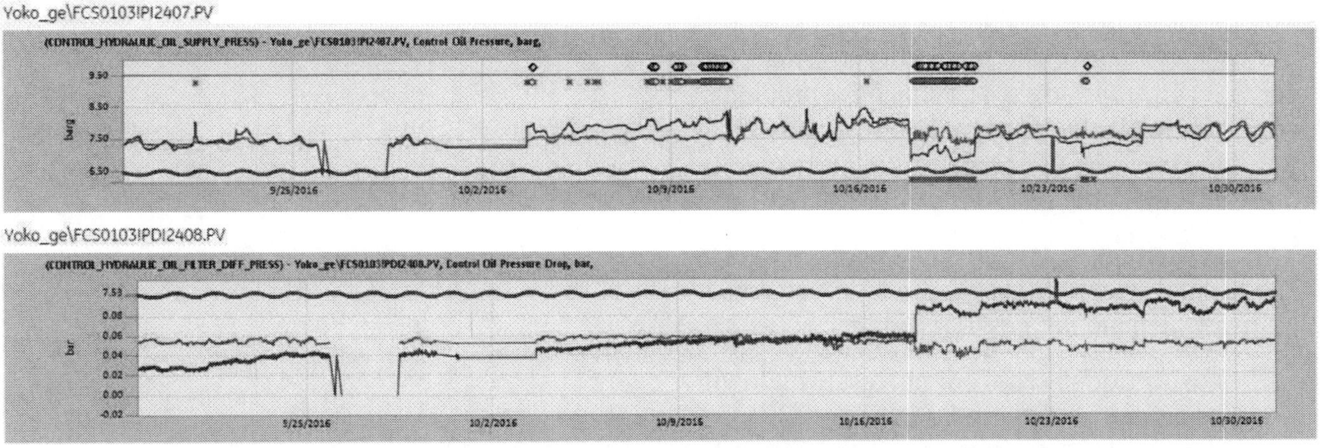

Figure 9. Control Oil Pressure Charts

Some months later a more significant issue was not detected by the system. A leak occurred in the bundle of a waste heat boiler leading to a loss of boiler feed water into the process gas, as the pressure of the water circuit is higher than the process side. The system did not detect any problem for more than a month after the leakage occurred.

The main cause of this miss was the wrong assets definition. They were originally designed with a strong focus on process adherence and a lack of attention to utilities.

The boiler feed water/steam circuit was split amongst several assets, and a critical tag that could have helped to detect the leakage was not included at all (i.e. process flow condensate in the Figure 10 which showed an increase due to the abnormal deviation of the water balance).

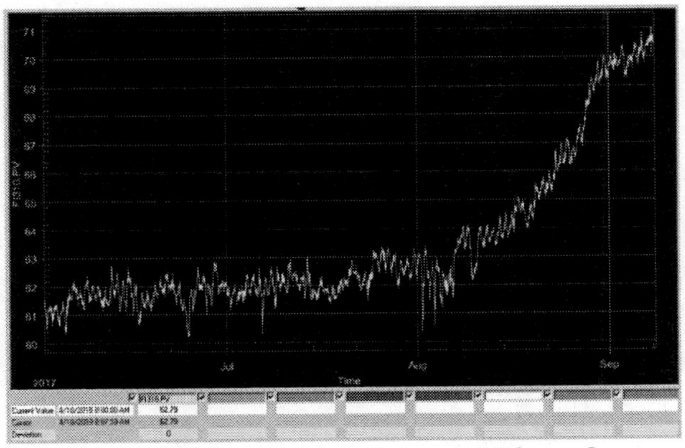

Figure 10. Process Condensate Flow chart

The corrective action was to redefine several assets, increasing the number of equipment included in each asset and reviewing which tags shall be used to identify such a failure. A new asset containing the boiler feed water and steam circuits, including several measurements on the process (inlet and outlet temperatures of heat exchanger, load of main compressor, etc.), has been defined.

Preliminary Results and Evaluation

Between September 2016 and March 2018, Yara received 109 advisories or cases and 20 of them were actionable cases. Yara used these actionable notifications to shift unplanned maintenance to planned, to reduce repair duration and to repair small problems before they became bigger problems. Figure 11 shows additional quick facts about the pilot project.

Pilot Project – Quick Facts	
# Rotating Assets	11
# Static Assets	11
# Sensors involved	970
# Models	65
# Weekly calls / reports	75
Total # of cases	109
# Actioned Cases	20

Figure 11. Pilot Quick Facts

The financial impact of such a potential failures has been evaluated, and the ability of SmartSignal to localize, prioritize and diagnose problems, weeks ahead of the standard alarm is a justification for extending the scope of the current solution and expanding it across additional affiliates.

Yara's adoption of the recommended processes and technologies will enable the following capabilities which are not possible today:

- An exception based work process which will provide earlier detection and diagnosis of a wider variety of equipment and process problems.
- Enable a proactive operations and maintenance culture, where a mainly reactive one exists today.
- A "Best Practice" remote monitoring capability which can be leveraged across all Yara Production assets.
- A strategy that leverages previous investments on condition based monitoring, people and processes.

Conclusion and Path Forward

Yara evaluations have demonstrated the value of the SmartSignal solution and found that equipment problems could be detected with enough advanced timing to optimize maintenance resources, reduce the risks and eliminate failures.

The strategy to expand predictive maintenance system to protect all Yara critical assets is under evaluation.

Acknowledgments

To the BHGE and GE Digital project team which show flexibility and commitment in the design and implementation phase.

Special thanks to Yara Sluiskil team for their time and efforts in the pilot phase of the project.

Reliability Improvement by Replacing the Vertical Process Gas Cooler with a Proven Design

The process gas cooler located downstream of the secondary reformer is one of the most critical pieces of equipment of an Ammonia plant. Many plants designed from the 1960's until the 1990's feature vertical boilers, with the primary process gas cooler as a water tube design and the secondary process gas cooler as a fire tube design. Both have several drawbacks which lead to excessive maintenance effort. For that reason, CASALE and ARVOS decided to adapt a well proven ARVOS product, which originally was designed for Ethylene and Gasification applications. Applying this design to Ammonia plants provides a more robust and trouble free vertical boiler.

Dr. Joerg Weidenfeller
ARVOS GmbH | SCHMIDTSCHE SCHACK

Francesco Baratto
Casale SA

Introduction

Many of the Kellogg Ammonia plants built between 1960 and 1990 are equipped with vertical process gas coolers (PGC) downstream of the secondary reformer. Reliability issues, damages and failures of the vertical PGCs are known to commonly occur. Since the PGC is one of the most critical pieces of equipment in an Ammonia plant, Casale SA and ARVOS GmbH SCHMIDTSCHE SCHACK teamed-up to overcome this situation.

Background

The typical Kellogg PGC arrangement downstream of the secondary reformer is shown in Figure 1. The effluent gas enters two parallel, primary PGCs at a temperature level around 1,000°C (1,832°F). The primary PGCs are connected to a vertical secondary PGC. The secondary PGC is equipped with an external bypass to control the gas outlet temperature.

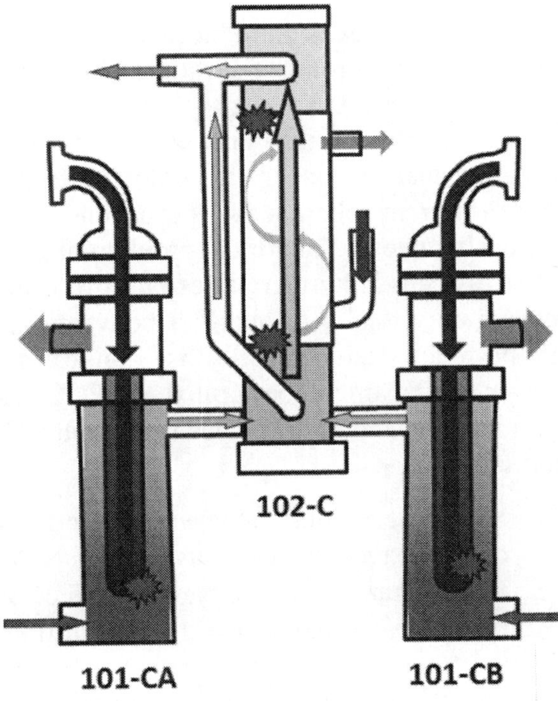

Figure 1: Typical Arrangement of PGCs

The primary PGCs (101-CA/CB) are vertical bayonet water tube type. The process gas enters the PGC at the bottom, and the flow is turned by baffles upwards in co-current flow with the boiler feed water (BFW). The shell is completely refractory lined inside and has a water jacket outside.

The secondary PGC (102-C) is a vertical fire tube design. The process gas enters the PGC at the bottom and flows upward in co-current flow with BFW.

It is reported in various conference papers that both PGCs have reliability issues and damages [1, 2, 3].

Reason for PGC failures

The primary PGCs (101-CA/CB) are equipped with the so-called bayonet tube as a heating surface, which is a tube in tube design. In general, the tube in tube design with boiler feed water flowing in the annular gap is an excellent solution for a heating surface. In this bayonet design, the feed water is introduced into the end cap of each tube through a coaxial open-ended inner tube and flows upwardly through the annular space between the inner and outer tube walls. Because the inner and outer tubes are held by different tube sheets, the bayonet-tube boiler design avoids the critical problem of tube sheet stress caused by thermal expansion of the tubes, which virtually disqualifies conventional shell-and-tube designs for this application. However, the bayonet tube applied at a PGC has several inherent reliability issues (see Figure 1), in particular:

- Sludge deposition in the end cap / pocket of the bayonets: deposits in the BFW agglomerate in the end cap and reduce or block the cooling capability. In addition, the cap is the section of the bayonet with highest heat flux, and, in consequence, this leads to local overheating. The situation gets worse when operating above the name plate capacity.
- Tube vibration: Since the process gas is turned by baffles, this always has the risk of flow induced tube vibrations and tube failure.

Damage to the tube bundles of these boilers is a leading cause of unexpected shut-downs in Ammonia plants (see Figure 2). In consequence, the boiler requires a high maintenance effort, and, frequently, the tube bundle has to be replaced completely. This requires a spare bundle which, on the one hand, has also to be maintained for the next replacement. On the other hand, spare inventory ties up capital with additional negative financial impact.

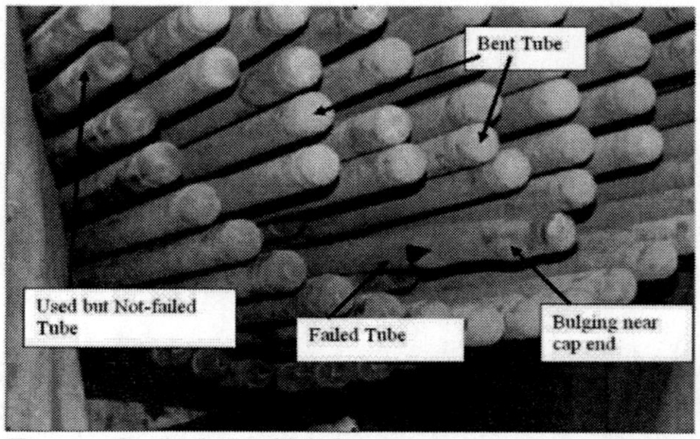

Figure 2: 101-CA/CB bayonet type typical failure (picture taken from [2])

The secondary PGC (102-C) is a standard shell and tube design in a vertical arrangement. This design principle has also inherent reliability issues due to:

- Sludge deposition at the lower tube sheet: deposits in the BFW agglomerate at lower tube sheet and reduce or block the local cooling capability. Despite the lower temperature level, this leads to overheating with failure of tube to tube sheet welding.
- Steam blanketing at the upper tube sheet: Due to mechanical reasons, riser nozzles cannot be located as close to the upper tube sheet as required. As a consequence, steam bubbles agglomerate which leads to reduced cooling of the

upper tube sheet and failure of tube to tube sheet welding.

A reliable design of a vertical PGC is always challenging. For that reason, SCHMIDTSCHE SCHACK's philosophy was not to develop a completely new design, but to apply and adapt a proven design to achieve a reliable solution. This design has been established for over 50 years in more than 7000 exchanger units in the Ethylene and Gasification industry.

Proven Design Solution

The basic element of this proven solution is the Schmidt´sche® double tube system (see Figure 3). It combines the advantages of fire and water tube design. The basic element is the double tube register. It comprises of a row of coaxial double tubes welded at each end to a horizontal header with an oval cross-section. A PGC is made up of a multiple array of these tube/header assemblies, with the headers welded together to effectively form hollow tube sheets.

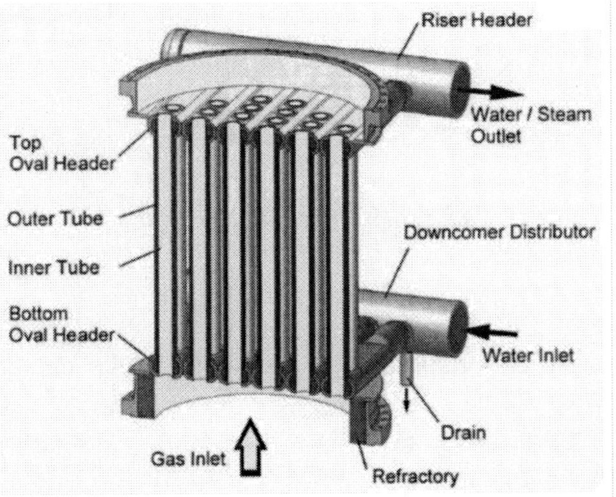

Figure 3: Schmidt'sche® double tube system

The water side of the exchanger operates in natural circulation. Water from the steam drum is fed through the downcomer piping to the bottom oval headers. The water then flows upwardly through the annulus between the inner and outer tubes where it is partially converted to steam, and into the upper headers which discharge into the riser manifold. Because the 'tube sheets' are cooled by the water/steam flowing through the tube headers, they do not have to be protected by a refractory layer.

The process gas flows through the inner tubes and is cooled by the water/steam mixture in the annulus between the inner and outer tubes.

Each process tube in the Schmidt'sche® Double Tube Exchanger thus has its own cooling element provided by the outer tube. Variations in the heat load on the individual tubes are automatically accommodated by the proportion of the water that is converted to steam.

Due to the clear defined flow path on the water and gas sides, there is no risk of flow induced vibrations.

Since the BFW flow from bottom to top is clearly defined, there is no risk of sludge settlement at the bottom or steam bubble agglomeration at the top. High water flow velocities with high turbulence intensively cool the oval headers (tube sheet) and each tube. In consequence, low wall thickness can be realized which leads to low wall and material temperatures. This results in low thermal stresses at the tubes and tube to tube sheet welding. In addition, the oval header as a main design feature enables a high degree of system elasticity for easy accommodation of differential thermal expansion between the inner and outer tube which also results in favourable stress patterns.

The Schmidt'sche® double tube system is ensured by full penetration and crevice free welds of the inner and outer tube. These features ensure a robust design. One of the main advantages is the absence of crevice corrosion due to water cooling at the weld position which keeps a low temperature at the welds (see Figure 4).

In case of need, the design still allows the possibility to both plug an inner tube or substitute inner/outer tubes.

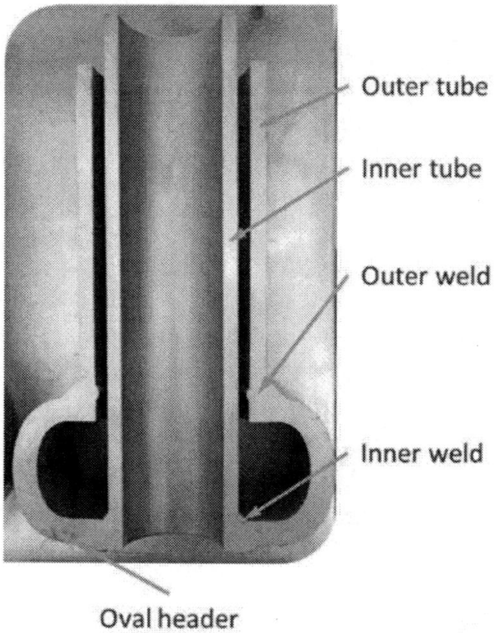

Figure 4: Double tube and oval header details

These features address the design drivers of reliability, availability and life time which have been demonstrated over 50 years in more than 7000 exchanger units in the Ethylene and Gasification industries.

Reference Applications

The majority of double tube systems are in the petrochemical industry in steam cracking processes for ethylene (olefin) production. The process gas is typically at an inlet temperature of 900°C (1650°F) at an inlet pressure up to 10 bar (145 psi). The water-steam system typically operates at 100 - 120 bar (1450 - 1740 psi). Due to upstream process fouling in the radiant section, these vertical heat exchangers are frequently operated in a regime of routine thermal cycles which is mechanically more severe than steady state operation.

In the last 20 years, approximately 19 units of double tube heat exchangers have also been installed downstream of the gasifier in coal, petcoke and biomass gasification processes. The particle and dust-laden process gas is typically at a temperature up to 900°C (1650°F) at 27 - 50 bar (400 - 725 psi), while the water-steam side typically operates at 50 - 115 bar (725 - 1670 psi). A comparison of ethylene and gasification with a typical ammonia application is shown in Table 1. The process conditions as well as size and weight of the double tube design replacing 101-CA/CB are within the existing fleet experience.

	Ethylene	Gasification	Ammonia 2000MTPD
Gas side (in / out)	900°C / 380°C (1650°F / 715°F)	900°C / 350°C (1650°F / 660°F)	~ 960°C / ~400°C (1760°F / 750°F)
	up to 10 bar (145psi)	50 bar (725 psi)	~30 bar (435 psi)
Water Side	100 – 120 bar (1450 - 1740 psi)	50 – 115 bar (725 – 1660 psi)	~105 bar (1520 psi)
Flow orientation	bottom → top	top → bottom	bottom → top
Length	10m – 17m (32 – 55 ft)	34m – 45m (111 – 147 ft)	13 - 15m (42 – 19 ft)
Weight	10 – 20 ton	300 – 620 ton	60 – 80 ton
Operation	Cyclic	Steady	Steady

Table 1: Schmidt'sche® Double Tube Fleet experience

Figure 5 shows the world largest PGC for a Gasification process. The double tube design was adapted to realize a PGC with an overall length of 45m (147 ft), a weight of 620 ton and a thermal power of 100MW (340 MMBTU/hr).

Figure 5: Double Tube system in coal gasification, 45m length, 620 ton weight

Gasification plants are even more problematic than Ammonia plants because of higher fouling, higher metal dusting and higher hydrogen attack due to higher partial pressures of hydrogen, carbon monoxide and carbon dioxide which is shown in Figure 6. This demonstrates that the syngas atmosphere is more severe in gasification compared to an ammonia application.

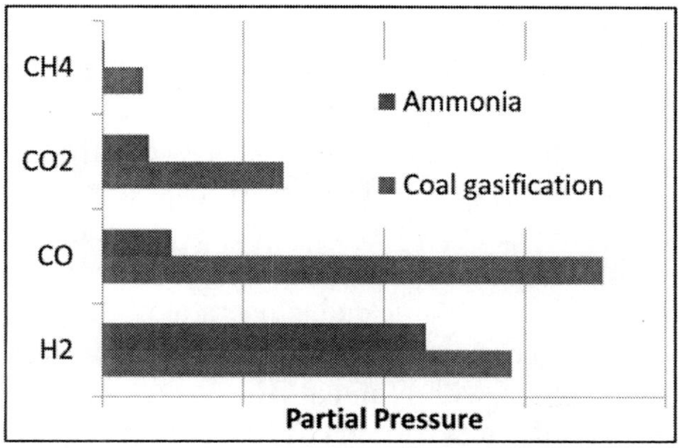

Figure 6: Comparison of gas compositions

Revamping Options

The boilers downstream of the secondary reformer (101-CA/CB and 102-C) can be replaced in three different scenarios:

- One by one: each of 101-CA/CB and 102-C can be replaced by a new PGC
- Two in one: both 101-CA/CB will be replaced by one new large 101-C. 102-C will be replaced separately. The existing by-pass for temperature control also remains.
- Three in one: all three boilers will be replaced by one large PGC including temperature control.

Comparing these revamping options, the highest customer benefit is achieved by a one by one replacement because of the following reasons:

- Usually the existing foundation inside the existing steel structure is suitable for the new PGCs (preliminary analysis can assess this issue); the installation of a new, large foundation will be avoided.
- No new plot space for the new PGC is required. In case of a new large PGC, new plot space has to be identified (which would reduce the available space for maintenance activities).

- The existing gas transfer lines can be used. In case of a new large PGC, the installation of a new long and large transfer line is necessary to connect the secondary reformer to the new large PGC which is cost intensive and could be a safety concern due to operation at high pressure and temperature.
- The existing water/steam piping can be used with minimal adjustments. In case of a new large PGC, this piping requires major extensions.
- For a 2000MTPD (2200STPD) scenario, the hardware costs (PGC, piping extension for water/steam and gas) for the one by one replacement are approximately 10% lower compared to the new large 101-C.

Figure 7 shows an example for the one by one replacement of 101-CA/CB. The new 101-CA/CB were designed for 2000 MTPD (2200 STPD) which already means a capacity upgrade from nameplate capacity (1360MTPD, 1500 STPD). The real arrangement of an existing plant was 3D laser scanned by Casale SA. The data was transferred into a CAD model. This approach ensures full transparency early in the design phase. Potential conflicts can be detected at an early project phase.

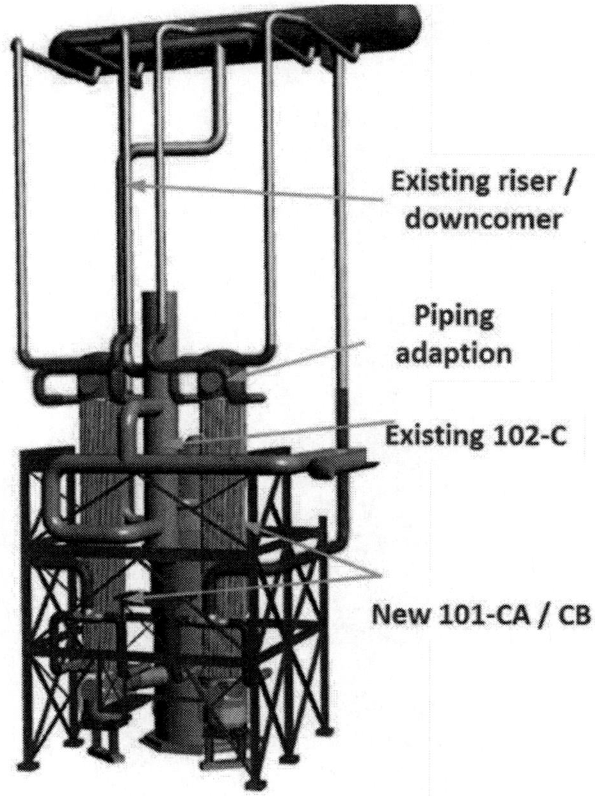

Figure 7: Use case for 101-CA/CB one by one replacement

The new PGCs fit into the existing steel structure as well as onto the existing foundation. The adjustments to the existing riser/downcomer piping are very limited (red / blue colored).

However, the double tube technology is flexible enough to support and enable all re-vamping scenarios depending on customer needs and requirements. Figure 8 shows an example for the 101-CA/CB replacement by a single PGC "new 101-C". The major extensions of water/steam piping as well as the gas transfer line can be seen clearly. Also the new PGC requires a new plot space outside of the existing steel structure.

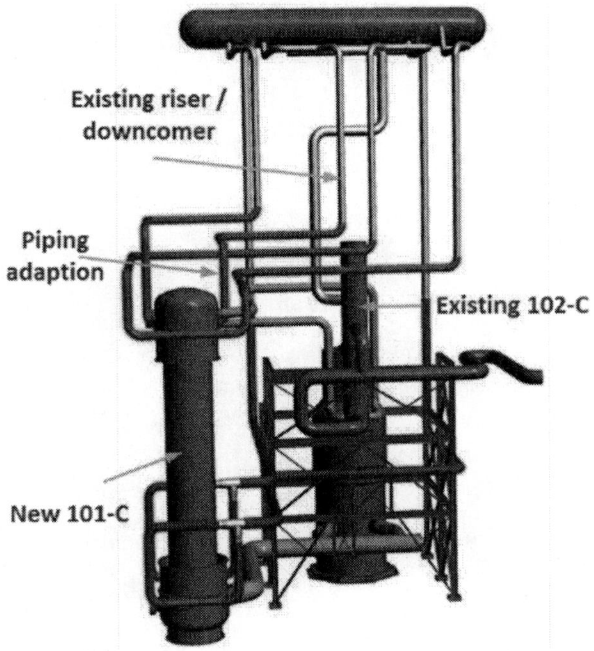

Figure 8: Use case for 101-CA/CB replacement by a single large PGC (New 101-C)

Conclusion

Many Ammonia plants built between 1960 and 1990 are equipped with vertical PGCs downstream of the secondary reformer. Those PGCs have reliability issues which lead to failures and damage of the PGC, higher maintenance and repair effort and more frequent bundle exchange.

A reliable design of a vertical PGC is a challenging task. A new design always has development and operational risks. For that reason, the adaption of a proven design minimizes uncertainties. The double tube design was originally developed for the ethylene industry and is established in more than 7000 exchanger units for over 50 years. It combines the advantages of a fire tube and water tube design. This design was successfully adapted to the gasification industry which has a more severe syngas atmosphere than ammonia synthesis with respect to hot gas corrosion and hydrogen attack. The process, operational and syngas conditions for Ethylene and Gasification envelope the Ammonia application downstream of the secondary reformer. Also, the PGC size for Ammonia plants is within the range of experience.

Different options are possible to replace the PGC in Ammonia plants. The double tube technology is flexible enough to enable all replacement scenarios, whereas the highest benefit is seen with a one by one replacement of the existing boilers with new ones.

The application of the proven double tube technology to PGCs in Ammonia provides a robust vertical PGC.

References

[1] I. Ahmad Gondal: Waste Heat Boiler (101-C) Leakages &Possible Causes. 50th Annual Safety in Ammonia Plants and Related Facilities Symposium, Ammonia Technical Manual Vol.46, 2005

[2]A. Din, S. Hassan and A. Ahmad: A Waste Heat Boiler Replacement – From Bayonet to Fire Tube Design. 53rd Annual Safety in Ammonia Plants and Related Facilities Symposium, Ammonia Technical Manual Vol.49, 2008

[3] S. Singh, M. Srivastava and M. Masood Ahmad: Failure of Waste Heat Boiler in a Kellogg Designed Ammonia Plant. 57th Annual Safety in Ammonia Plants and Related Facilities Symposium, Ammonia Technical Manual Vol.53, 2012

The Successful Debottlenecking of the OCI Beaumont Complex

The OCI Beaumont facility is an integrated methanol and ammonia complex that was built in the late 1960's and revamped in the 1980's with Lurgi's parallel steam raising methanol converters. In 2012, the OCI Beaumont facility was restarted after being moth-balled in 2004. On restart, the plant exceeded design rates and energy consumption for the case without CO2 addition. In 2015, a debottleneck project was successfully implemented to increase methanol and ammonia production by 25% and 12% respectively. This paper discusses the safety measures taken during shutdown to preserve methanol synthesis catalyst while performing upgrade works in the high-pressure synthesis loop; and the successful restart and recovery from a start-up incident that posed great risk to Clariant's high performing MegaMax® methanol synthesis catalyst, already close to 3 years on-stream. Also discussed are the upgrades made to the facility to increase capacity while considering optimum operating conditions for extended catalyst life.

Brandan Rossi
OCI Beaumont, Beaumont, TX

Michelle Anderson
Clariant, Louisville, KY

Introduction

The OCI Beaumont Complex produces both methanol and ammonia. The front end of the unit converts natural gas to syngas which is suitable for methanol production (similar to an ammonia front end, except without CO Shift Converters and CO_2 removal since CO and CO_2 are required for methanol production). There is a methanol synthesis loop and an ammonia synthesis loop which utilizes recovered hydrogen from the methanol plant along with additional required hydrogen and nitrogen over the fence. The integrated facility was restarted in 2012 after being mothballed by the previous owner in 2004. Following the restart, OCI studied the feasibility of debottlenecking the plant with the target of increasing methanol and ammonia capacity by 25% and 12% respectively. The project also included environmental upgrades and energy improvement initiatives. The study showed that there were no limitations at the methanol converter which was loaded with Clariant's high activity MegaMax® 800 catalyst.

As part of planning for the debottleneck shutdown and restart, careful attention was given to preserve the high activity Clariant MegaMax® 800 methanol synthesis catalyst shown in Figure 1. Similar to other copper based syngas catalysts such as low temperature shift (LTS) catalyst, methanol synthesis catalyst will react exothermically with air potentially causing a high temperature excursion that can affect catalyst performance or worse lead to catalyst and/or vessel failure.

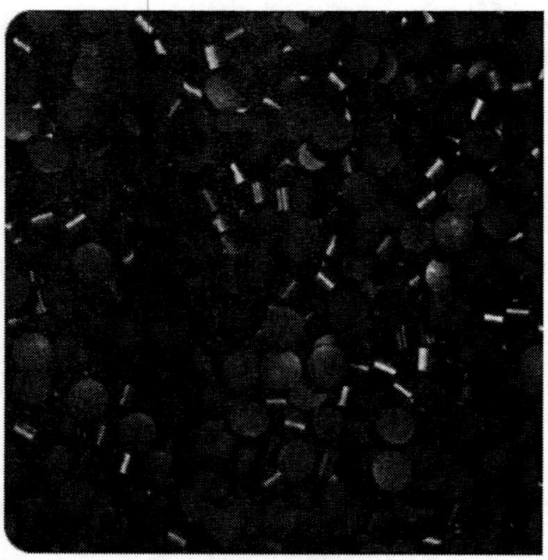

Figure 1. MegaMax® 800 6x4 mm Tablets Cu/Zn Methanol Synthesis Catalyst

On restart, all operations performance targets were met or exceeded, and the debottlenecking was deemed successful.

Project Feasibility

From 2012 to 2015 prior to the debottlenecking, the plant synthesis loop showed consistent, high performance. Through technical collaboration with Clariant, the MegaMax® 800 catalyst was operated at lower than design temperature which significantly extended its life by reducing the effects of typical sintering that similarly affects LTS catalyst.

The proven success of Clariant's installed high activity MegaMax® 800 methanol synthesis catalyst was a paramount component in the feasibility of the debottlenecking project. Clariant projected that the installed catalyst activity and volume was sufficient to exceed the targeted methanol production. This was a substantial savings in terms of capital expenditure required for debottlenecking. Based on the high conversion rates at low temperature, the projected remaining life of the catalyst showed that replacement of the catalyst was not needed for the debottlenecking project further improving its feasibility. Figure 2 shows the performance of the MegaMax® 800 catalyst at low operating temperature up to the debottlenecking project. The converters operated at 18°C (32°F) lower than design temperature yet achieved over 98% loop carbon conversion.

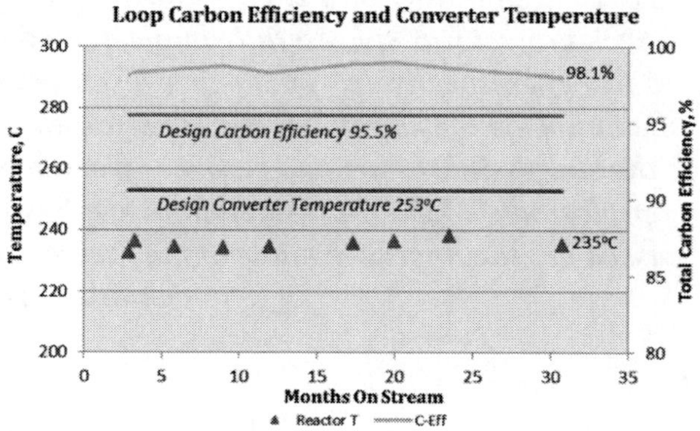

Figure 2. Total Carbon Conversion and Converter Operating Temperature from 2012 to 2015

Operating the converters at low temperature raised some concern of the possibility of wax formation and deposition on the downstream cooling heat exchangers. This had the potential to increase loop pressure drop, reduce heat exchanger efficiency and thereby reduce overall plant efficiency and yield. Such low operating temperatures are not typical within the industry for that reason, but Clariant was confident that the catalyst components were of a high quality to safely operate in this region. Concerns of wax formation were eased over time as there were no operating conditions that suggested that it was occurring. In fact, after three years with MegaMax® 800 catalyst in service, inspection of the loop cooling heat exchanger tubes showed negligible wax build-up. Figure 3 shows a picture of the findings.

Figure 3. Picture of OCI Loop Heat Exchanger after 3 years in-service at low temperature converter operation

Debottlenecking Strategy

To meet the Debottleneck Project targets, the following modifications were made:

- Installation of a pre-reformer and preheater to reduce the load of the primary reformer
- Installation of a deep desulfurization vessel to protect the pre-reformer catalyst from sulphur poisoning
- Replacement of the reformer tubes with larger ID, thinner walled tubes to limit the increase in pressure drop as well as increase heat transfer efficiency
- Installation of a feed gas saturator to meet increased steam demand while lowering plant waste water
- Upgrade of the methanol and ammonia synthesis loop compressors
- Increase the cooling capacity in the methanol and ammonia synthesis loops.
- Replaced specific trays in the Distillation unit

In addition to the capacity increase modifications, the following environmental improvements were made to meet the Texas Commission on Environment Quality (TCEQ) permits for the project:

- Installation of a selective catalytic reduction unit (SCR) in the reformer flue gas stream to reduce NOx emissions
- Removal of flue gas liquid fuel burners in the reformer convection section
- Reduction of plant wastewater by utilization in the feed gas saturator.
- Installation of a new reformed gas flare system. All frontend process vents were rerouted from atmosphere to the flare header.

Schematics of the highlighted modifications is provided in Figures 4a and 4b.

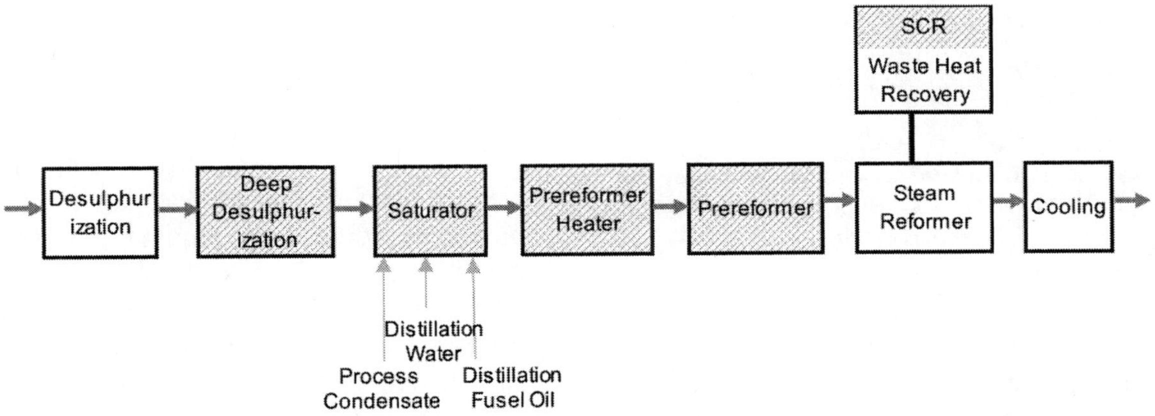

Figure 4a. Front End Debottleneck Schematic

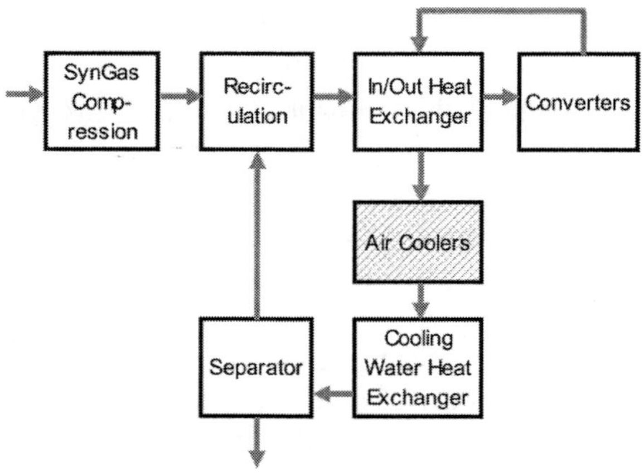

Figure 4b. Back End Debottleneck Schematic

With the additional rates to the methanol synthesis loop, air coolers were installed to increase the cooling capacity and achieve effective separation of crude methanol from the converter effluent. These air coolers are shown in Figure 5.

Figure 5. Air Coolers Installed in the Methanol Synthesis Loop

To achieve increased capacity at the front end, the bottlenecks identified were the pressure drop limitation of the front end based on existing compression capability, lack of steam availability, and primary reformer furnace duty.

By installing the pre-reformer, the total increase in duty required at the primary reformer was reduced. The replacement of the primary reformer tubes, already close to 10 years on-stream, with a larger ID thinner wall design allowed increased heatflux to the process gas and a lower pressure drop increase.

The feed gas saturator provided the additional steam requirements to achieve the re-designed steam to carbon ratio. Additionally, it reduced CO_2 emissions of the plant as the process condensate from the front end cooling section was re-routed from the CO_2 stripper directly into the process. Dehydrator Column tails water along with Fusel Oil, which is a mixture of water and higher alcohols drawn off from the distillation unit, was also re-routed to the feed gas saturator which reduced the plant's net wastewater flow to its bio-oxidation pond.

The addition of the selective catalytic reduction (SCR) equipment was necessary to reduce the NOx emissions from the primary reformer and new pre-reformer fired heater to meet the necessary permit levels.

Shutdown Planning – Synthesis Catalyst

The preservation of the methanol and ammonia synthesis catalysts were of utmost importance to ensure production and energy targets were met post shutdown as well as ensuring that the remaining life of the MegaMax® 800 catalyst would allow high productivity until the next planned outage some 3-4 years later. As the ammonia converter loop required no work that posed a risk to the installed catalyst, nitrogen blanketing and isolation sufficiently protected it. By contrast, the copper-based methanol synthesis catalyst required a special procedure to isolate the catalyst as the work required in the synthesis loop was done under atmospheric conditions.

With guidance from Clariant, the threat to the MegaMax® 800 catalyst during and after the shutdown was carefully assessed, namely:

- MegaMax® 800 reactivity. Exposing this catalyst to air would lead to spontaneous self-heating with the potential for uncontrolled high temperature excursion. High temperature could lead to catalyst fusing, complete deactivation and even vessel damage.
- MegaMax® 800 performance. The copper based catalyst is extremely sensitive to small levels of oxygen which will oxidize the catalyst leading to localized heating that can sinter the small copper crystallites which give it high activity. Additionally, the physical integrity of the catalyst can be weakened by reducing the crush strength through an oxidation/reduction process. Reduced crush strength could lead to increased bed pressure drop during restart and over time.
- MegaMax® 800 impurity poisoning. Any work conducted in the loop which exposes piping and equipment metallurgy to the environment has the potential to form loose, scaly rust which on restart can deposit at the top of the catalyst bed. Iron catalyzes Fischer-Tropsch reactions which promote the formation of by-products such as higher hydrocarbons and wax. These by-products can reduce the loop efficiency and plant yield by adding additional load on the distillation unit, fouling the cooling heat exchangers and reducing the loop separator efficiency.

A typical modern methanol synthesis loop operates at around 100 bar (1450 psi) with the large diameter process piping fully welded without gasketed flanges to prevent leakage. However, the OCI methanol synthesis loop, consisting of two Lurgi-designed steam generating converters in parallel, is designed with two inlet and two outlet 16 inch flanges for each converter (Figure 6). Thus isolation of the converters required eight, 16 inch ½-inch thick slip blinds. In collaboration with Clariant, the blinding and un-blinding procedure was written.

Figure 6. Parallel Steam Generating Methanol Converters at OCI Beaumont

Blinding and Shutdown

To minimize the risk of air exposure to the methanol synthesis catalyst, the following steps were established following normal shutdown:

1. The steam side was emptied and pressurized with nitrogen.
2. The loop was depressured to 1-2 psi
3. The two converters were isolated from each other via hand valves in the loop.
4. Blinds were installed one at a time to avoid any chimney effects that can suck air through the catalyst
5. All four blinds on the first converter were installed and the converter pressured to 1-2 psig with nitrogen before proceeding to the second converter.
6. To ensure positive pressure was maintained in the converter, a nitrogen hose was fitted to a 2" valve at the bottom of the converter and a water seal assembly was connected at the 1" sample connection at the converter inlet piping. The water seal assembly consisted of ½" tubing submersed 2-3 inches deep into

water filled in a 50 gallon drum. Before installing a blind, positive nitrogen flow was established via the nitrogen hose and vented at the water seal. The bubbling rate at the water seal was monitored continuously by a dedicated operator and nitrogen flow was adjusted as necessary when a flange was opened.

7. The loop piping and equipment were opened up only after all blinds were successfully installed and the converters were at 1-2 psi with nitrogen.

During the shutdown the nitrogen pressure at the DCS was monitored as well as the limited temperature measurements around the converter.

Following the installation of the air cooler in the synthesis loop and the inspection and cleaning of the water cooled heat exchanger and the separator, special care was taken to ensure that the piping and equipment were free from rust and loose iron deposits.

Blind Removal

The process for pulling the blinds required special precaution as the inlet piping and associated equipment was a separate system than the outlet piping and associated equipment with the blinds installed. These systems had to be treated separately and precautions taken to avoid introducing air from these systems to the converter. The following steps were followed:

1. The inlet and outlet loop piping and equipment systems were pressure purged with nitrogen to 70 psi at least 4 times each until oxygen levels were less than 0.5% by volume.
2. In order to ensure positive pressure in the converter, the water seal was again installed and nitrogen flow was established from the bottom of the converter and vented at the inlet using the wet seal assembly.
3. One blind was pulled at a time to avoid chimney effects and sucking air through the catalyst.
4. A dedicated operator was stationed at the water seal to monitor the bubbling rate. Caution was taken when removing a blind and the nitrogen flow was increased as necessary.
5. The process was repeated on the other converter.
6. The loop was finally pressurized to 70 psig with nitrogen to check for leaks.

Plant Restart Incident

The start-up of the front end commenced in April 2015 and required the commissioning of the various new equipment items. On April 21st, 2015, the methanol synthesis loop was in start-up with the catalyst at temperature. While increasing the pressure with make-up gas, the gasket inlet the west converter blew out. Immediate action was taken to depressure and purge gas from the loop. Clariant was contacted and all precautions were again put in place to protect the active MegaMax® 800 catalyst while the eight inlet and outlet converter piping gaskets were replaced with new ones.

Performance Post Debottleneck

On April 22nd methanol production commenced and plant rates were increased over a period of time while the plant conditions were optimized. During this period, Clariant was consulted on the manner in which to optimize the loop considering the need to increase the converter steam drum header pressure sufficiently for process steam feed to the front end. It was understood during the project design phase that this would be necessary and would effectively increase the converter operating temperature by 7°C (13°F) post shutdown. By May 23, 2015, the target production of 2500 MTPD (2750 STPD) was achieved.

Further optimization resulted in a maximum production of 2550 MTPD (2810 STPD) being achieved.

Changes to some key performance indicators of the plant performance following the debottleneck project are shown in Table 1 below:

Parameter	Post Debottleneck
Methanol Production Increase	550 MTPD (606 STPD) 28 %
Ammonia Production Increase	165 MTPD (182 STPD) 20 %
Energy Utilization Reduction Natural Gas Energy/Ton Methanol	2 %
Reduction of Import Hydrogen Flow/Ton NH3	22 %
Fluegas NOx reduction	90 %
Reduction in Wastewater effluent flow to BioOxidation Pond	66 %

Table 1. Changes in Plant Key Performance Indicators Post-Debottleneck

Loop and MegaMax® 800 Performance

Despite the high risk that the active MegaMax® 800 was exposed to air during the shutdown and during the restart incident, which had the potential to impact its activity as well as physical integrity, the activity of the catalyst remained unchanged even at the higher space velocity through the catalyst (i.e. higher process flow/volume of catalyst). Further, the pressure drop remained unchanged, and, in fact, at over 5 ½ years in service at the time of writing, the pressure drop is still at the start of run value.

Some key operating parameters are given in Table 2:

Parameter	Pre-Project	Post-Project
Make-up Gas Stoichiometric Number	3.00	2.92
Recycle Ratio	4.9	3.7
Converter Operating Temperature	235°C (455°F)	242°C (468°F)

Table 2. Key Loop Operating Parameters

For methanol synthesis, the following chemical reactions occur:

$$CO + H_2 \leftrightarrow CH_3OH$$
$$CO_2 + H_2 \leftrightarrow CH_3OH + H_2O$$

The stoichiometric number (SN) of the make-up gas is a measure of the amount of carbon monoxide and carbon dioxide relative to hydrogen in the process gas and is given by the equation:

$$SN = \frac{(H_2 - CO_2)}{(CO + CO_2)}$$

A make-up gas composition with SN close to 2.05 is optimum for methanol synthesis.

By the debottlenecking modifications made in the front end, the make-up gas SN was reduced with more CO and CO_2 relative to hydrogen which technically improves loop performance and efficiency as less purging is required. With this improvement, the lower recycle ratio was compensated.

Figure 7 shows the performance of the loop from start of run to December 2017 and identifies the unchanged loop efficiency and the increase in operating temperature of the converters post-debottleneck. It also clearly shows the stability of the MegaMax® 800 catalyst from start of run to over 5 ½ years on-stream.

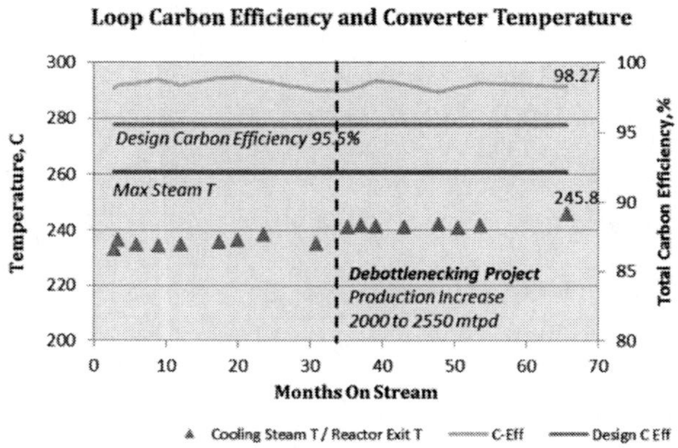

Figure 7. Loop performance from SOR to December 2017

As Figure 7 shows, the operating temperature of the MegaMax® 800 catalyst was increased in a step change fashion following the debottleneck project. Similar to LTS catalyst, optimization of the methanol synthesis catalyst over time is achieved by strategically increasing the operating temperature in increments to improve conversion but being mindful that large increases can prematurely reduce the life of the catalyst.

The by-product make at OCI continues to be low with the MegaMax® 800 high selectivity catalyst. This has significant energy saving benefits in the distillation section and thus improves the energy utilization of the complex. Laboratory analysis of the crude methanol from the OCI synthesis loop shows low levels of all by-products in comparison to plant guarantee figures. Table 3 lists the key byproducts and what percent of the guarantee figure is present for each.

By-product in Crude Methanol	Percent of Plant Guarantee Value (5 Years On-stream)
Ethanol	53%
Total Higher Alcohols	34%
Di-methyl ether (DME)	62%
Methyl Formate	53%
Acetone	Not Detected

Table 3. By-Product Make in Crude Methanol

Based on the stability of the catalyst performance post-debottlenecking, it is clear that the well planned procedures and actions taken to protect the catalyst during the shutdown and during the start-up incident proved to be a success.

Plant Projection

OCI Beaumont continues to look at ways in which production can be increased. Currently an expansion project is under review. The timing of such project is dependent on the remaining life of the methanol synthesis catalyst. Based on current catalyst performance and the maximum limitation of the catalyst operating temperature set by mechanical design, Clariant continues to guide OCI Beaumont for planning purposes. As an example, Clariant provides routine catalyst evaluations with guidelines on optimization for temperature increases and remaining life projections. An example of such projection is shown in Figure 8 below.

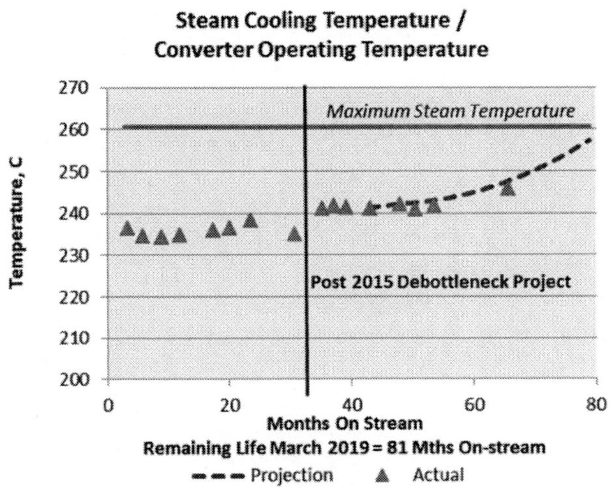

Figure 8. Converter Operating Temperature Projection by Clariant

The proven performance of Clariant's MegaMax® 800 catalyst has shown that through collaborative partnership a catalyst change out cycle of 6+ years is possible. This long life is a significant financial benefit to a producer like OCI when catalyst cost, shutdown costs and lost production is considered.

Conclusion

In line with OCI's strategy of growth into a global producer of fertilizers and industrial chemicals, the acquisition, restart, and debottleneck of the OCI Beaumont complex has made the facility into one of the world's largest merchant methanol producers.

The debottleneck project of 2015 achieved a significant increase in methanol and ammonia production capacity, and at the same time, complied with all permitted emission levels.

With the successful implementation and results of the debottlenecking project, OCI continues to look at opportunities for further expansion.

Acknowledgments

The authors wish to acknowledge OCI N.V. and OCI Beaumont management on their vision and leadership in successfully executing the debottleneck project. Additionally, we wish to acknowledge the contributions of the technical specialists, operations and maintenance teams whose commitment and dedication to the complex is invaluable.

New Alloy Materials for Reformer Outlet Systems

This paper describes the work performed on the development and characterization of new alloy materials used in steam methane reformer furnaces (SMR) outlet systems. New material developments allow outlet components to operate at more severe conditions without the need of increasing the wall thickness of these parts. In existing designs, reduction of wall thickness is possible while decreasing thermal stresses in such components. Selected results from improved materials, including results of creep and accelerated ageing tests carried out, are presented. Additionally, finite element case studies on existing outlet designs showing the benefits of new materials are also described in this paper.

Pablo Cardín
Schmidt+Clemens Group

Pedro Imízcoz
Schmidt+Clemens Group

Iosu Mutilva
Schmidt+Clemens Group

Dietline Jakobi
Schmidt+Clemens Group

Introduction

Outlet systems are a main part in steam methane reformer furnaces (SMR). Although these elements are typically located outside the reformer firebox, they operate at severe conditions up to the metallurgical limits. Traditionally, centrifugally cast low carbon materials are used instead of wrought components [1]. The most commonly used material is alloy GX10NiCrNb32-20 or ASTM CT 15C (20Cr-32Ni-Nb), commercially named as Centralloy® G 4859 for the Schmidt + Clemens Group.

This paper will discuss new material developments done by the Schmidt+Clemens Group in outlet manifold components. New materials with superior ductility after ageing and better creep strength have been developed and placed into service under the most stringent conditions of steam methane reformer furnaces.

Results of different comparison tests performed between newly developed materials and previous ones, including case studies and finite element analysis (FEA) simulation on real outlet components, will show the benefit of these improved materials.

Objective

The main objective of this paper is to show new material improvements in outlet components for steam methane reformer furnaces. These components are operated with a process gas temperature in the range of 800°C [1472°F] to 950°C [1742°F] at a pressure of approximately 40 bar [580 psi]. Consequently, component lifetime is consumed by the combined effects of creep and alternating thermal and mechanical stresses (creep ductility exhaustion) [1]. Oxidation/corrosion processes typically have a lower impact on material damage mechanisms.

While significant research efforts have been made in improving alloys used for steam reformer catalyst tubes, no significant efforts were devoted in improving outlet system materials over the last 30 years.

The main requirements for these alloys are an excellent material ductility after ageing and an adequate creep resistance. Several studies have related ductility reduction with the transformation in service of niobium carbides (NbC) into intermetallic phases like G-phase ($Ni_{16}Nb_6Si_7$) and η'-phase (Nb_3Ni_2Si) [2, 3, 4]. Additionally, it has been suggested that such intermetallic phases offer less inhibiting dislocation movement as compared with niobium carbides, leading to a lower material creep resistance [2].

Chemical composition optimization has a strong potential in inhibiting/reducing the presence of such phases. Schmidt+Clemens has been working in areas like optimum Si content, controlled Nb/C ratios and micro additions, developing superior materials for outlet header components.

Material Description

Outlet manifold materials differ depending on reformer designer preferences. Typically, low temperature systems having design temperatures below 829°C [1525°F] can either use 800 HT or cast equivalent 20Cr-32Ni, while those designs targeted for higher temperatures would use cast materials 20Cr-32Ni almost exclusively [5].

20Cr-32Ni alloy material supplied by the Schmidt+Clemens Group is named Centralloy® G4859. This alloy is a low carbon alloy designed to maintain good ductility after ageing and a good creep resistance in the range of 800°C [1472°F] to 1000°C [1832°F]. Niobium is used as carbide forming element (20Cr-32Ni-Nb). A substantial benefit of the cast alloy with Nb is that the creep strength is about 50% higher than the wrought alloy (800 HT) with Al and Ti [6].

Two materials have been considered as a potential replacement of alloy 20Cr-32Ni-Nb. Alloy material Centralloy® G4859 Micro (micro alloyed version of standard 20Cr-32Ni-Nb) and Centralloy® H 101 Micro (low carbon version of standard reformer tube material 25Cr-35Ni-NbTi). Nominal composition of these materials is represented in Table 1.

Composition wt.%	C	Si	Mn	Cr	Ni	Nb	Ti/Zr	Fe
Centralloy® G4859	0.1	1	1.5	20	32	1	-	Balance
Centralloy® G4859 Micro	0.1	1	1.5	20	32	1	Additions	Balance
Centralloy® H101 Micro	0.13	0.5	0.5	25	37	0.5	Additions	Balance

Table 1. Typical composition of Schmidt+Clemens low carbon alloy materials for outlet components.

Centralloy® G4859 Micro

A logical improvement of 20Cr-32Ni-Nb is the development of a micro alloy version. This material is commercially known as Centralloy® G4859 Micro. Addition of small quantities of elements like titanium and composition optimization has enhanced significantly material performance. Like standard material G4859, this alloy has a maximum operating temperature of 1000°C [1832°F].

Centralloy® G4859 Micro was available in the market since 2011. Currently more than 102 outlet manifold arms and 2950 outlet reducers have been installed in service, together with more than 40 T-pieces and transition cones. Performance of these components in operation has been regarded as excellent.

Centralloy® H 101 Micro

Low carbon versions of HP-Nb (25Cr-35Ni-Nb) were originally designed for operation under severe corrosion conditions only requiring a moderate ductility. These conditions are typical in steam cracker applications.

Nevertheless, these materials were described in the literature as a potential replacement of 20Cr-32Ni-Nb materials [6, 7], especially in designs requiring higher operational temperatures.

Unfortunately, lower ductility after ageing performance compared with 20Cr-32Ni-Nb has been regarded as the main drawback. Therefore, material replacement in outlet manifold components was not advisable [6, 7].

Schmidt+Clemens has designed a micro alloyed version of this low carbon material, with an optimized chemical composition, capable of enhancing creep resistance and ductility after ageing over existing 20Cr-32Ni-Nb materials. Furthermore, higher chromium and nickel contents would allow outlet components to be designed for higher temperatures up to 1050°C [1922°F].

First reference in reformer outlet components was supplied in 2016.

Characterization and Results

Schmidt + Clemens did an extensive characterization of tensile properties and creep performance. Additionally, significant efforts were made on material microstructural characterization and accelerated ageing tests in order to study the driving forces regarding material ductility exhaustion.

Tensile Tests

Room temperature (RT) and high temperature tensile tests were performed on these three materials. 0.2% Yield Strength, Ultimate Tensile Strength and elongation to rupture (l=5d) properties were measured.

Results for 0.2% Yield Strength are shown in Figure 1. Addition of micro alloying elements has improved material yield strength in comparison with alloy G4859. Figure 2 shows the comparison of Ultimate Tensile Strength (UTS) for these materials. No relevant differences can be seen between G4859 and G4859 Micro, while H101 Micro UTS values seem to be slightly inferior.

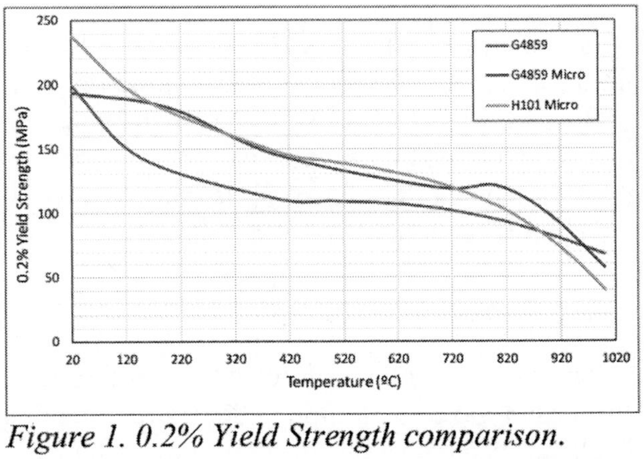

Figure 1. 0.2% Yield Strength comparison.

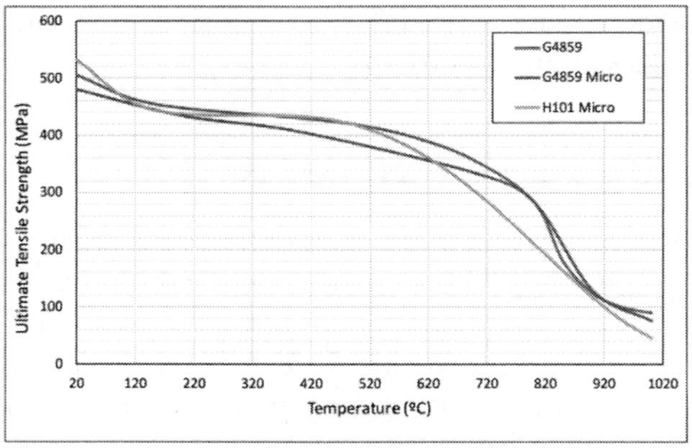

Figure 2. Ultimate Tensile Strength comparison.

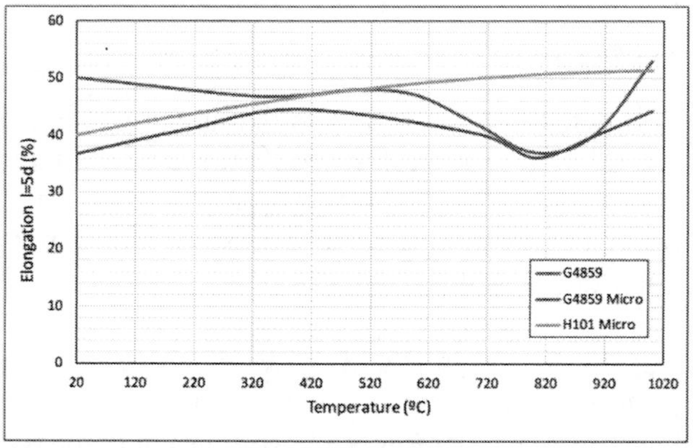

Figure 3. Elongation to rupture comparison.

Elongation to rupture values are shown in Figure 3. Results indicate a superior ductility of G4859 at low temperatures, while H101 Micro has a better ductility at higher temperatures. G4859 Micro alloy has a slightly lower ductility.

It is important to remark the relevant reduction on ductility seen for alloy G4859 in the range of 700-900ºC [1292ºF-1652ºF]. Such behavior might be related to the stability of brittle intermetallic phases like G-phase [8], leading to a localized embrittlement of the alloy material.

Creep Tests

Creep resistance is one of the major requirements for these type of materials. Creep strengthening is accomplished by the formation of carbides in the microstructure [2,8].

Material creep resistance is typically shown in Larson Miller curves (parametric stress rupture strength). Minimum parametric stress to rupture values for a life time of 100,000 hr is represented in Figure 4.

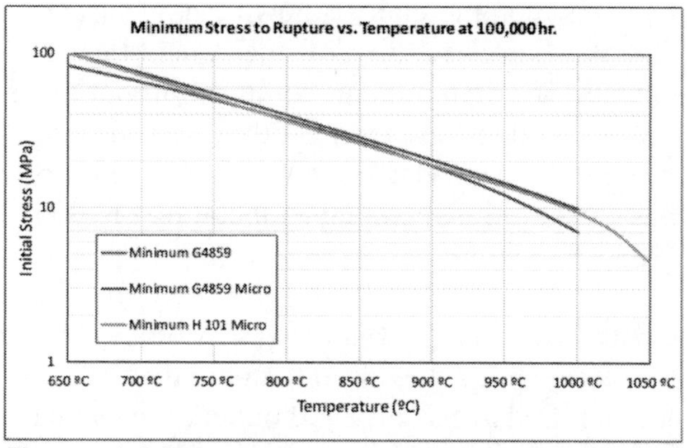

Figure 4. Minimum creep resistance comparison.

Both micro-alloyed materials have a superior creep resistance compared with G4859. Small additions of strongly carbide forming elements lead to the formation of stronger and more evenly distributed secondary carbides in the austenitic matrix [8].

Ductility after ageing

Ductility after ageing is a major requirement for these type of materials. Material samples are aged in a furnace (controlled temperature and time of ageing), and later aged samples are tensile tested at room temperature to measure elongation to rupture. Ageing conditions and test results are shown in Figure 5.

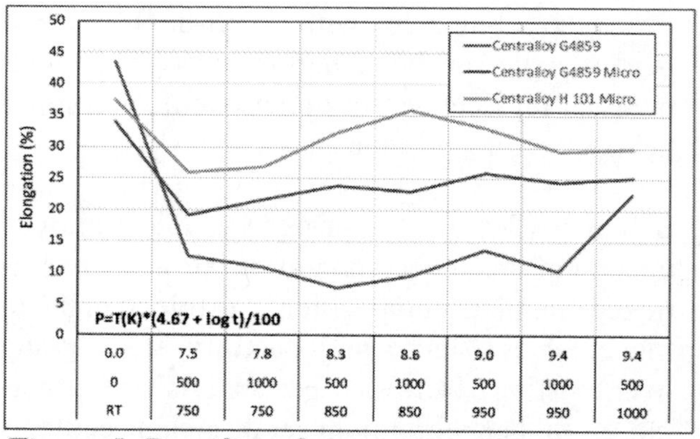

Figure 5. Ductility after ageing comparison.

Considering all three materials, superior ductility after ageing performance is achieved with alloys H101 Micro and G4859 Micro in comparison with alloy G4859. Such superior behavior might be explained by the formation of more thermodynamically stable primary carbides and due to delay/suppression of intermetallic phase formation [9].

Microscopy Inspection

Microstructure typically consists of austenitic dendrites surrounded by carbides in the interdendritic region. In addition to this network of primary carbides formed during material solidification, fine secondary carbides precipitates form in the austenitic matrix at service temperature [8].

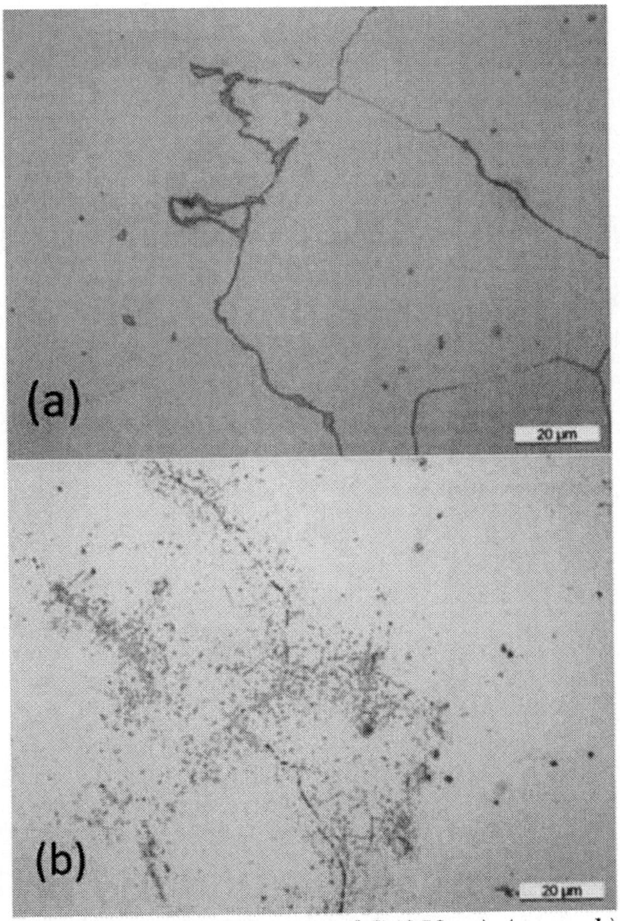

Figure 6. Microstructure of G4859 -a) As cast b) Aged at 850°C -1000hr.

Results for "as cast" and aged G4859 are presented in Figure 6. Carbide precipitation is almost continuous in interdendritic areas, while the aged sample has a secondary carbide precipitation around interdendritic zones.

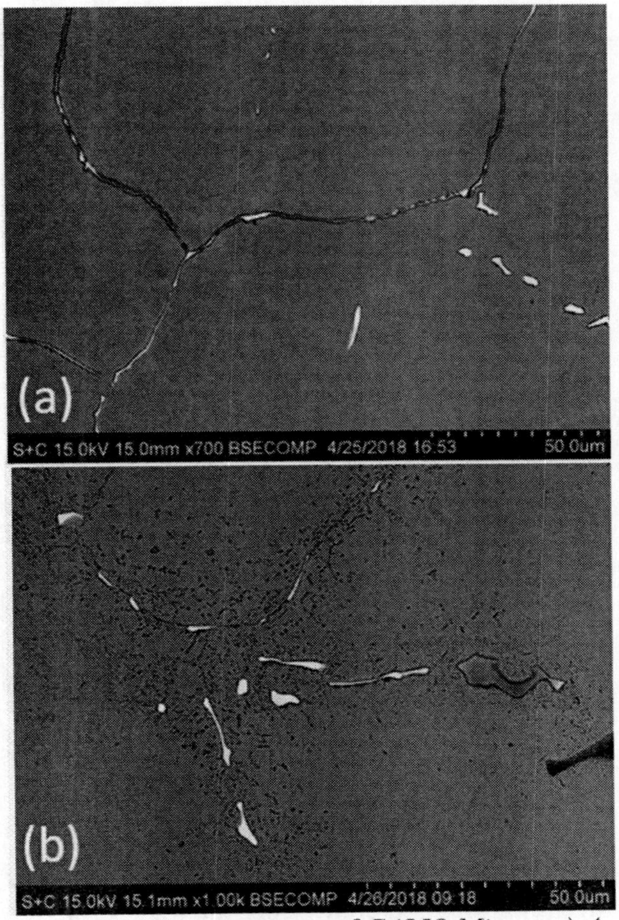

Figure 7. Microstructure of G4859 Micro -a) As cast b) Aged at 850°C -1000hr.

Figure 7 shows similar results for material G4859 Micro. Titanium carbides provide some disruption on the continuous interdendritic carbide precipitation which enhances creep material resistance.

Microstructure of H101 Micro samples is shown in Figure 8. Carbide structures are similar to those seen in previous samples. Nevertheless, austenitic matrix is more stable due to the higher content in Ni and Cr.

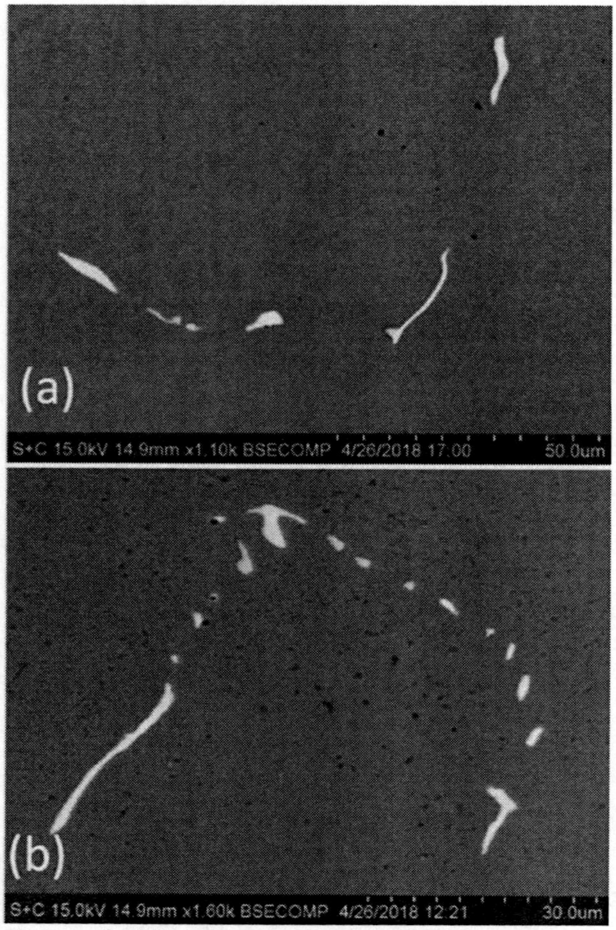

Figure 8. Microstructure of H101 Micro -a) As cast b) Aged at 850°C -1000hr.

No G-phase was detected in aged G4859 Micro and H101 Micro. Addition of elements like Ti are inhibiting and controlling the transformation of niobium carbides (TiNb)C into G-phase [3,4].

Effect of Material Selection on Expected Life

Lifetime calculations of these components are determined by the creep rupture strength of alloy material. In outlet components, the normal design method is the standard ASME B31.3, for a typical design time of 100,000 hr [10]. Based on design data (temperature, pressure, tube OD/ID), ASME B31.3 provides the stress thickness (ts), which is the minimum required wall thickness for a 100,000 hr design time:

$$t_s = \frac{P(d + 2 \cdot c)}{2[S \cdot E - P(1 - Y)]}$$

where:

P= design pressure (MPa).
d= inside diameter (mm).
c= additional thickness for corrosion/erosion (mm).
E= quality factor.
Y= coefficient.
S= 67% of average/80% of minimum stress to rupture at design temperature.

ASME B31.3 includes several quality factors, coefficients which are tabulated [10]. Examples of wall thickness calculations are presented in Table 2. Selection of an alloy material with a higher creep resistance would reduce minimum wall thickness of the component.

Unfortunately, this model does not consider the effect of cycling and extra wall thickness allowance is required to consider such cyclic operation [5].

The main benefits of wall thickness reduction would be: weight reduction (cost reduction) and increased resistance to thermal shock.

Alternatively, maintaining original wall thickness with the same design and operating conditions, more creep resistant materials can substantially increase the lifetime of such components as indicated in Table 2.

Centralloy®	P (MPa/psi)	T (°C/°F)	ID (mm)	OD (mm)	ts (mm)	Design time (hr)
G4859	2.94/426.4	940/1724	160	213.8	26.9	100000
G4859 Micro	2.94/426.4	940/1724	160	207.1	23.6	100000
H101 Micro	2.94/426.4	940/1724	160	211.8	25.9	100000
G4859 Micro	2.94/426.4	940/1724	160	213.8	26.9	216000
H101 Micro	2.94/426.4	940/1724	160	213.8	26.9	128000

Table 2. ASME B31.3 Calculations. Effect of alloy material over wall thickness of component.

FE Modelling of Outlet Manifolds

Finite Element Analysis (FEA), provides the opportunity of testing real components simulating the continuous operation in the furnace. Accuracy of these FEA models depend mainly on the quality of the creep model, boundary conditions, geometrical models and correct meshing.

Creep modeling is clearly the most complicated issue. Manufacturers like Schmidt+Clemens provide stress to rupture values parametrized in models like Larson-Miller. Such stress to rupture data is not adequate for dynamic modeling where creep elongation and stress relaxation behavior information is required.

Therefore, having a good creep behavior modeling is imperative in providing good FEA results.

FEA Model Description

ANSYS v. R.19.0 FEA software was used for these outlet header simulations. One standard outlet system was selected, with three supports for the outlet manifold tube arms. Location of these supports was based on existing designs (Figure 9).

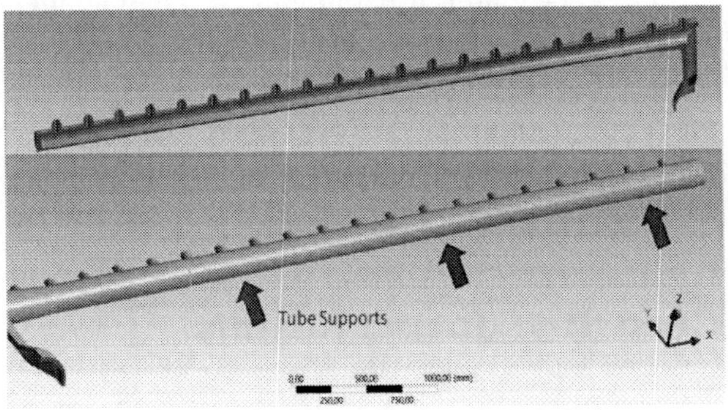

Figure 9. FEA Model and support position.

Model meshing comprises 2,170,628 nodes and 1,440,966 elements. Quadratic (solid 187) mesh elements were selected.

As boundary conditions for the model, fixed support at the bottom part of the cone was considered. A frictionless support was selected for the outlet header arm supports (allowing longitudinal displacement of header tubes).

As in typical header systems, self-weight of the system was compensated in the weldolets. Effect of gravity was also considered in the model.

As a first approach for material creep modelling, ANSYS creep model 10 (Norton equation) was selected. Norton is a power law model, only valid for the secondary creep stage. Nevertheless, stress relaxation behavior is considered in this model.

$$\dot{\varepsilon}_{cr} = C_1 \sigma^{C_2} e^{-C_3/T}$$

where:

$\dot{\varepsilon}_{cr}$= material creep rate (mm/(mm s)).

σ = material stress (MPa).

T= temperature (°C).

C_1, C_2, C_3= material parameters to be evaluated.

Regarding outlet header design conditions, homogeneous internal pressure of 2.5 MPa [391.5 psi] and temperature of 930°C [1706°F] were selected.

Case Studies and Results

Two simulation steps were considered for all three materials selected (Centralloy® G4859, G4859 Micro and H101 Micro)

First step considers a static structural simulation of just 1 second, to evaluate stress/strain distribution on the outlet header. Second step is a 1-year simulation of the part under creep regime.

Wall thickness of the component was estimated using ASME B31.3 calculations for alloy material G4859 considering a design time of 100,000 hr.

Loading Effect

Some surprising results were obtained from the static structural simulation. Even considering that reformer tube counterweight system was compensating all weight stresses over the material, significant part deflection was observed on header tube arms due to the thermal expansion of the transfer line as illustrated in Figure 10.

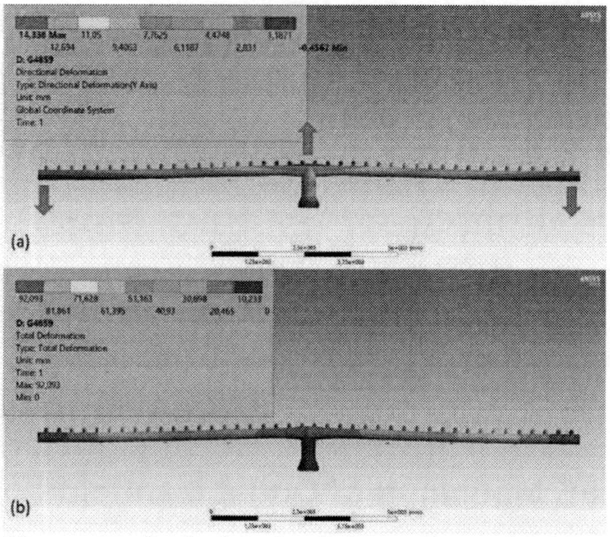

Figure 10. Outlet header deformation – a) Deformation Y-axis b) Total deformation.

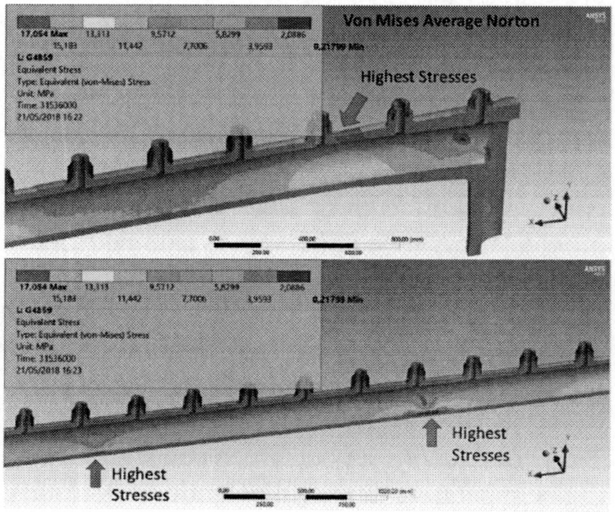

Figure 11. Average Norton (Von Mises) stress distribution in outlet header, alloy G 4859.

Effect of Material Selection

Average Norton stress distribution (Von Mises) plots for material G 4859 are presented in Figure 11 as examples of creep stress distribution in these materials. All material results are showing that the outlet header tube support location has a strong impact on the stress distribution in the component. Supported areas and T-piece/arm tube zones have the highest stresses in these models. Therefore, these parts are the most susceptible to failure due to creep damage.

Life time predictions can be applied into the model considering the Larson Miller curves for each material. Most stressed portions of the outlet header are determining the lifetime of the component in operation.

Time to rupture on the T-piece/arm tube zone for each material is represented in Figure 12. Minimum stress to rupture values are considered. According to this simulation, material G4859 will have a time to rupture of 133,000 hr in such area. On the other hand, more creep resistant material like G4859 Micro (255,000 hr) and H101 Micro (220,000 hr) will have a significantly higher time to rupture in operation.

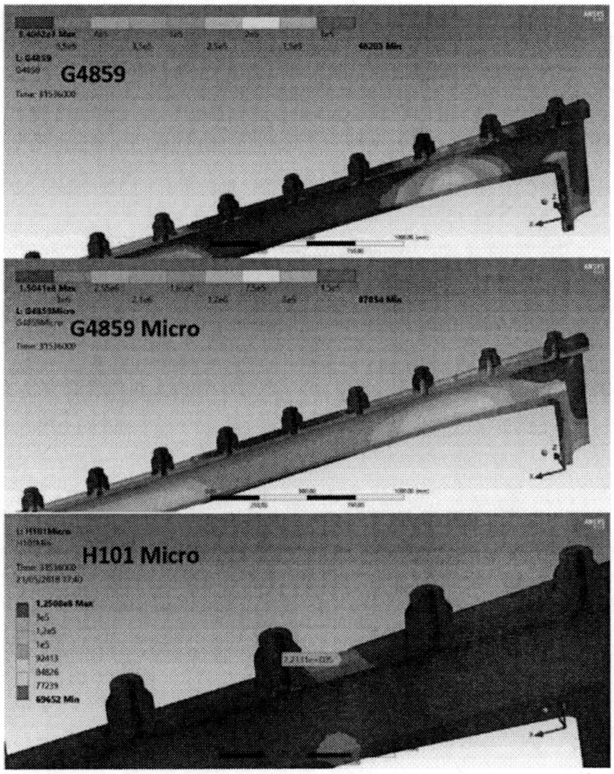

Figure 12. Time to rupture T-piece/arm tube zone for each material (minimum stress to rupture).

Lack of maintenance or inadequate alignment of header components (like arm supports) might add additional stresses on the part and, consequently, reduce dramatically the lifetime of the outer header. Figure 13 shows the effect of overstressing of the cast parts in material H101 Micro produced by the support. In this case time to rupture decreases to a value of just 81,403 hr.

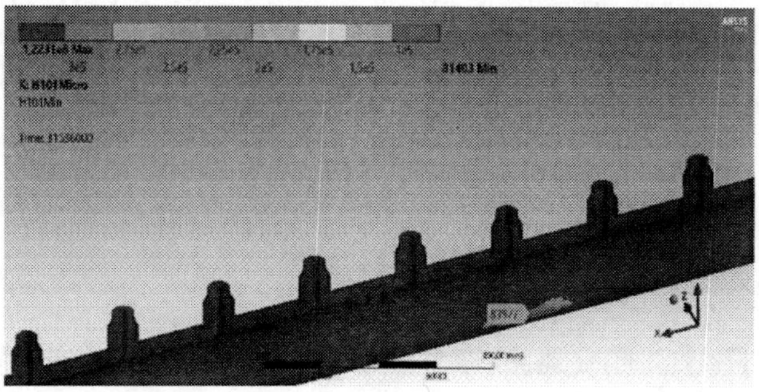

Figure 13. Time to rupture header arm support area for material H101 Micro (minimum stress to rupture).

Results of these simulation efforts show that material selection is the most critical part in ensuring that the outlet system is able to achieve the required design time.

Also, components of the entire header system like tube supports, can also have a significant impact in the outlet header performance/ life if they are not working adequately.

Conclusion

New Schmidt+Clemens materials are available for outlet header components as alternatives to standard material GX10NiCrNb32-20 or ASTM CT 15C (20Cr-32Ni-Nb).

These materials named as Centralloy® G4859 Micro and Centralloy® H101 Micro can significantly improve creep resistance and ductility after ageing over the standard alloy 20Cr-32Ni-Nb.

The main advantage for plant operators are wall thickness reduction of outlet header components, with benefits in weight reduction and increased resistance to thermal shock.

Alternatively, maintaining original wall thickness, lifetime of such components can be substantially increased keeping the same design and operating conditions.

Finite element analysis (FEA) can be carried out on outlet header systems, but the model accuracy strongly depends on the quality of the creep model, model design and part meshing. Creep modeling with existing data (Larson Miller plots) is especially complicated and requires additional efforts in developing a model with enough accuracy.

Analysis results indicate that new materials are improving time to rupture compared with standard 20Cr-32Ni-Nb materials. Nevertheless, inadequate operation or maintenance of outlet header components can lead to overstressed zones that can lead to early component ruptures in operation, even if the entire header is securely designed according corresponding standards.

References

1. *Improving Reformer Pigtail Reliability.* Nitrogen and Syngas 2014, Paper 34.
2. *High Temperature Aging Characteristics of 20Cr32Ni1Nb Castings.* Hoffman, J.J. s.l.: NACE Corrosion, 2000, Paper. 512.
3. Almeida, L.H., et al., Material Characterization, 2003, 49.
4. *Effects of Si Content on the Microstructure of HP-Mod Austenitic Steels.* Ibañez, R.A.P., de Almeida Soares, G.D. at al.s.l.: Materials Characterization, 1993, 30.
5. *Reformer Outlet System Reliability, Design and Repair*. Barnett, D. and Price, J. s.l.: Nitrogen and Syngas 2013, Paper 324.
6. *State-of-the-Art of Materials and Inspection Strategies for Reformer Tubes and Outlet Components.* Gommans, R., Jakobi, D., Jiménez, J.L. s.l.: 46th Annual Safety in Ammonia Plants and Related Facilities Symposium 2001.
7. *Experiences with Nb Containing Cast Materials for Fitting in Steam Reformer Furnace.* Shibasaki, T., Mohri, T., and Takemura, K. s.l.: 46th Annual Safety in Ammonia Plants and Related Facilities Symposium 2001.
8. *Centricast Materials for High Temperature Service. Jakobi, D., Huber, J. s.l.:* Nitrogen and Syngas 2010, Paper 307.
9. *Use of Titanium and Zirconium in Centrifugally Cast Heat Resistant Steel.* Caballero, F.G., Imízcoz, at al., s.l.: Materials Science and Technology 2007, Vol. 26 No. 5.
10. ASME Code of Process Piping, ASME B31.3-2016.

Management of Hot Spots in Refractory Lined High Temperature Equipment

Refractory lined equipment is typically used in high temperature processes in the chemical processing industry. Degradation of refractory over time commonly leads to development of hot spots at the pressure boundary which may present a serious risk to the structural integrity of the equipment and can even lead to a catastrophic failure. This paper discusses a successful experience of managing hot spots developed in refractory lined high temperature syngas piping in a steam methane reforming (SMR) plant. Level 3 fitness for service (FFS) analysis was performed to define a series of operating limits. Through rigorous temperature monitoring, in-situ inspection of critical dimensions and application of external cooling, the plant has been successfully operated under these conditions for an extended time period before the piping was replaced.

K. Xu, A. Alexander, T. S. Varma, M. DeLoatch and J. Armstrong
Praxair

Introduction

The outlet piping system of a steam methane reformer (SMR) plant gathers the reformed syngas from SMR catalyst tubes inside a furnace and transports the hot syngas to the process gas boiler (PGB). For hydrogen production, the reformed syngas typically has a temperature exceeding 870°C (1600°F) at the catalyst tube exit. One of the common outlet piping system designs is refractory lined piping where the pressure boundary can be constructed with low alloy steels such as carbon steel or Cr-Mo steels. This type of design is known as a cold outlet system. The refractory is designed with the following considerations:

1. The low alloy steel piping must stay below a certain temperature limit to prevent from high temperature creep and high temperature hydrogen attack (HTHA);
2. The temperature of the low alloy steel piping must also be greater than the dew point of the syngas to prevent carbonic acid induced aqueous corrosion.

Under normal conditions, such design provides adequate reliability for long term operation. However, it is not uncommon that premature degradation of the refractory occurs even during the early stage of plant operation possibly due to inferior quality or installation issues. The degradation can have a significant impact on the performance of the refractory, and as a result, hot spots can develop at the pressure boundary. Depending on the temperature, size, quantity, and location of the hot spots, the outlet piping system may be subjected to various failure mechanisms that present a significant risk to the structural integrity of the outlet system. Continuing operation with the presence of hot spots must be validated by fitness for service analysis. This technical paper shares a successful experience of managing hot spots developed in the outlet piping of an SMR plant. Through engineering stress analysis, rigorous monitoring, inspection and external cooling, the plant was successfully operated for an extended time period until the replacement of the outlet piping system.

Outlet Piping System Design and Operating Anomalies

The subject outlet piping system is part of the SMR plant that produces 8.75 tons/hour (82 MMSCFD) of hydrogen. The piping system consists of a transfer line that connects six headers to a PGB. Each header is connected to 39 catalyst tubes inside the SMR furnace. A schematic illustration of the outlet piping system is shown in Figure 1.

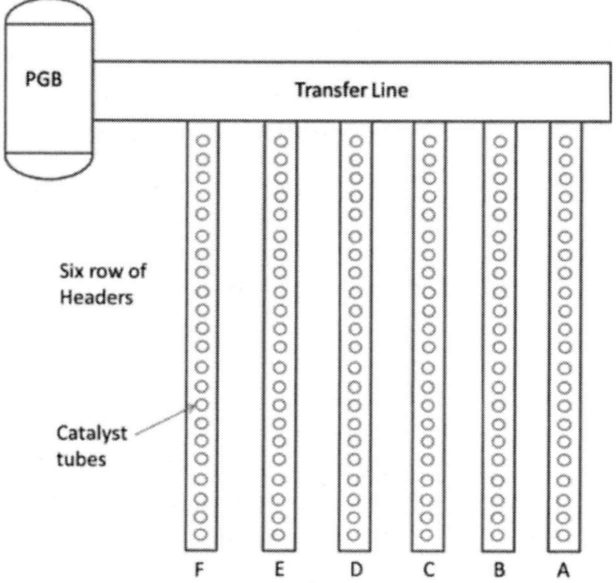

Figure 1. Illustration of outlet piping system

The piping is anchored at the PGB and the transfer line is allowed to grow away from the PGB to accommodate thermal expansion during operation. Similarly, the headers are allowed to grow away from the transfer line. The headers are also designed to support the catalyst tubes.

The piping system is designed to ASME B31.3 and is constructed with refractory lined ASME SA-387 Grade 11 Class 2 which is a 1.25 Cr 0.5 Mo alloy steel. The B31.3 design temperature is 400°C (750°F). The metal temperature of the piping is 200°C (390°F) at the normal operating condition. The temperature limit for thermal expansion design is 260°C (500°F).

During operation, numerous hot spots were detected, primarily at the catalyst tube to header joints but also the header to transfer line joints. Typical hot spots at the tube and header joints are shown in Figure 2. The temperature at the hot spots was up to 340°C (644°F), which is still below B31.3 design temperature, but significantly exceeded the normal operating temperature of 200°C (390°F) and the thermal expansion temperature limit.

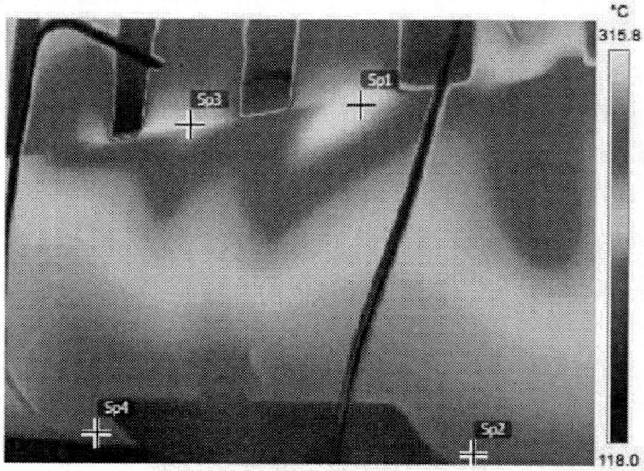

(a) Thermal image

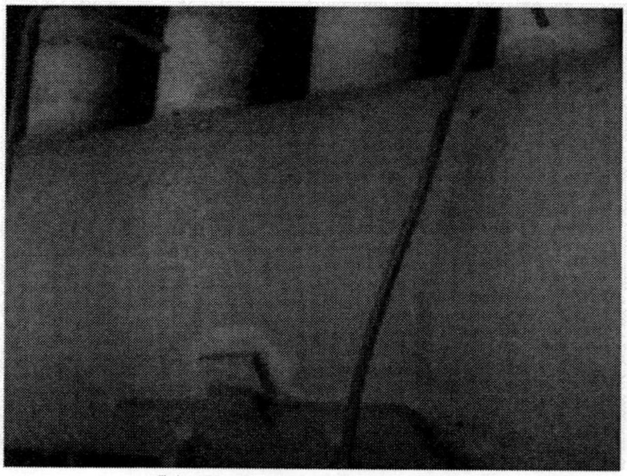

(b) Corresponding piping

Figure 2. Representative hot spots on header

As an immediate action, external cooling was applied by blowing air on the hot spots. The objective was to prevent further temperature excursion and to maintain the temperature at the hot spots below the B31.3 design temperature. Thermal imaging temperature measurement was deployed to constantly monitor the temperature variation.

Since the temperature was below the B31.3 design temperature, the risk of imminent failure due to high temperature failure mechanisms such as creep or HTHA was insignificant. However, due to the large quantity of hot spots, and the high temperature gradient at the hot spots, concerns were raised about possible high piping stress due to uneven temperature distribution, high localized stress at hot spots, and the ability to maintain thermal expansion and contraction at different plant rates. Therefore, extensive fitness for service analyses were performed to establish the operating limit of the SMR plant.

Fitness for Service Analysis

The unexpected temperature excursion obviously exceeded the original piping design scope. A Level 3 fitness for service study as defined in API-579[1] was conducted to investigate the structural integrity of the outlet piping system with the presence of hot spots and to define the temperature limits at the hot spots. The following factors were considered:

1. Thermal expansion limits of the global piping system
2. Global piping stresses in accordance with ASME B31.3[2]
3. Piping thermal stresses due to hot spots
4. Local stresses at hot spots in accordance with ASME Section VIII Division 2[3]

In the Level 3 fitness for service study, piping stress and flexibility analysis of a three-dimensional (3D) piping model and finite element analysis (FEA) of localized stresses at a hot spot were performed. The analytical calculations were performed by PAI Engineering.

Piping stress and flexibility analysis

A 3D piping model was established using Bentley AutoPIPE® that included the transfer line, six rows of outlet headers and catalyst tubes as shown in Figure 3. Refractory lining was also included in the model because of its significant influence on the piping stiffness. The global system stresses were calculated in accordance with ASME B31.3 with considerations of a range of friction coefficients at the piping supports. The results indicated that the highest piping stresses were located at Header A due to bending of the transfer pipe from thermal expansion. The calculated stresses are tabulated in Table 1.

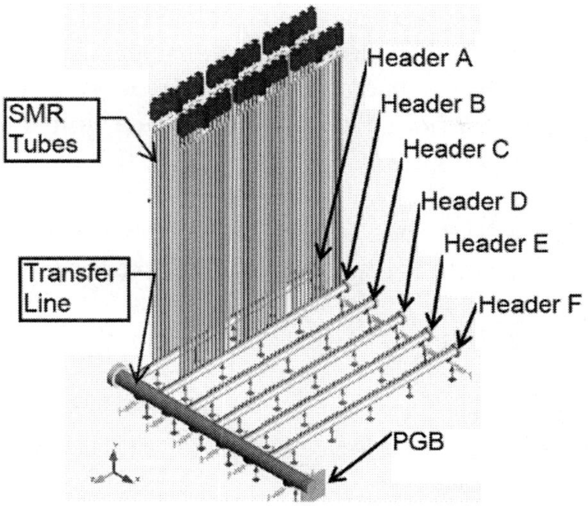

Figure 3. 3D piping model for piping stress and flexibility analysis

Table 1. Piping stress analysis results

Average pipe wall temperature	Calculated Stress Ratio		
	μ=0	μ=0.05	μ=0.2
200°C (392°F)	0.60	0.63	0.78
220°C (428°F)	0.68	0.70	0.84
260°C (500°F)	0.84	0.89	0.96

With a very conservative friction coefficient of 0.2, the piping stress was found to be below the B31.3 allowable stress (stress ratio less than 1) at 260°C (500°F). Based on the calculations, a maximum average pipe wall temperature of 260°C (500°F) was established. The temperature limit agreed with the design limit for thermal expansion. This temperature served as the base temperature for the FEA hot spots analysis.

Uneven average temperatures at the top and bottom of the headers were also taken into account in the piping stress calculation. At the observed overall temperature gradient, it was found that the effect of thermal bowing was not significant. However, the thermal bowing may affect the vertical piping supports and limit their design capability for thermal motion. Periodic inspections were recommended to ensure the performance of the piping support.

FEA analysis of hot spots

The impact of hot spots on localized stresses was investigated by extensive FEA analysis. The following five temperature profile cases were considered for hot spots at tube to header joints as shown in Figure 4:

Case 1: Hot spot around nozzle circumference with hot spot on adjacent nozzle at top

Case 2: Hot spot located at the side of the nozzle edge

Case 3: Hot spot located at the top of the nozzle edge

Case 4: Adjacent nozzles with hot spots around the nozzle circumference

Case 5: Adjacent nozzles with hot spots at the side edge

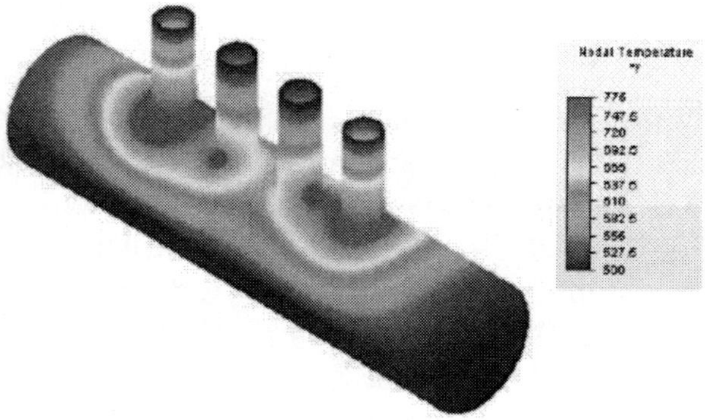

(a) Case 1

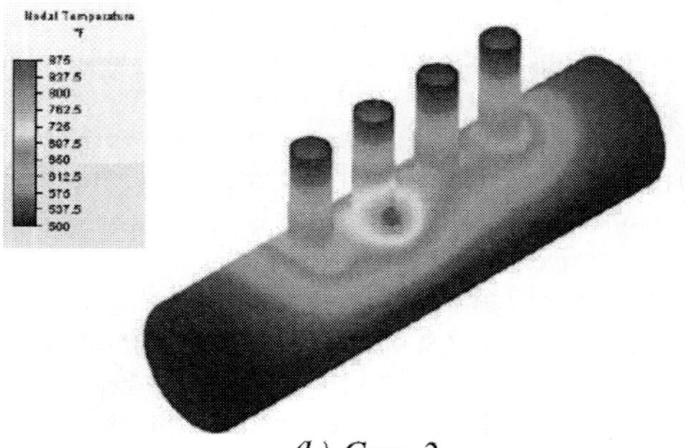

(b) Case 2

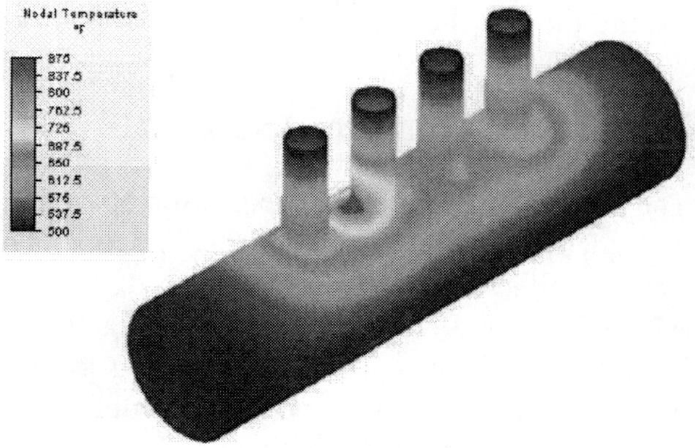

(c) Case 3

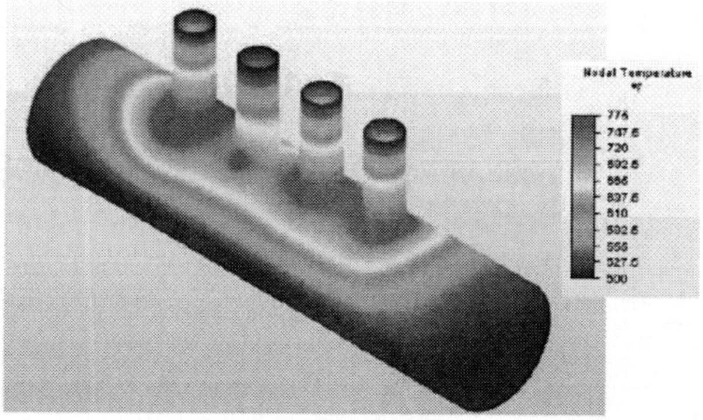

(d) Case 4

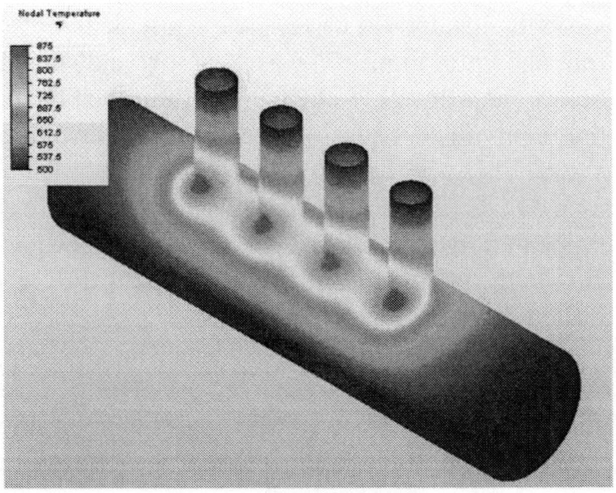

(e) Case 5

Figure 4. Temperature cases at SMR tube and header joints in FEA analysis

Similarly, two temperature profile cases were considered for hot spots at the header to transfer line joints as follows and shown in Figure 5:

Case 1: Hot spot near top of the transfer pipe

Case 2: Hot spot near reinforcement pad weld

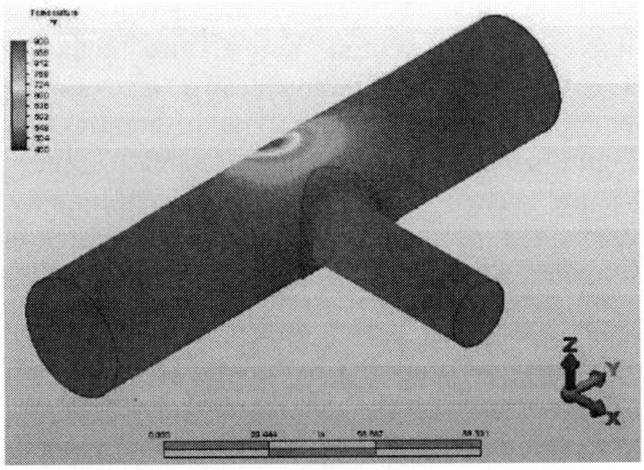

(a) Case 1

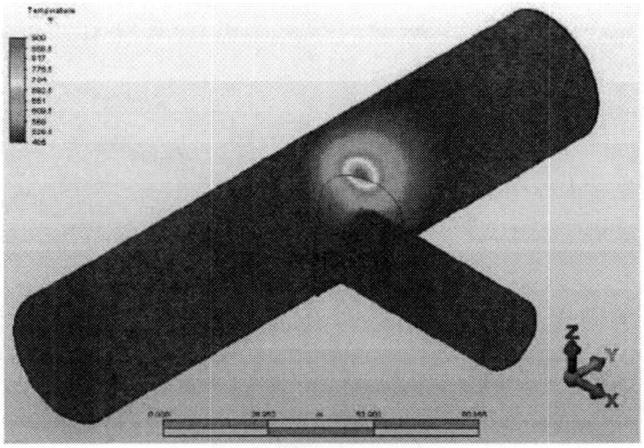

(b) Case 2

Figure 5. Temperature cases at header to transfer line joint in FEA analysis

In the FEA analysis, the lateral and vertical translational constraints were applied to one end of the header and transfer line such that the piping is anchored on one end and is allowed to move axially on the other end. The whole pipe section is allowed to move radially. The refractory lining was not included in the FEA model since the additional loads from the refractory were already considered in the piping analysis.

The local stresses were calculated at hot spots with the operating stress from the internal pressure. Based on the thermal imaging temperature measurements, the maximum hot spot temperature varied from 300°C (500°F) to 482°C (900°F). The average piping temperature was fixed at 260°C (500°F) at the bottom of the pipe. The calculated Von Mises equivalent stress was compared with the allowable stress determined based on primary stress plus secondary stress (S_{ps}) in accordance with ASME Section VIII Division 2 [2].

Figures 6 and 7 show the FEA stress analysis results for the header and transfer line respectively. The equivalent stresses are plotted against the hot spot temperature in five temperature profile cases. The allowable stress (S_{ps}) is also plotted as a comparison which shows a slight decrease with the hot spot temperature and a sharp decrease at about 454°C (850°F). On the header, the maximum temperature limited by the hoop stress is 473°C (885°F) based on B31.3 stress calculations.

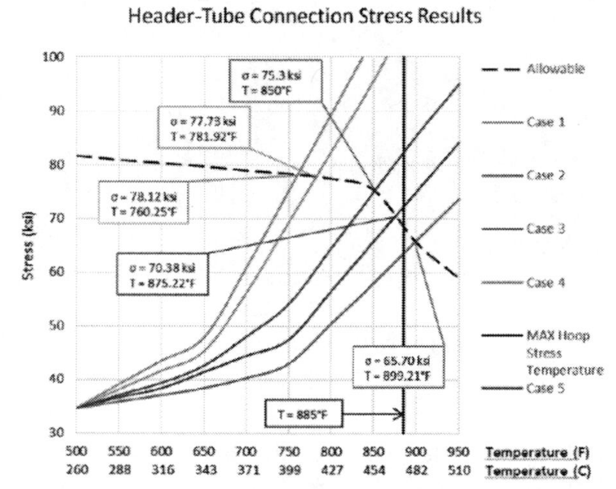

Figure 6. FEA stress analysis results on hot spot at tube to header joints

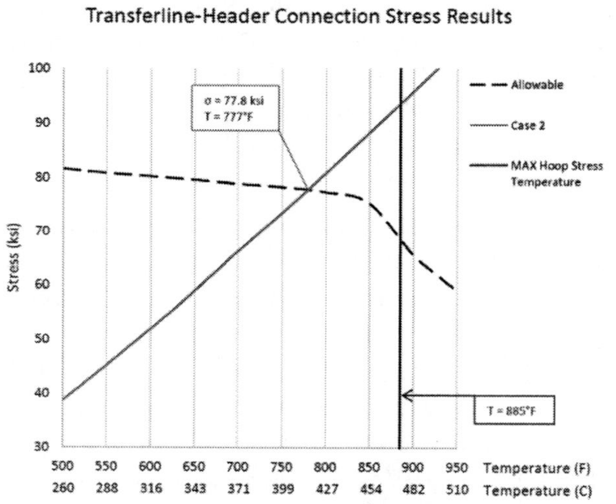

Figure 7. FEA stress analysis results on hot spot at header to transfer line joint

In all temperature cases except Case 2 for the tube to header joint, the equivalent stresses exceeded the allowable stress at a lower temperature than the B31.3 temperature limit. In the worst case (Case 4), the equivalent stress at the hot spot exceeded S_{ps} at 404°C (760°F) which represents a temperature gradient of 260°F. The maximum stress is located at the edge of the tube branch connection to the header. Therefore, Case 4 is the control case for the header. On the transfer line, the B31.3 temperature limit is the same as on the header. In Case 1, the equivalent stress (not shown in Figure 7) was very low and stayed below the allowable stress. On the other hand, the equivalent stress exceeded the allowable stress when the hot spot temperature is greater than 414°C (777°F) in Case 2. The highest stress is located at the weld of the reinforcement pad next to the hot spot. Therefore, Case 2 is the control case for the transfer line.

A sensitivity analysis was also performed to study the size effect of the hot spots on the stresses. The results indicated that the maximum local stress is dictated by the temperature gradient from the hot spot to the rest of the pipe rather than by the absolute temperature at the hot spots. Based on the hot spot size measurement and FEA analysis, the critical local stress can be represented by the differential temperature over 380mm (15in) distance between the hot spot and the bottom of the pipe.

In the actual hot spot situation, the temperature profile closely resembled Case 2 and Case 5 on the headers; and Case 1 on the transfer line. Based on the FEA analysis results, a conservative maximum differential temperature equal to 156°C (280°F) over 380mm (15in) distance at the hot spot was established. This differential temperature also agreed with the worst case scenario on the transfer line. The external cooling was controlled to maintain the maximum differential temperature throughout the outlet piping system.

Monitoring and Inspection

Once the operating temperature limit was established, a rigorous monitoring and inspection procedure was implemented in the plant. The monitoring and inspection matrix included:

- Twice a day temperature measurement at hot spots.
- Thermal imaging documentation of hot spots 2 - 3 times every month.
- Piping support inspection every two days

- Thermal expansion gaping inspection every two days

As an example, Figures 8 to 10 provide representative average header temperature measurement, the differential temperature measurement and thermal expansion of the headers respectively. The inspection frequency was increased whenever the plant experienced rapid load changes to meet customer demand.

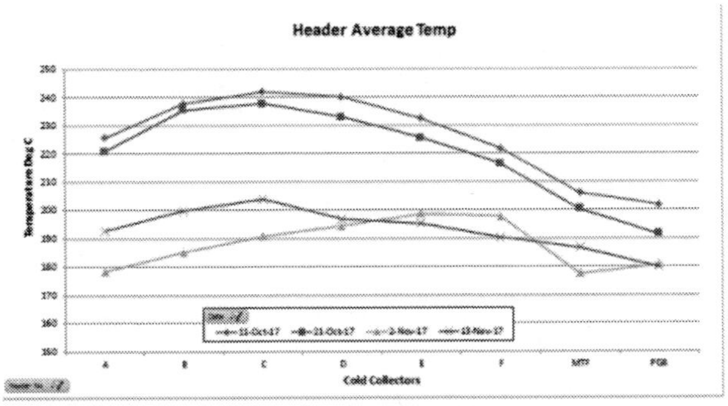

Figure 8. Header average temperature

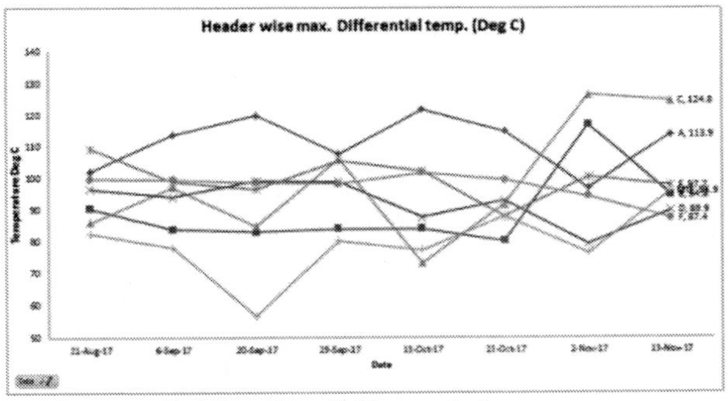

Figure 9. Header differential temperature

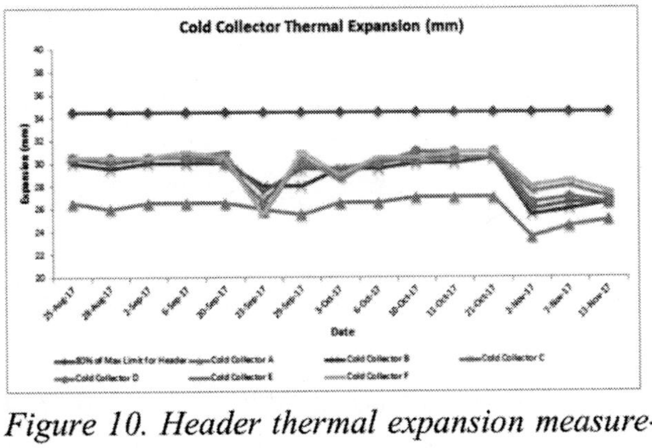

Figure 10. Header thermal expansion measurement

The inspection results indicated that the average header temperature was maintained below 240°C (464°F) in the worst case, which is lower than the 260°C (500°F) limit. The differential temperature between the hot spot and the bottom of the pipe was maintained below 125°C (224°F) in the worst case, below the 156°C (280°F) limit. The thermal expansion was maintained at 30mm (1.2in) in the worst case, below 80% of the design thermal expansion limit. In addition, the inspection of piping supports showed no indications of piping lift or restrictions of thermal motion. The outlet piping system has been successfully managed to operate within the safe design margin with the presence of significant hot spots.

Conclusions

Significant hot spots were detected on the refractory lined outlet piping system of an SMR plant. Extensive fitness for service analyses were performed to study the impact of hot spots on the structural integrity of the piping system and to establish the safe operating limits. External cooling was applied to control the piping temperature. Rigorous temperature monitoring and inspection procedures were implemented to ensure the plant was operated within the safe limits established by above analysis. The SMR plant was successfully operated over a year with the presence of hot spots on the outlet piping system until the refractory lined piping was replaced.

The successful experience demonstrated that the state of the art engineering analysis, robust monitoring and inspection procedures and a strong culture of operational discipline are essential components for safe plant operation in case an unexpected situation arises.

Acknowledgments

The authors are grateful to Praxair Inc. Management for the support and approval of publication of this work.

References

1. API-579/ASME FFS-1, "Fitness for Service", 2016
2. ASME B31.3-2016, "Process Piping", ASME, 2016
3. ASME Boiler and Pressure Vessel Code, Section VIII Rules for Construction of Pressure Vessels, Division 2 Alternative Rules, 2017

An Engineered Approach to Drying Refractory Concrete for the Ammonia and Petrochemical Industries

The drying or rapid heating of refractory and ceramics used in the ammonia and methanol industry can be problematic for industry. Despite many advances, the industry still has little choice but to follow supplier "rules of thumb" or "trial and error" when drying or heating refractory.

Using advanced engineering simulation techniques and testing installed refractory samples means drying schedules can be optimized and the most economical and safest schedule can be achieved.

Greg Palmer, James Millard and Juan Cobos
Palmer Technologies Pty Ltd - Australia

Hayden Williams
Ballance Agri-Nutrients – New Zealand

Introduction

Despite the advances in engineering analysis techniques, the drying and heating of refractory concrete after plant maintenance still relies on material supplier schedules which are historical rule based or based on testing of small refractory samples. The science of predicting water removal in porous concrete/ceramics under process conditions is complex. While the use of historical "rule based" schedules has sufficed, it has been at the expense of industry, with costs either in failures or lost production. The end result is that manufacturers' schedules for refractory drying remain highly conservative.

Other techniques have been tried such as using phosphoric acid refractory (water with a low pH) or various additives to try and improve water removal. Unfortunately, such non-engineered schedules are not risk free and do not add economic value to industry.

These risks have been compounded by the developments of high performance concrete (HPC), self-compacted concretes (SCC), low cement refractory concrete (LCC), ultra-low cement refractory concrete (ULCC) and no cement refractory concrete (NCC); which is achieved by selecting the particle size distribution to improve packing and replacing cement with ultra-fine particles such as silica fume and hydrate-able alumina.

There are two drivers for spalling of concrete - thermal strain caused by rapid heating and internal pore pressure during the heating of "wet" concrete. The problem of predicting safe heating rates for refractory concrete and the mechanism that causes spalling during drying is still debatable and a number of possible reasons exist.

However, using computational engineering techniques is a very efficient and reliable method of analyzing and optimizing drying schedules.

Industry Practice

Many industrial plants undertake refractory drying or heating by placing gas burners in process vessels and controlling the burners to heat the concrete at a prescribed rate. This rate, as mentioned, is based on historical heating rates or the heating of small samples heated in laboratory ovens. Both of these approaches have limitations which can over predict or under predict heating schedules.

Understanding the effects of different heating rates on "wet" concrete structures is of primary importance to engineers and industry. This is because heating of concrete and refractory materials can result in serious problems, particularly if an explosive spalling event occurs. If an explosive spalling event occurs, projectiles of reasonable mass (1 - 10 kg, 2 – 22 lbs) can be thrust violently over many meters. While surface spalling is less violent, the extent of damage can still be severe and, in both cases, repairs will be required resulting in significant costs to industry.

In the syngas reforming processes, which operate at high pressures and temperatures, typically 775°C to 1000°C (1427°F to 1832°F), it is necessary to line process vessels such as auto-thermal reforming reactors and transfer lines with refractory. Historically, reactors were lined with refractory brick. However, with the advent of refractory concretes in the mid 1950's, these linings were progressively changed to refractory concretes due to their lower price and faster installation times. This lead to numerous cases where the concrete could explode during heat-up due the build-up of steam pressure.

The refractory lining is necessary to prevent the steel shell from overheating, but it must be stable under process conditions. To achieve this, a high thermal resistance (insulation) refractory layer is installed next to the shell which is protected from the process gas by a low thermal resistance (dense) refractory layer or a (Inconel) metal liner. However, the use of metal liners introduces the risk of buckling and collapse, which can also be catastrophic.

The shift from brick lined systems to "wet" concrete systems, while having tremendous economic benefits, could also cause serious reliability issues.

Asset owners had to rely on refractory supply companies. Since the computational techniques were not available, heating schedules were developed by "trial and error." When these "trial and error" values failed, a schedule would become even more conservative.

Concrete Explosion During Heating

The mechanisms proposed for concrete explosive failure [4] are: Localized constraint by a colder zone of a heated plate which results in compression of the heated surface, pressure shear due to internal vapour pressure (moisture clog), hydraulic spalling and frictional forces.

Hydraulic spalling occurs when the expansion of water in a saturated pore is higher than the surrounding volume. If a closed pore is filled with water, then the expansion of heated water will cause hydraulic pressure to rise very quickly fracturing the surrounding material.

Spalling can occur when the material's tensile or shear stress is exceeded. Both these stresses can be estimated as they are related to compressive stress, f'_c.

Explosive failure of concrete is a particularly serious case and for this to occur there must be a sufficient instantaneous energy release to fracture the concrete and thrust a mass (perhaps 1 -2 kgs, 2.2 - 4.2lbs) through space over several metres. Figure 1 is of an exploded refractory panel for a LCC refractory concrete which occurred during dry-out.

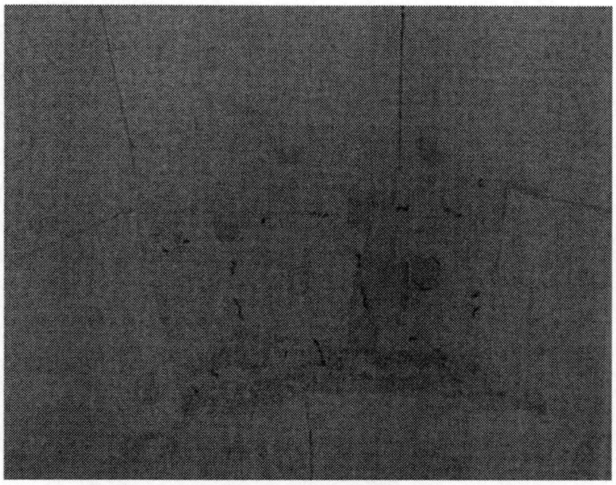

Figure 1. View of an exploded refractory panel which occurred during dry out

A quick check of the strength to thrust a projectile through space can be estimated from ballistic theory. This shows that a mass of approximately 1.1kg (2.4lb) can travel a distance of approx. 35m (115 ft) at a velocity of 17 ms^{-1} (56 ft/sec). The pressure required is approximately 1.8MPa (260 psi). However, there must be sufficient propellant available to ensure a pressure wave continues and a crack doesn't just instantaneously relieve pressure.

Thus, it seems quite realistic that vapour pressure and pressurized water have sufficient energy to cause an explosion spalling event.

Laboratory testing of concrete samples show it doesn't readily replicate what happens at the plant level. For example, Moore *et al* [3] studied the dehydration of 60% w/w alumina monolithic with 11% water. The panels were 600x600x150mm (2'x2'x6"), and were subject to very rapid heating and hold periods up to a max temperature of 600°C (1112°F). It was concluded that spalling was controlled in part by pore pressure which in turn is a function of the material's permeability.

Figure 2 shows the damage to concrete cylinders tested by Noumowe *et al* [4]. It shows the explosion failure of two concrete materials, one normal strength concrete with polypropylene fibres and a lightweight concrete without polypropylene fibres. During testing the lightweight concrete samples exploded between 290° and 430°C (554 and 806°F) and the normal strength concrete (2kg/m^3, 0.4lb/ft2) did not explode.

Figure 2. Explosive spalling of high performance concrete

Experiments were undertaken by Crowley and Johnson [5] who reported that curing temperature plays a critical role in the explosive tendency of refractory concretes. They found that concretes, which were inclined to explosively spall did not explode when they were cured above 21°C (70°F) before placing them in a furnace preheated to 870°C (1600°F).

Other researchers found: 1. Cement dehydration appears to have little effect on the drying profile and drying time for heating rates above 10°C/min (18°F/min), but explosive spalling occurred in bodies heated above 20°C/min (36°F/min) [6] and 2. Rapid heating does not cause damage to high alumina refractory but rapid cooling does, and that combined rapid heating and cooling will also cause refractory damage [7].

Thus, the research shows that laboratory testing of refractory concrete samples is variable and does not adequately simulate in-situ large refractory concrete sections in industrial processes.

Put simply, drying simulations using small laboratory samples do not replicate what will happen in larger scale industrial applications.

Water Migration

The removal of water from refractory concretes during heating occurs by diffusion and steam pressure. As the temperature of the material starts to increase towards the cold face, water will start to evaporate and move through the pores towards the heated surface due to pressure and humidity differences. However, this movement will be restricted by capillary size and presence of liquid water in the unheated side of the concrete, which will create a barrier for the steam to flow as water will be more resistant to flow due to higher viscosity. This liquid water in equilibrium with the steam is known as the “moisture clog” zone [8].

As steam pressure builds, water will be progressively pushed towards the cold surface where it will exit the concrete if unsealed. This common water migration has been reported by Jansson [1] and it is a very important factor to understand and consider during determination of refractory dry-outs. Not only the movement of liquid water and steam vapour can be aided by vapour pressure gradients, but also the thermal expansion of water can aid in this process. Water expansion can cause very high pressures if the constraining volume restricts water movement. In the case of open pores, water could be restrained in the circumferential direction, but it is the relief of pressure in the axial direction that causes the water to be “pumped” though the porous network.

Movement restrictions can cause excessive pressure to build up within a refractory concrete to the point where explosive spalling will occur. Thus, understanding material permeability, its changes during curing and heating is critical to predict water migration and steam pressure development.

Figure 3 shows the water content in both the hotface and insulation layers during heating. The results show there can be a small accumulation of water near the pore pressure peak. Figure 3 also shows that water is pushed to the shell side and the insulation becomes totally saturated adjacent to the shell. However, the most important issue during the hold period is that there is almost no change in the material water content. In fact, it is not until the pore pressure starts to develop that there is a decrease in the water content.

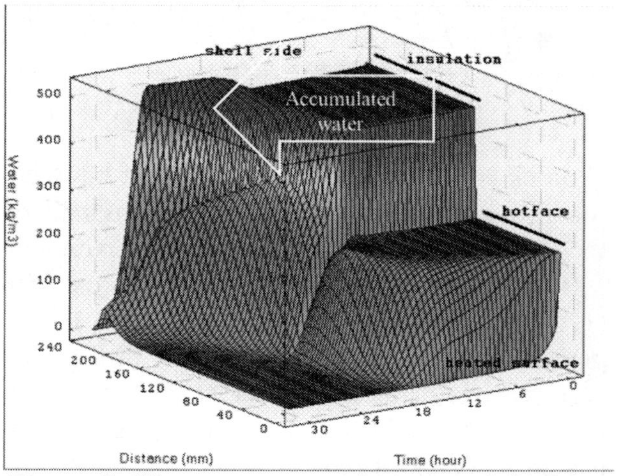

Figure 3. Water content with time and thickness – 50°C/hr with 10hr hold at 125°C (vertical axis is water content per m3, red greatest and blue least amount of water/m³)

Pre-weld and Post-weld Heat-Treatments

Welding of refractory lined steel sections for pressurized vessels such as transfer lines is critical. Drying the refractory concrete in these sections prior to being installed in-situ is not always practical, which means that welding needs to be completed on sections with undried refractory, which can be challenging.

Pre-weld heat treatments of the steel components are usually done at low temperatures (around 150°C or 300°F) for several hours, but they are designed only to prepare the steel for welding. What happens in practice is that once the welding process starts, the heat generated heats up the refractory to temperatures above 150°C (300°F) which will generate steam from the water contained in the undried refractory. This steam makes the welding process very difficult and can contribute to defects in the final weld. What can be done in this case is to complement the pre-weld heat treatment to remove just enough water to the point where generation of steam during the welding process is eliminated. This can be done in a controlled manner through simulation and has been proven in practice

After welding is completed then the finished weld will be post-weld heat treated by heating up to approx. 650°C (1200°F). This treatment needs to be complemented with a dry-out schedule to avoid damage to the refractory concrete. During the heat treatment, the water in the still wet refractory will be "pumped" or evaporated into steam and will move along the steel refractory interface or the porous refractory by diffusion and pressure differentials, just as if it was heated from the inside, but will move in opposite direction. What is important is that steam pressure is controlled, while the water and steam leave the refractory through the front and sides of the lining, to prevent high pressure peaks and ensure steam spalling will not occur.

A visualization of a simulation result for a post-weld heat treatment is shown in Figure 4. A redistribution of water away from the externally heated shell can be observed in (a). This re-distribution of water to the inner volume of the insulation layer must be taken into account during a subsequent refractory dry-out performed from the inside of the vessel. Dried refractory adjacent to the heated shell is shown in (b) at 4 hours into the heat treatment.

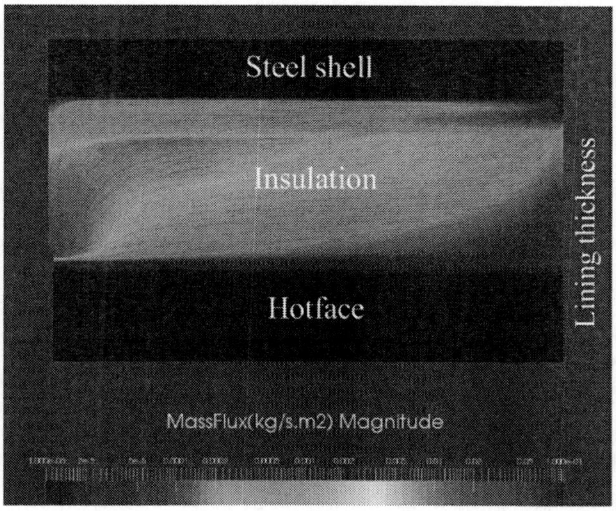

(a)

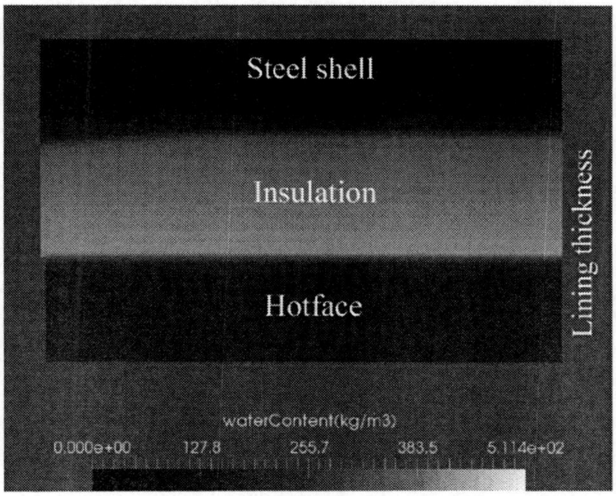

(b)

Figure 4. Visualization of water flux (a) and water content (b), throughout lining cross-section 4 hours into a post-weld heat treatment.

Predicting Concrete Dry-outs

Heat and mass transfer in concrete can be described by the water mass flux rate due to the pressure gradient and the water released during dehydration. Using numerical techniques, the governing equations for heat and mass transfer in concrete can be written to calculate the water mass flux rate due to the pressure gradient, the water released during dehydration, the heat flux rate due the heating of concrete, water (including heat of evaporation) plus heating by convective movement of water.

To overcome historical problems, namely explosive spalling, during dry-out manufacturer's heating schedules have been very slow and lengthy hold periods have been applied. These schedules have been on the assumption that hold periods lower the risk of explosion and improves drying. However, by using engineering techniques more reliable results can be achieved and optimal drying schedules prepared.

The numerical technique allows for multiple material types with different cement content, water content, permeability and thermal conductivity to be simulated.

Gas Reformer Heating Schedules – Case Study

Ballance Agri-Nutrients has always used the manufacturer's specifications for refractory dry-out in their ammonia plant. Typically, a refractory dry-out to normal operating temperature would take about 60 hours, see Figure 5.

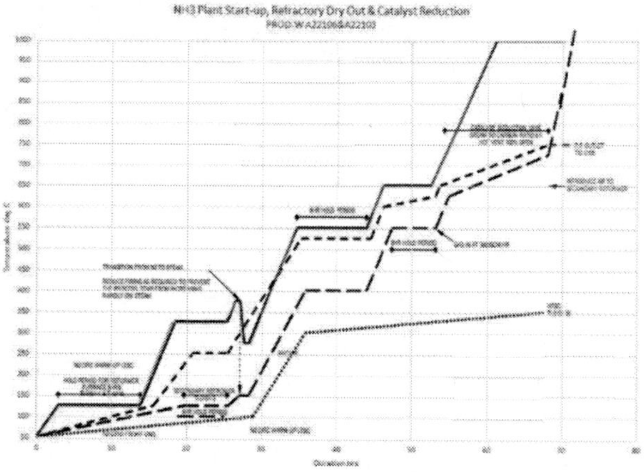

Figure 5. Ballance Agri-Nutrients Dry-out previous dryout schedules

During the 2015 ammonia producer's symposium, Palmer Technologies presented a methodology for working out customized curves for specific plants. Ballance engaged Palmers' to develop a curve for their syngas reformer and ancillary equipment customized to their turnaround repair scope. The software predicted a safe heating rate, which almost halved the startup duration, from the previous heat-up schedule, whilst avoiding adverse pressure and thermo-mechanical stress.

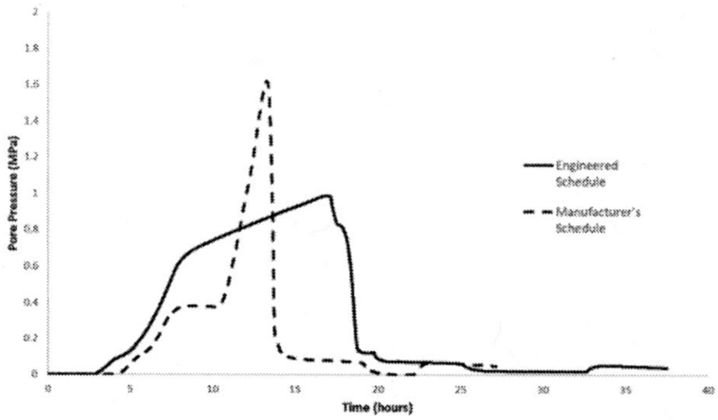

Figure 6. Pore pressure vs time for manufacturers and our dry-out curve.

As shown in Figure 6, the engineered schedule substantially lowered the risk of an explosive event when compared to the manufacturer's schedule, controlling the peak pore pressure at a lower value and minimizing the sensitivity to minor temperature divergences and process fluctuations.

Ballance will look to use this technology in the future to correlate turnaround repair scope with a customized reformer startup curve.

Installed Quality

One important consideration is the quality of the installed material. The quality is very important as under process condition it behaves very differently. The installation of the lining is critically important, it cannot be thought of just another piece of concrete, i.e., "near enough is good enough".

Refractory is commonly installed by casting (pouring) or gunning (spraying), both of which require addition of water when mixing. To ensure the desired material properties are achieved, it is very important to control quality during placement, as increasing water content to improve workability will decrease strength and increases porosity. The other problem of increased water content is concrete shrinkage during drying causing cracks.

Conclusions

From a review of the literature the following was concluded:

- Accurate measurement of permeability is essential.
- The addition of polymer fibre can reduce explosive spalling risk but cannot eliminate it.
- Maximum pore pressure is highly dependent on heating rate.
- Explosive spalling results when the peak pore vapour pressure initiates a crack and the sudden release of the superheated liquid water into steam acts as a propellant.
- Size effects prevent naïve testing of small samples in laboratory settings.

We have established that with a detailed simulation of the heat and mass transport during drying, the effects of process-refractory interaction can be accurately resolved.

It has been found that for the analyzed lining configurations, the existing manufacturer heating schedules produce uncontrolled pore pressure build up, with high sensitivity to variations in temperature.

The 3D model of heat mass transfer has been applied to industrial gas reformers to optimize the heating schedule, improving operational up-time and decreasing exposure to the risk of catastrophic spalling events.

References

1. Jansson R., Material Properties Related to Fire Spalling of Concrete, Report TVBM-3143, Lund Institute of Technology, 2013
2. Zhao J., Fire-Induced Spalling Modelling of High-Performance Concrete, PhD Thesis, University Delft, 2012
3. Moore R.E., Smith J.D., Sander T.P. and Severin N., *Dewatering of Monolithic Refractory Castables: Experimental and Practical Experience*, UNITECR 97
4. Noumowe A. N., Siddique R. and Debicki G., “Permeability of high-performance concrete subjected to elevated temperature (600°C)” Construction and Building Materials 23, 2009
5. Crowley, M.S.; Johnson, R.C.: Guidelines for installing and drying refractory concrete linings in petroleum and petrochemical units. Amer Ceram. Soc. Bull. 51 (1972) [3] 226-230
6. Innocentini, M.D.M.; et al.: Drying behaviour of refractory castables. Amer. Ceram. Soc. Bull. 80 (2001) [11] 47-57
7. Pereira, A.H.A.; et al.: Elastic modulus, dampening and modulus of rupture changes in refractory castable due to thermal shock damage. 52nd Int. Coll. On Refractories, Aachen, Germany, 2009
8. Velez, M.; Headrick, W.L.; Karakus, M.; Moore, R.E.: Dewatering Castable. 39th Symposium on Refractories. St. Louis, April 2003

Asset Integrity Management Programs for Ammonia Plant Facilities

Effectively managing the integrity of key assets in ammonia plant facilities is essential to maintaining high levels of both personnel and process safety as well as ensuring that economic benefits are obtained from the investment in those facilities. This paper reviews the essential elements of an asset integrity management program that were addressed in a recent audit of several world-class ammonia plants that typically produce more than 500,000 metric tons of ammonia per year. Methods for judging the effectiveness of an asset integrity management program are reviewed.

Carl E. Jaske
HSI GROUP, INC.

Steven J. Weichel
DNV GL USA, INC.

Michiel P. H. Brongers
DNV GL USA, INC.

Introduction

Most ammonia plants use a hydrocarbon feedstock, typically natural gas, to produce hydrogen by means of steam methane reforming. Hydrogen is then combined with nitrogen to produce ammonia by means of the Haber-Bosch process. After sulfur is removed from the feed gas, the gas is preheated before being mixed with steam and fed to the primary reformer. The gas from the primary reformer is mixed with air in the secondary reformer to produce the synthesis gas. This gas then flows through shift reactors to produce carbon dioxide, which is removed from the mixture before the methanation process followed by ammonia synthesis.

The processes described above require critical equipment, including piping, pressure vessels, boilers, tanks, compressors, turbines, valves, instrumentation, structures, and electrical cabling. This includes equipment that operates at low temperatures and other equipment that operates at very high temperatures. Some of the equipment also is exposed to corrosive operating conditions. To maintain and assure reliable operation of such equipment, world-class ammonia plants typically employ asset integrity management (AIM) programs.

In a previous paper [1], the authors presented an overview of important features of an AIM program for ammonia and methanol plants. The present paper focusses on the essential elements of

AIM programs that were included in a recent national facilities audit/benchmarking study of 30 companies in Trinidad & Tobago's energy sector [2]. This audit covered rating both the organization's AIM systems and the application of those systems to equipment in the facility.

What is asset integrity management? Asset integrity has two components: (1) technical integrity is the management of barriers, which are typically non-physical such as operating limits, and (2) process safety is the management of hazards to prevent major incidents. Performing these two activities together is then referred to as asset integrity management. When properly implemented, AIM enables a company to use an asset efficiently while safeguarding life and environment.

In the framework of AIM, it is important to consider the difference between process safety and personal safety. Figure 1 compares these two aspects of safety in terms of risk. Process safety incidents usually have a lower frequency and higher consequences than personal safety. Process safety is not as intuitive as personal safety because people often lack personal experience with it. Since high consequences and potentially massive and lasting impacts are considered unacceptable, process safety necessitates highly reliable controls. Thus, managing process hazards is not directly linked to managing personal health and safety.

Nomenclature

AIM Asset Integrity Management
E&I Electrical & Instrumentation
FFS Fitness for Service
HSEQ Health, Safety, Environment, and Quality
KPIs Key Performance Indicators
LOC Loss of Containment
MOC Management of Change
PHA Process Hazard Analysis
PSM Process Safety Management
PSV Pressure Safety Valve
QA Quality Assurance
RBI Risk Based Inspection.

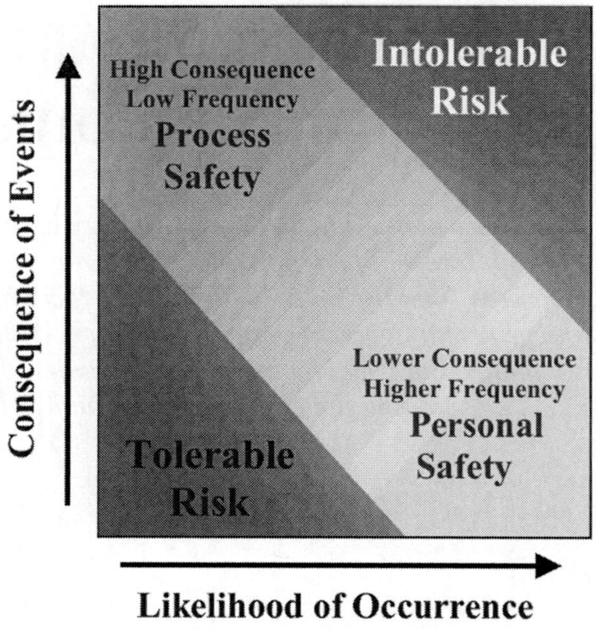

Figure 1. Comparison of Risk for Process Safety with Risk for Personal Safety.

Topics Covered in Audit

The following twelve topics were reviewed and rated in the audit of AIM systems in ammonia plants. The highlighted ones are discussed in this paper.

- Leadership
- Planning and administration
- Risk management
- Training & competence
- Contractor management
- Communication & promotion
- Project management
- Asset management
- Change management
- Learning from events
- Emergency response
- Results & review

Application of AIM to the following eight categories of equipment was rated in the audit of ammonia plants:

- Electrical & instrumentation

- Facility piping
- Fired heaters & boilers
- Pressure vessels
- Pressure safety valves
- Rotating equipment
- Tanks
- Structures

For the most part, the ammonia plants that were audited were rated above the average of the 30 energy companies that were included in the overall audit. These plants and others that were above average were typically part of large multi-national organizations with well developed process and personnel safety cultures.

The twelve audit topics do not stand alone, but tie together to form an AIM plan. As will be discussed later, ISO 55000 – Asset Integrity Management [3] defines 24 different topics that should be contained in a complete AIM plan. Important aspects are discussed in the remaining sections of this paper, organized under the headings Leadership, Risk Management, Management of Change, and Emergency Response.

Leadership

Commitment and buy-in from upper management all the way down to the personnel working daily in the plant are needed for a truly effective and successful AIM program. The audit showed that the ammonia plants have leadership that defines their organizations' vision, purpose, goals, policies, and values. These were typically aligned with those of other internal and external stakeholders. The leaders of these organizations also identified the major risks of the ammonia production business and each of the audited plants had established a formal program for process safety management (PSM).

In our audit of the ammonia plants' management systems, we found that various levels of management from team foremen to the plant manager to corporate management typically supported AIM programs. Some plants held quarterly "town hall" style meetings to keep workers up to date. Additionally, most plants used newsletters, electronic communication, bulletin boards, and employee recognition to keep personnel informed. Some companies encouraged staff to attend industry meetings and conferences that address safety and integrity management.

Risk Management

The companies included in the audit employed a systematic approach to managing their exposure to intrinsic risk by control of major accident hazards. An example of a risk management approach is provided in ISO 31000 and 31010 [4, 5]. Intrinsic risk scores remain fairly static and change only with significant changes at the site, such as inventories and processes, local population, or surrounding environment. Intrinsic risk can be assessed using safety hazard grouping and environment hazard grouping in accordance with the SEVESO III Directive on the Control of Major Accident Hazards (COMAH) approach [6], as illustrated in Figure 2.

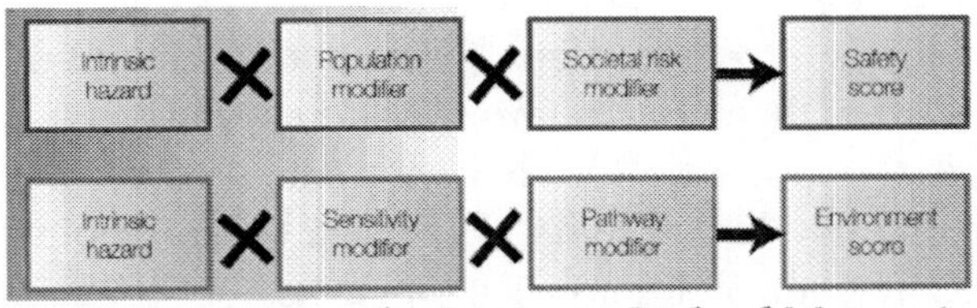

Figure 2. Rating the Intrinsic Risk of Major Accident Hazards.

Company leadership teams were committed to a defined risk management policy that engaged stakeholders and aligned with the culture of the organization. The organizations provided the resources needed to carry out the risk management program, measure its effectiveness, and make improvements.

Both internal and external factors were used to define the types of risks that the organization considered, the processes that they used to manage risk, and the types of risk criteria that they considered. These criteria included the types of consequences that can occur and what events can cause them, measures of likelihood of events, how risk levels are determined, and levels of risk acceptability.

Typical risk management activities include the following steps:

- Identification
- Analysis
- Evaluation

Risk identification consists of determining the risk sources, their causes, and their potential impacts, including if knock-on effects are possible. A knock-on effect is a secondary or incidental effect that is indirectly caused by an event. The main tasks of risk analysis are determining the magnitude of potential consequences and the likelihood of those consequences. The risks are evaluated by comparing them with the established criteria, so controls can be selected to result in risks that are in accordance with the risk management policy.

The ammonia plants used process hazard analysis (PHA) techniques to identify and evaluate the various process safety risks in their facilities. Most of them used a formal risk-based inspection (RBI) approach for their pressure vessels and piping systems. Some of the plants also applied RBI to their ammonia storage tanks. RBI was typically not applied to the plant structures or electrical and instrumentation systems because formal RBI procedures have not been developed for these applications. However, the plants did have inspection programs for electrical and instrumentation, with emphasis on equipment related to safety. Inspection of structures was often a lower priority and identified as an area of improvement.

Risk control strategies typically included the following four options: (1) terminate, (2) treat, (3) transfer, or (4) tolerate.

1. Terminating is the elimination or avoidance of risk by deciding not to undertake a particular activity or course of action.
2. Treating the risk can include taking actions to modify its likelihood and/or altering its consequences.
3. Transferring risk is sharing the risk with other parties. Examples include using a joint venture and using insurance.
4. Tolerating a risk is accepting both its likelihood and potential consequences.

The selection and implementation of control strategies are reviewed and modified if needed periodically as part of the AIM program.

The ammonia plants employ a hierarchy of controls to manage the risks.

1. Engineering or design controls are the first choice to eliminate risks where possible.
2. Administrative controls including procedures, rules, work permits and warning signs are the next choice to mitigate risk.
3. Personal and environmental protective equipment are the last line of defense.

Materials and products are identified, labeled, stored, and inspected to ensure that their quality is properly controlled. Process controls are maintained within critical parameters as part of process safety management.

The benchmarking study revealed that several ammonia plants replaced their converter vessel because a potential problem had been identified with a major weld seam. The plant management and owners decided that they could not accept the risk of this large, high pressure vessel failing even though the cost of replacement was quite high.

Management of Change

Effective management of change (MOC) processes are essential component of AIM programs. Two main types of change occur in operating ammonia facilities: organizational and engineering.

Management of Organizational Change

Organizational change is a constant in modern facilities and should be controlled with an effective change management process. The process should include mechanisms for effectively retaining critical skills and capabilities and for addressing the risks associated with changing the structure of the organization.

The audit of ammonia plants revealed that most of them have some type of system to make sure that staff with the proper skill levels fill the organizational positions related to AIM. However, such systems were not normally part of their formal MOC processes.

Management of Engineering Change

Effective engineering change management provides a coherent, systematic, and simple mechanism for identifying and controlling hazards throughout the process of change with emphasis on the transition phase. Well implemented, MOC ensures that the health, safety, and environment of the facility and its personnel are not compromised by inadequate evaluation of hazards, threats, and other potential undesired events related to change. MOC also ensures that the intended benefits of the change are fully realized as planned.

The audit showed that the ammonia plants had a documented, formal process for managing engineering changes. This MOC process typically is tracked and documented through a computerized database system to ensure tracking, signoff, and retention of records. Temporary changes were found to go through the same process and have limited lifetimes. In such cases, they were required to be replaced by a permanent change at the earliest opportunity or no later than the next turnaround.

Communication of Change

Good communication is essential for effective management of change. In an ever-changing ammonia plant workplace, effective communication is critical both to inform and motivate personnel. Good communication is more than telling; it should be an interactive process of giving and getting understanding. Promotional campaigns and a variety of communication channels should be used to promote process safety as well as health, safety, environment, and quality (HSEQ) improvement in a fresh and interesting way. Exceptional group and individual performance should be identified and widely communicated to reinforce positive behaviors.

The audited ammonia facilities required approvals of changes from all affected groups. Changes affecting safety and asset integrity were reviewed at daily, weekly, and monthly staff meetings. Staff were recognized for outstanding performance and for identifying problems or potential safety issues that led to the need for change.

Learning from Events

Learning from events is critical to continual improvement in safety, environmental and business performance. An effective system for learning from events can transform undesired events into opportunities for improvement. Managers should strive to create a no-blame culture to foster high levels of event reporting. Personnel should be encouraged to report near-misses because they offer the greatest number of learning opportunities. All events should be assessed for risk, considering what did happen and what could have happened (actual vs. potential consequence), and investigated appropriately. This effort should involve all relevant personnel. Investigations must uncover the basic (underlying) causes of events before determining the necessary corrective and preventive actions. These actions should be tracked to completion and the results should be communicated to all necessary stakeholders.

Besides learning from events within their own facilities, personnel can learn from events that occur elsewhere in the ammonia industry. This includes neighboring plants in an industrial estate, other plants within the organization, and similar plants within the global ammonia industry. An especially important source of relevant industry information are the publications from this annual symposium on safety in ammonia plants [7]. In 2005, Pattabathula et al. [8] reviewed 290 incidents that had been reported in the first 49 symposia. In 2015, Pattabathula and Richardson [9] updated this review to summarize the 86 incidents that had been reported in the next 10 symposia. In the current study, we identified 14 more relevant incidents from papers published in the 2015, 2016, and 2017 symposia [10-21].

Based on the descriptions of the 390 incidents in the referenced papers, we used our judgment to estimate a level of safety risk for each one based on the following four categories:

- Low: likely to result in no or slight injury
- Medium Low: likely to result in minor injury and affect work performance
- Medium High: likely to result in severe injury or disability
- High: likely to result in loss of life

Our results are summarized in Table 1.

Risk Category	Total	Percentage
Low	14	3.6%
Medium Low	222	56.9%
Medium High	140	35.9%
High	14	3.6%
Total	390	100%

Table 1. Summary of estimated safety risk for incidents reported in previous ammonia safety symposia.

The high and medium high levels of safety risk made up 154 or 39.5% of the incidents. Plant management and engineering staff should review the reports of such incidents and make sure that the issues that caused them have been properly addressed in their AIM program. They also should consider sharing incidents from their plant by participating in safety-oriented conferences and symposia. In fact, many of the organizations included in the AIM audit participate in the AIChE symposia.

The results of our audit showed that the ammonia plants were open to employee feedback. There typically was a blame-free culture in the organizations. Reporting of safety related events, including near-misses, was encouraged. Most of the plants applied root-cause analysis in their investigation of incidents. Some plants sent staff to corporate meetings to share learnings with other plants in their global organization. In a few cases, the plant personnel participated in local or international conferences where events related to AIM were discussed.

Emergency Response

Effective emergency preparedness requires planning and practicing in advance so that, in the event of an emergency, the harm to people, to the environment, and to the business is minimized. Potential emergencies need to be identified and categorized across the spectrum of risk including the extreme cases, such as low-likelihood, high-consequence events. The level of consequence impact should be determined for each scenario that is considered. Plans should then be developed to respond to each of these emergencies. Systems for emergency communication should be established and technical systems, for example, fire protection and emergency power. Teams of experienced personnel should be established to execute the emergency plans; their competence should be ensured through regular training, drills, and exercises. Finally, adequate first aid and medical support should be available if required.

The plants in the AIM audit had developed emergency response plans that included training on-site personnel to respond to emergencies, working with neighboring plants, and coordinating with local emergency units. Various scenarios for incidents were postulated and drills were held on a rotating basis to practice addressing them.

IMPLEMENTATION OF PROGRAM

Principles, requirements, and guidelines for implementation of an AIM program are provided in ISO standards [3, 22, 23]. Figure 3 shows the relationship among the key terms used in asset management while Figure 4 illustrates the key elements of an assessment management system as presented in ISO 55000 [3].

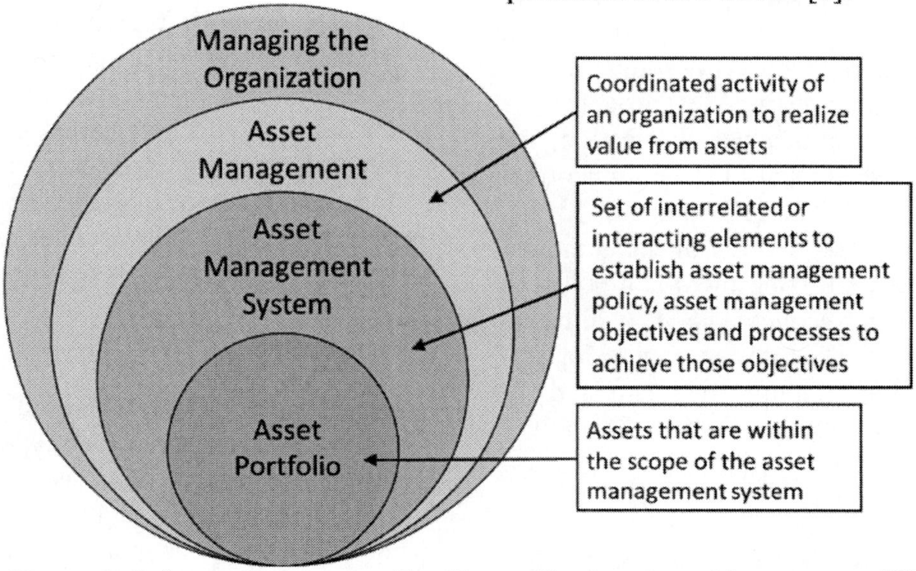

Figure 3. Relationship Among Key Terms Used in Asset Management [3]

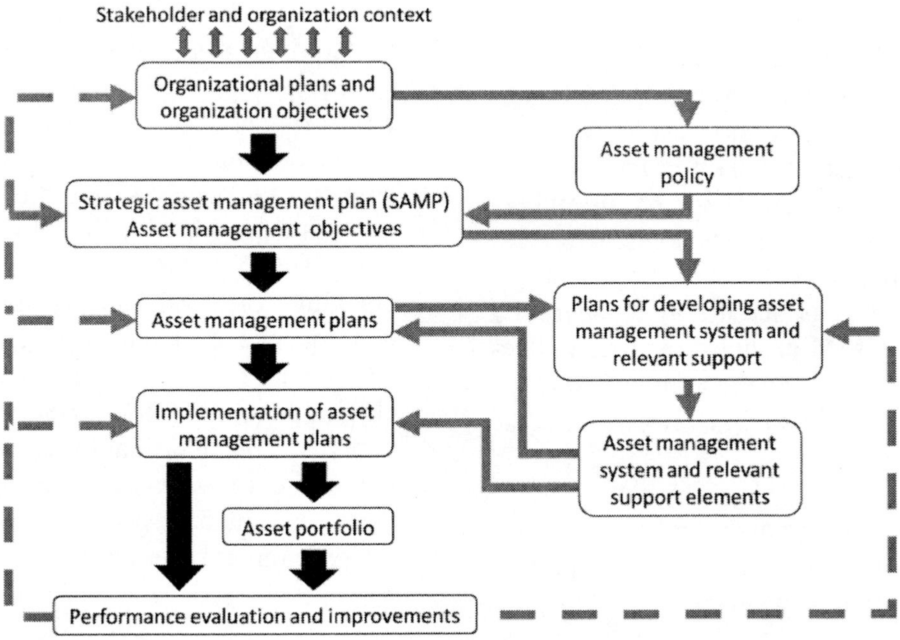

Figure 4. Relationship Among the Key Elements of an Asset Management System [3].

Asset management is part of managing the organization and includes the management system as well as the asset portfolio. An AIM program is an important part of asset management. The system includes a continuous improvement process where results are evaluated and fed back into various aspects of the system as required.

Process safety management [24] includes many of the same items as asset integrity management and is typically part of an AIM program. Important items to include in AIM program are quality assurance, maintenance, inspection, fitness-for-service assessment, and equipment repair and replacement. The ISO standards [3, 22, 23] were not referenced by any of the ammonia plants in the audit. Utilizing them is a potential opportunity for improvement. PSM, quality assurance, maintenance, inspection, fitness-for-service assessment, and equipment repair and replacement were used by the plants in the audit.

Quality Assurance

To control asset integrity, quality assurance (QA) is an element in all aspects of an AIM program and should consider the equipment lifecycle. High quality fabrication and installation of the equipment in ammonia plants begins with robust designs based on comprehensive standards and specifications that carry through the procurement, fabrication, transportation and delivery and installation stages. Quality maintenance, inspection, engineering assessment, and repair and replacement procedures need to be implemented to ensure that the initial condition of the equipment is not significantly degraded during its operating life. An effective AIM program requires collection, integration, and application of a large amount of data. Continual QA reviews of these data are essential.

Many of the plants in the audit had an ISO 9001 certified quality management system as part of their asset integrity management.

Maintenance

Strategic and preventative maintenance of pressure vessels, piping, tanks, rotating equipment, and electrical equipment is essential for effective management of asset integrity. This work includes following manufacturers' recommendations as well as accounting for local conditions within the facility. Particular attention should be given to equipment that is subject to time dependent degradation, such as corrosion, aging, fatigue, and high-temperature creep. Maintenance schedules should be established, followed, and adjusted based on experience and inspection results. Maintenance of safety critical equipment, such as pressure safety valves and process control instrumentation, is particularly important.

Inspection

Inspection and monitoring of equipment and structures in ammonia are essential to maintaining the integrity of those assets. Risk based inspection (RBI) following the methods of API RP 580 [25] or similar guidelines is used by many ammonia plants to effectively plan and implement inspection of pressure vessels, piping, and tanks. RBI methodology helps operators select and apply effective inspection techniques to critical equipment to manage safety risk.

Release of ammonia is a significant risk in ammonia operations. Facilities that apply RBI to ammonia tanks may also further evaluate this risk by considering the likelihood of loss of containment (LOC) resulting from stress-corrosion cracking, corrosion, and structural failure and the consequence of plant personnel and the local community being exposed to ammonia vapor, including accounting for containment and isolation systems. This additional risk assessment should include comprehensive consequence analysis that uses a dedicated dispersion model across a range of release scenarios.

Fitness for Service Assessment

If inspection reveals defects or degradation in equipment, its fitness for service (FFS) must be evaluated to make sure that it can safely and reliably operate for a desired future period. API 579-1/ASME FFS-1 [26] or a similar standard is used to perform FFS assessments of possible damage identified in pressure vessels, piping, and tanks in ammonia plants. The results of a FFS assessment indicate if operation of the equipment as-is can continue or if a repair or replacement is required. Repairs must be made following applicable standards and regulations. Replacement equipment should meet or exceed the requirements of the original equipment that is being replaced.

Summary

Asset integrity management is critical to maintaining the safe and reliable operation of ammonia plants. As reviewed in this paper, many plants in the ammonia industry have established AIM programs. It also was pointed out that industry standards are available for effectively implementing asset management.

References

1. Jaske, C. E., Weichel, S. J., and Brongers, M. P. H., 2017, "Essential Elements of an Asset Integrity Management Program for Ammonia and Methanol Plants," Paper PVP2017-65977, Proceedings of the ASME 2017 Pressure Vessels and Piping Conference, ASME, New York.
2. National Facilities Integrity Audit & Beyond Workshop, 2017 Trinidad & Tobago Energy Conference, Hyatt Regency Port of Spain, January 25, 2017.
3. ISO 55000, 2014, "Asset management – Overview, principles and terminology," ISO, Geneva.
4. ISO 31000, 2009, "Risk management – Principles and guidelines," ISO, Geneva.
5. ISO 31010, 2009, "Risk management – Risk assessment techniques," ISO, Geneva.
6. SEVESO III Directive, 1 June 2015, Control of Major Accident Hazards (COMAH).
7. AIChE Technical Manual Collection, 1956 to 2017, Proceedings of Annual Safety in Ammonia Plants & Related Facilities Symposium, AIChE, New York.
8. Pattabathula, V., Rani, B., and Timbres, D. H., 2005, "The AIChE Ammonia Safety Symposium 50 Years of Shared Experiences," Proceedings of 50th Annual Safety in Ammonia Plants & Related Facilities Symposium, AIChE, New York, pp. 12-54.
9. Pattabathula, V. and Richardson, J., 2015, "Sixty Years of History of the AIChE Ammonia Safety Symposium," Proceedings of 60th Annual Safety in Ammonia Plants & Related Facilities Symposium, AIChE, New York, pp. 1-26.
10. Orooji, A., Hosseininia, S., and Mahvelati, E. A., 2015, "Catastrophic Explosion in the CO2 Removal Unit of an Ammonia Plant," Proceedings of 60th Annual Safety in Ammonia Plants & Related Facilities Symposium, AIChE, New York, pp. 105-113.
11. Aliabadi, N., John McGrath, J., Keen, D Pattabathula, V., and Firth, D., 2016, "Failure of Secondary Reformer Dome and Its Repair," Proceedings of 61st Annual Safety in Ammonia Plants & Related Facilities Symposium, AIChE, New York, pp. 41-49.
12. Stoffels, J., 2016, "Failures in Aging Ammonia Plants," Proceedings of 61st Annual Safety in Ammonia Plants & Related Facilities Symposium, AIChE, New York, pp. 51-62.
13. Balgobin, K. and Babwah, R., 2016, "Failure of Level Bridle in Benfield Service," Proceedings of 61st Annual Safety in Ammonia Plants & Related Facilities Symposium, AIChE, New York, pp. 71-79.

14. AbdAl-Wahab, I. and Abdel-Rahman, H., 2016, "Syngas Compressor Hydrodynamic Thrust Bearing Failure," Proceedings of 61st Annual Safety in Ammonia Plants & Related Facilities Symposium, AIChE, New York, pp. 103-112.
15. Majeed, R. and Ali, S. 2016, "Ammonia Converter Reaction Loss Followed by Fire Incident," Proceedings of 61st Annual Safety in Ammonia Plants & Related Facilities Symposium, AIChE, New York, pp. 113-118.
16. Bosselaar, R., 2016, "Severe Cracks on a New Duplex Syngas Loop Cooler," Proceedings of 61st Annual Safety in Ammonia Plants & Related Facilities Symposium, AIChE, New York, pp. 123-132.
17. Orooji, A., Hosseininia, S., and Mahvelati, E. A., 2016, "Ammonia Plants Compressor Room Fire; Prevention and Remedies," Proceedings of 61st Annual Safety in Ammonia Plants & Related Facilities Symposium, AIChE, New York, pp. 133-140.
18. Singh, I. and Mehta, S., 2017, "BFW Preheater Exchanger Leak and Smooth Recovery of LTS Catalyst after Wetting,", Proceedings of 62nd Annual Safety in Ammonia Plants & Related Facilities Symposium, AIChE, New York, 39-48.
19. Sinha, R. and Lalchan, N., 2017, "HP Syngas Compressor Failure Due to Reverse Rotation," Proceedings of 62nd Annual Safety in Ammonia Plants & Related Facilities Symposium, AIChE, New York, 49-60.
20. Turi, I., 2017, "Process Air Compressor HP Stage Rotor Damage Due to Reverse Rotation," Proceedings of 62nd Annual Safety in Ammonia Plants & Related Facilities Symposium, AIChE, New York, 79-84.
21. Ramjattan, G. and Widrig, J. R., 2017, "Failure of a Riser Transition Assembly on a Primary Reformer," Proceedings of 62nd Annual Safety in Ammonia Plants & Related Facilities Symposium, AIChE, New York, 199-207.
22. ISO 55001, 2014, "Asset management – Management systems – Requirements," ISO, Geneva.
23. ISO 55002, 2014, "Asset management – Management systems – Guidelines for the application of ISO 55001," ISO, Geneva.
24. OSHA 1910.119, 2017, "Process safety management of highly hazardous chemicals," Title 29 – Labor, U.S. Code of Federal Regulations, Washington, D.C.
25. API Recommended Practice 580, 2016, "Risk-based Inspection," Third Edition, American Petroleum Institute, Washington, DC.
26. API 579-1/ASME FFS-1, 2016, "Fitness-For-Service," American Petroleum Institute, Washington, DC.

Fitness-for-Service Case Studies in HTHA Equipment

Improvements in Non-Destructive Examination (NDE) and Fitness-for-Service (FFS) have been developed in the area of High Temperature Hydrogen Attack (HTHA). The NDE advances have greatly increased the accuracy of and confidence in HTHA detection and sizing and have therefore allowed FFS evaluations of equipment that were not technically feasible just a few years ago. This paper will deal with case studies involving inspection planning using flaw and detection tolerance, remaining life dependent on inspection intervals, Minimum Pressurization Temperature (MPT) considerations, Post Weld Heat Treatment (PWHT) considerations in pressure vessel repairs of aged HTHA equipment and other operational concerns for Pressure Vessel equipment via API 579.

Michael Nugent, Trace P. Silfies, Paul J. Kowalski, Phillip E. Prueter, P.E.
The Equity Engineering Group, Inc.

Introduction

The purpose of this paper is to highlight the evolving application of Fitness-For-Service (FFS) evaluations for equipment in HTHA service. Although HTHA has been a known problem for over 65 years, the publication of API's 8th edition of RP 941[1] has promoted ammonia producers to reconsider that there are now two carbon steel (CS) Nelson curves (Figure 1), distinguished by whether the equipment has been PWHT or not. In addition, as a result of recent HTHA failures below the Carbon steel curve, the U.S. Chemical Safety and Hazard Investigation Board (CSB) has recommended re-validation of components currently operating in hydrogen service [2].

HTHA risk re-assessment is required for a good deal of carbon steel (and 0.5Mo steel) equipment in hydrogen service now that the curves have changed. Due to the removal of the old C-0.5Mo Nelson Curve, historical operation of this equipment above the newly recommended PWHT CS Nelson curve has been observed. If a vessel has been in operation for more than 30 years, the lifetime far exceeds what is recommended by the API RP 941 incubation curves.

What the 8th edition of RP 941 did not address was the ongoing refinement of the non-destructive examination (NDE) modalities (methods) which were unchanged from the previous edition (Tables D1 and D2). Reliable and repeatable NDE techniques are now gaining more widespread acceptance and confidence due to an ongoing Performance Demonstration Initiative and the emergence of a proposed API 579-1/ASME FFS-1 *Fitness-For-Service* (API 579) approach to HTHA damage evaluation [3-7]. This paper will present a brief overview of these NDE initiatives and focus on some FFS considerations not directly addressed in the API 579 Level 1 and Level 2 equipment evaluations.

Background of NDE

Non-destructive examination (NDE) has historically been used to ensure the quality of new fabrication and for assessing the quality of service-aged equipment. Although there are published recommended inspection techniques for HTHA, many owner-users have found varying results on the same piece of equipment by both the same and different techniques. In addition, some of the field reported data for the 'Nelson curves' are formulated with no underlying laboratory data to support the field experience.

Since 2012, the Equity Engineering Group (E^2G) HTHA joint industry project (JIP) has collected over 200 samples of in-service HTHA damage. The initial 25 samples were examined by six major NDE service providers using multiple NDE methods. The samples were destructively examined to verify decisions made based on the NDE. Since then, the results of the phase 1 round robin were reviewed by multiple NDE subject matter experts (SMEs) to identify the deficiencies and to develop a new scanning and sizing methodology.

These findings have been packaged into a NDE Performance Demonstration Initiative (PDI) where highly skilled inspectors are being experienced on the latest equipment (with Original Equipment Manufacturer (OEM) representatives present) on ex-service damaged components. This has provided a higher level of confidence in these inspections and greatly improved repeatability, reliability and sizing. These inspectors have been qualified on an extensive catalogue of incipient to advanced HTHA ex-service damage in a variety of forms, shapes, and thicknesses. The PDI contains in-depth understanding of HTHA damage manifestation, advanced ultrasonic testing (UT) applications, the fine-tuning of equipment parameters, and data acquisition from a variety of thicknesses and states of material degradation.

Data analysis consisted of many forms of raw ultrasonic data, including time of flight diffraction (ToFD), phased array ultrasonic testing (PAUT), and backscatter data. Damage samples of HTHA – ranging from colonies of microscopic damage to through-wall cracks – were inspected. The characterization of material degradation is a critical part of this PDI. Sources of false positives (e.g., material inclusions, stepwise hydrogen induced cracking (HIC) damage, and weld back cladding issues) are discussed. The PDI involves hands-on testing, as well as discussion of the theoretical approach and offline data analysis tutorials.

Historically, a common method for HTHA inspection has been advanced ultrasonic backscatter technique (AUBT)/velocity ratio measurements. However, this is a straight beam UT method, and it is unable to interrogate weldments effectively, as shown for a HTHA crack in Figure 2.

The main concept of ToFD is imaging weak ultrasonic energy that radiates from flaw extremities, as shown in Figure 3. This phenomenon is known as diffraction. Low amplitude diffracted waves are best received using a simple pitch and catch probe arrangement. Typical damages are identified and characterized. The advantage of ToFD is the potential data acquisition speed over other computerized ultrasonic inspection systems. The main benefit of high speed data acquisition is that it allows rapid screening of the weld and the heat affected zone (HAZ) to identify potential sites of HTHA. Detection of HTHA using ToFD has been found to require careful selection of inspection parameters, including transducers, filters, and contrast palette. Failure to optimize these can result in missing early stage HTHA damage.

Background of Fitness-For-Service (FFS)

As a result of the ongoing JIP, an enhanced tandem methodology (HTHA FFS combined with NDE) has been created to assist owner-users in lifecycle management of components in HTHA service. This methodology has been made possible because of the increased confidence owner-users can now have in NDE results. By transitioning to ToFD, qualified NDE contractors are now able to reliably locate HTHA damage, if present, and are also repeatedly capable of accurately sizing any indications found. By combining the to-be-released HTHA FFS evaluations with NDE results, a complementary methodology was created that incorporates both the HTHA FFS results and NDE results in the final "output" for remaining life. The research and model creation behind the HTHA FFS rules have been successfully benchmarked against the Nelson Curves, ensuring accurate predictions.

The scope has ranged from a risk ranking of a fleet of plants to the individual ranking of components or their specific nozzles. The model application is ongoing and has been used for quantitative remaining life predictions to be used in either corporate risk planning or API type risk-based inspection [8-11].

Inspection Intervals using Flaw Tolerance on Inspection

The lifecycle management methodology starts with identifying equipment that is in HTHA service. The FFS evaluations accept past and future operating conditions, equipment geometry, material of construction, and stress states as inputs into the model. The model outputs a crack growth rate dependent on an initial assumed flaw size as well as insight onto when the crack growth will become unstable. The crack growth model was developed with the knowledge gained from a plethora of prior studies dating as far back as 1938 as well as current ongoing JIP HTHA crack growth testing [12-18]. Using the output from the model a critical flaw size (flaw tolerance) can now be created.

The flaw tolerance of the vessel serves as the smallest flaw that must be identified during inspection. A flaw smaller than the critical flaw size will be noted and monitored with future inspections, however a flaw at the critical flaw size or larger requires action. Action could include weld repairs, if feasible, although weld repairs on vessels with accumulated HTHA damage have historically been shown to be difficult and often not successful. In order to be 100% confident the vessel does not contain a flaw size outside of the flaw tolerance, the entire vessel must be inspected. However, this is often not feasible due to timing and economic issues. New-found understanding of HTHA as a result of the ongoing JIP can help guide inspection locations to the areas in the vessel "most likely" to contain HTHA damage. These locations include areas of highest stress (nozzles, structural discontinuities, etc.) as well as areas that are higher in temperature relative to the rest of the vessel (i.e., outlet nozzle versus inlet nozzle).

Following inspection, a "baseline" has now been created for the equipment, assuming a flaw larger than the critical flaw size was not found. The limiting flaw size (largest flaw) that was discovered by NDE is now compared to the HTHA model outputs, using the crack growth rate predictions, a remaining life is created and a "half-life" (half the projected remaining life) approach is utilized. The required inspection interval is now driven by the new half-life and the equipment can be safely managed into the future.

Consider a Methanator reactor constructed of C-0.5Mo that experiences a short-term operation excursion of approximately 620°C (1150°F) for 3 hours while the Hydrogen partial pressure remained the same at 17.5 bara (250 psia). The enhanced methodology would screen this equipment against the PWHT CS Nelson curve (now the recommended curve per API RP 941), while also accounting for the temperature excursion and likely stress states in the vessel. The enhanced methodology would determine how much life has been exhausted during this excursion as well as likely flaw sizes that may have resulted. This vessel would be evaluated per HTHA concerns as well as pure creep concerns at this high of a temperature. The FFS methodology would likely predict that rapid methane formation occurred, prompting a requirement for inspection.

Minimum Pressurization Temperature Considerations

As a part of the enhanced FFS methodology for HTHA considerations, Minimum Pressurization Temperature (MPT) curves are also provided for different reference flaw sizes, away from and near structural discontinuities. These curves are generated using the fracture mechanics-based approach outlined in Welding Research Council (WRC) Bulletin 562 and References and provide acceptable pressure and temperature combinations to avoid brittle fracture during start-up or shut-down (Figure 4) [19-22]. As long as operation occurs below the curve (to the right of MPT curve), brittle fracture will be avoided. Note the different curves per flaw size; this can help in determining a flaw tolerance. If, for example, a t/4 (25% of wall thickness) flaw MPT curve is too limiting during start-up, then the vessel would have to be inspected for flaws of this size to ensure they do not exist. Additionally, as the material experiences HTHA damage, it generally becomes embrittled. Therefore, a decreased toughness relative to that of virgin material is used in this assessment. By combining material toughness with stress intensity factors (flaw size), an acceptable MPT curve can be created using elastic-plastic fracture mechanics consistent with API 579 and WRC Bulletin 562. Owner-users will often plot actual start-up / shut-down data (minute data) onto this plot to compare how past operation aligns with the recommended curves.

An alternate strategy that is occasionally utilized should too limiting of a flaw size be found, is to attempt repair of the vessel. This can be as elementary as grinding out the flaw if geometry permits, or the strategy can become complex with attempting to fill the flaw with weld build up or install a patch. However, welding on damaged equipment can prove challenging.

Post Weld Heat Treatment Considerations in Pressure Vessel Repair of Aged HTHA Equipment

Weld repairs of vessels in Hydrogen service have often proved difficult, either from delayed Heat Affected Zone (HAZ) cracking of the weld or due to the requirement for PWHT of the repair weld to minimize residual stresses[23,24]. However, PWHT of the entire vessel is often impractical. Local PWHT of the repair weld also includes challenges from local thermal stresses that could potentially lead to further HTHA damage. Hau found HTHA damage propagation could occur during the short time scale (only a few hours) of PWHT despite a prior bake-out treatment [25]. If a local PWHT needs to be performed, special consideration must be given and detailed engineering analysis must be carried out to "bullseye" heat treatments (concentric circles heating elements resulting in local thermal gradients) to ensure local distortion is avoided (see Figure 5 and Figure 6) [26,27,28]. Heating/Cooling rates, thermal gradients, soak temperature/width must all be carefully engineered and controlled in order to prevent local distortion or detrimental residual stresses.

An Application of HTHA FFS Methodology

Often in practice, successful management of an asset requires an interdisciplinary effort that includes understanding of the interactions between HTHA, NDE findings, fracture mechanics, and MPT curves. Owner-users have chosen to create unique solutions to complex problems encountered in the field.

A C-0.5Mo reactor with 300 series SS cladding was evaluated using the enhanced tandem methodology (HTHA FFS + NDE) due to concern over past operation. This reactor operated above 315°C (600°F) at approximately 13.8 bara (200 psia) H_2 which placed it above the PWHT carbon steel nelson curve, the recommended screening curve per API 941, 8th Edition. Additionally, this vessel had been operating for over 40 years which outlives the current API 941 dataset. The approximately 9 years of operating conditions provided were modeled by creating a time-scaled operating histogram for reactor temperature and hydrogen partial pressure. Sensitivity studies were performed for a number of variables including future operating temperature, clad versus non- clad (if cladding were compromised), stress state (near or away from structural discontinuity), among others. The HTHA model results showed this vessel to have a high likelihood of cracks present rather than volumetric damage. Therefore, the inspection planning and inspection intervals were created from predicted HTHA crack growth rates along with inspection findings and utilizing the half-life approach per the limiting flaw size found. Additionally, MPT curves were created for a number of different flaw sizes to ensure safe startup and shutdown.

An inspection plan for the JIP ToFD technique was created. The re-pad on the outside of the nozzle prevented inspection from the OD while the cladding on the shell and loose liner in the nozzle bore presented challenges as well. This inspection plan presented inspection techniques optimizing for interrogating the welds of interest, ensuring any HTHA damage, if present, would be discovered. A few of the variables were optimized such as included probe spacing, probe angle, orientation, and offset versus single scans.

During inspection, a t/3 (33% of wall thickness) flaw was found at a nozzle to head weld (high stress location). Accounting for future operating conditions and the predicted HTHA crack growth rates from the HTHA model, this vessel was given a remaining life of 10 years, pushing the inspection interval to 5 years from this inspection. However, this flaw size was found to be too limiting per MPT concerns. Process restrictions prevented safe operation of the vessel below the t/3 MPT envelope during the start up or shut down. This was overcome by installing steam tracing around the reactor (this kept the reactor hot enough to withstand pressurization during startups and shutdowns while minimizing the risk for brittle fracture).

Conclusions

Historically, HTHA damage has been difficult to detect repeatedly and accurately with NDE. Recent enhancements in NDE techniques and a rigorous Performance Demonstration Initiative have brought a great deal of confidence to NDE for HTHA. Eventually, these methods will likely find their way into API documentation, and the NDE Table D1 and D2 from API 941 8th edition will be retired.

With these methods, FFS can now be considered to develop life cycle management strategies for fixed equipment to better manage the risk associated with operating HTHA-prone pressure vessels and piping components. The FFS approach has been provided to API for inclusion in the API 579 standard and will go through additional review for widespread applications. Even after the vetting and review of these FFS "rules," the operation and repair of HTHA-serviced equipment will remain a challenge due to PWHT and brittle fracture (MPT) concerns. With this new technology, the Asset Owner-user now has a wider selection of equipment prioritization techniques and fitness-for-service methodologies than ever before.

Figures

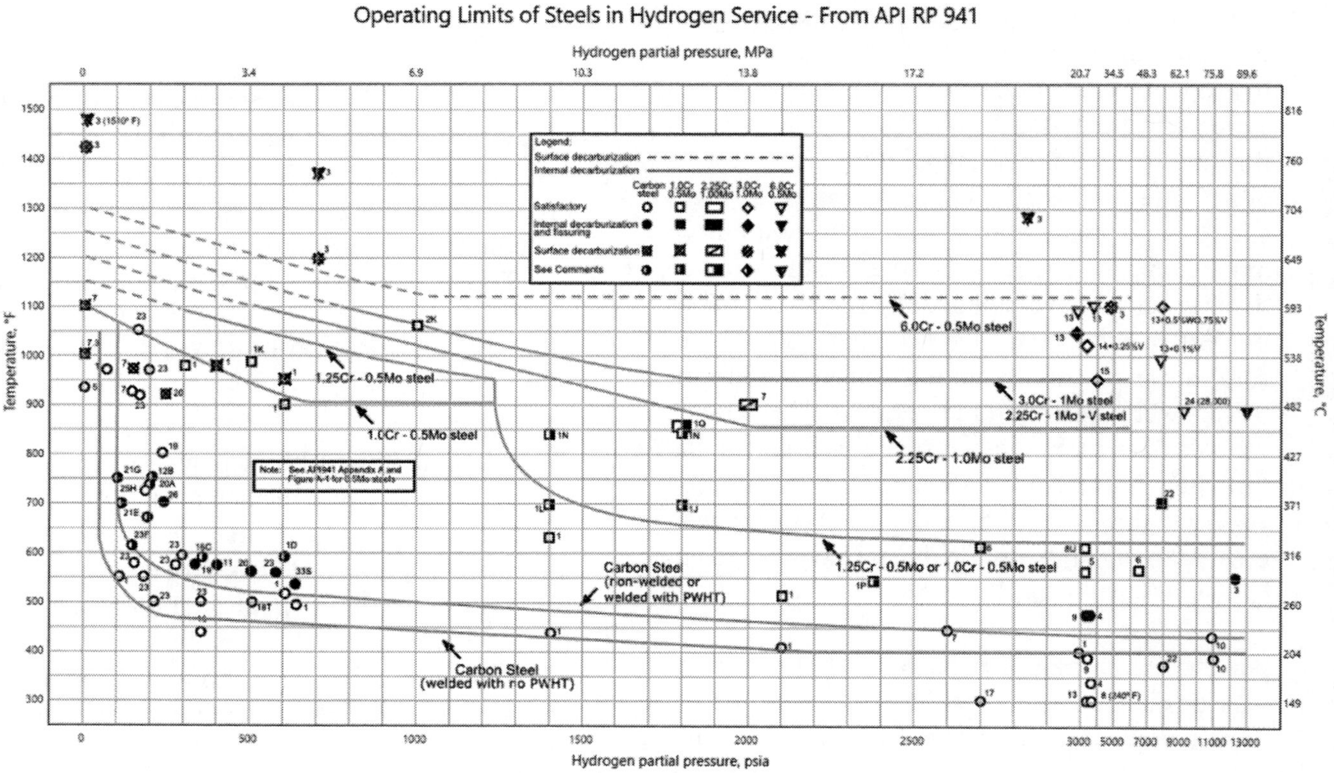

Figure 1. RPI 941, 8th Edition Nelson Curves [1]

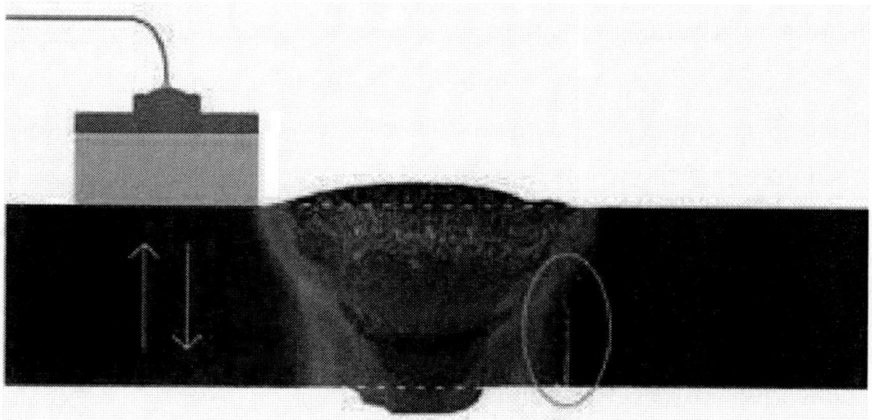

Figure 2. Straight beam AUBT may not detect HTHA damage either aligned with or 'covered' by the weld cap.

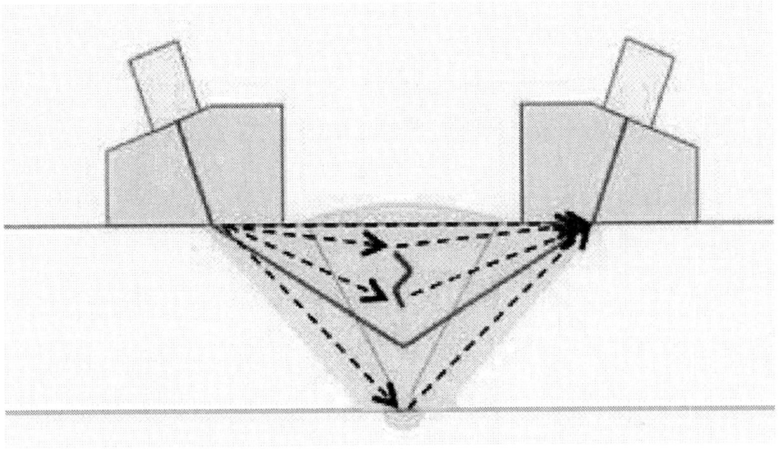

Figure 3. Basic ToFD arrangement.

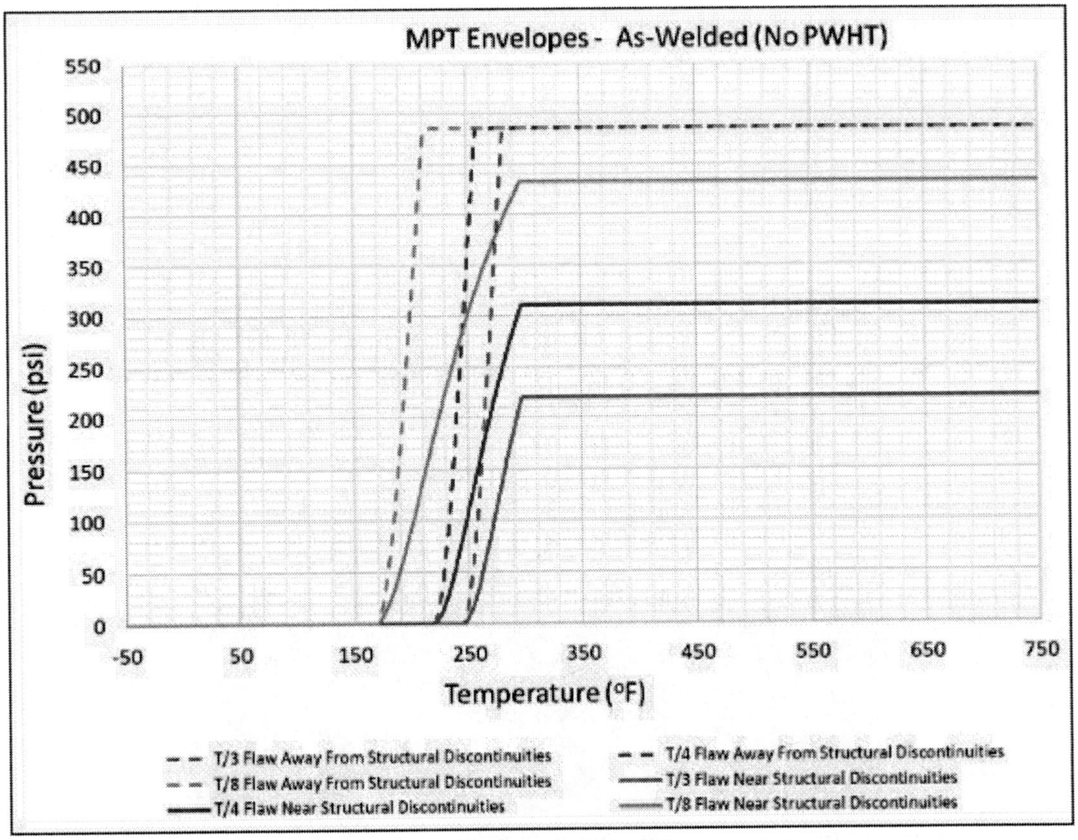

Figure 4. Example of MPT curves.

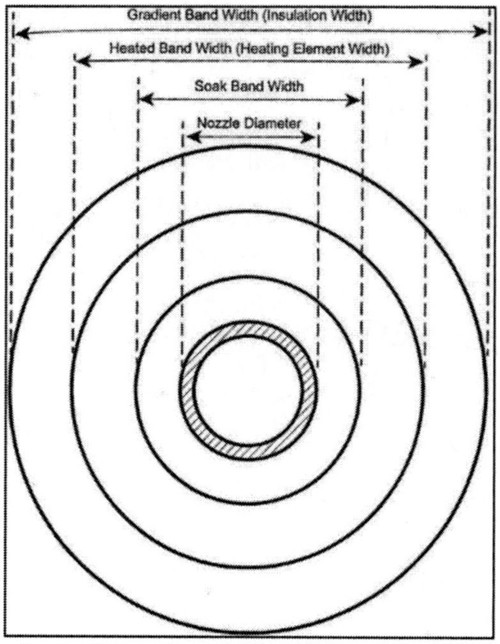

Figure 5. Example of a bullseye local PWHT configuration around a nozzle [29].

Figure 6. Example of local distortion due to improper local PWHT [26].

References

1. API RP 941, Steels for Hydrogen Service at Elevated Temperatures and Pressures in Petroleum Refineries and Petrochemical Plants, eighth edition, American Petroleum Institute, February 2016.

2. "CSB Safety Alert: Preventing High Temperature Hydrogen Attack (HTHA)," The U.S. Chemical Safety and Hazard Investigation Board (CSB), Aug. 2016, www.csb.gov/assets/1/20/safety_alert_htha_final.pdf?15689.

3. API, 2016. API 579-1/ASME FFS-1 Fitness-For-Service. The American Petroleum Institute, Washington D.C.

4. M. Nugent, T. Armitt, T. Silfies, J. Dobis, "State-of-the Art Prediction and NDE Techniques for High Temperature Hydrogen Attack in Piping and Equipment Paper 43" 7th Biennial Inspection Summit, Galveston Bay TX, 2017

5. M. Nugent, T Armitt "HTHA IN PETROCHEM EQUIPMENT:PREDICTION, PRIORITIZATION ANDNDE METHODS FOR DETECTION" International Chemical & Petroleum Industry Inspection Technology (ICPIIT), Galveston Bay TX, 2017

6. T. Armitt, M. Nugent, T. Silfies "HTHA Detection, Characterization and Verification" ASNT Fall meeting, 2017

7. T. Armitt, M. Nugent, T. Silfies "State-of-the Art Prediction and NDE Techniques for High Temperature Hydrogen Attack In Piping and Equipment Paper 407" 1st WCCM London UK, 2017.

8. M. Nugent, J. Dobis, T. Silfies and T. Armitt, "High Temperature Hydrogen Attack: Historic Review, Current Status and Future Perspective," in IPEIA, Banff, 2016.

9. M. Nugent, T. Armitt, J. Dobis and T. Silfies, "Advances in High Temperature Hydrogen Attack (HTHA) Non-Destructive Examination, Modeling, and Prediction, in Petrochemical and in Refinery Applications," in NACE CORROSION, New Orleans, 2017.

10. T. Silfies, J. D. Dobis and M. Nugent, "Looking behind the curtain of API RP 941 High Temperature Hydrogen Attack (HTHA) Data," in Corrosion 2016, Vancouver, BC, Canada, 2016.

11. M. Nugent, T. Silfies, P. Kowalski, and N. Sutton, "Recent Applications of Evaluations of Equipment in HTHA Service" in Corrosion 2018, Phoenix, AZ, USA, 2018

12. Materials Property Council, "MolyHi Joint Industry Project".

13. G. Manna, Factors Limiting the In-Service Lifetime of CrMoV Steels and Weldments of Pressure Equipment, Swansea: University of Wales, Swansea, 2004.

14. P. Shewmon, L. C. Chen and M. Prager, "Stress Assissted Hydrogen Attack Cracking in 2.25Cr-1Mo Steels At Elevated Temperatures," WRC Bulletin 527: Practical Aspcts of Hydrogen Attack, pp. 79-110, 2009.

15. P. Shewmon and Y. H. Xue, "Effect of High-Pressure Hydrogen on Crack Grwoth in Carbon Steel," Metallurgical Transactions A, vol. 22A, pp. 2703-2707, November 1991.

16. L. C. Weiner, "Kinetics and Mechanism of Hydrogen Attack of Steel," Corrosion - NACE, vol. 17, pp. 110-115, 1961.

17. F. K. Naumann, "Influence of Alloy Elements in Steel Upon Resistance to Hydrogen Under High Pressures," Technieche Nittellungen Krupp, Zorechungeberichte, vol. 1, no. 12, pp. 223-234, 1938.

18. M. McKimpson and P. G. Shewmon, "Initial Hydrogen Attack Kinetics in a Carbon Steel," Metallurgical Transactions A, May 1981.

19. Osage, D.A., Spring, D.W., Anderson T.L., Kummari, S.R., Prueter, P.E., and Wallin, K., "Welding Research Council Bulletin 562: Recommendations for Establishing the Minimum Pressurization Temperature (MPT) for Equipment." 2017, Welding Research Council, New York.

20. Prueter, P.E., Kummari, S.R., and Osage, D.A., 2017. "A Fracture Mechanics-Based Approach to Establish Minimum Pressurization Temperature Envelopes". In 2017 ASME Pressure Vessels and Piping Conference, July 16-20, 2017, Waikoloa, Hawaii, United States, PVP2017-65805.

21. Prueter, P.E., Kummari, S.R., and Osage, D.A., "Developing Life-Cycle Management Strategies and Minimum Pressurization Temperatures for Heavy-Walled Hydroprocessing Reactors." In 21st Annual International Pressure Equipment Integrity Association (IPEIA) Conference. March 1 – 3, 2017, Banff, Alberta, Canada.

22. Prueter, P.E. and Kummari, S.R., "A Fracture Mechanics-Based Minimum Pressurization Temperature Assessment on a 2-1/4 Cr – 1 Mo Hydrotreater Reactor using Finite Element Analysis." NACE CTW16 State-of-the-Art Session on MPT/Brittle Fracture, September 2016, Houston, Texas.

23. Prueter, P.E., Smith, K., Macejko, B., and Shipley, K.S., "Commentary on Recent Changes in ASME B31.3 Post Weld Heat Treatment Requirements and the Effectiveness of Weld Preheat." In 2017 ASME Pressure Vessels and Piping Conference, July 16-20, 2017, Waikoloa, Hawaii, United States, PVP2017-65859.

24. Prueter, P.E., "Observations on Flaw Tolerance and Brittle Fracture: The Effect of Post Weld Heat Treatment." January/February 2018 issue of Inspectioneering Journal.

25. J. L. Hau, M. Ledezma and U. A. Yepez, "Material Damage Implications of Post Weld Heat Treating H2 Blisters in Carbon Steel," in Corrosion 2001, Houston, TX, 2001.

26. Prueter, P.E. and Macejko, B., "Establishing Recommended Guidance for Local Post Weld Heat Treatment Configurations Based on Thermal-Mechanical Finite Element Analysis." Proceedings of the 2016 ASME Pressure Vessels and Piping Conference, July 17-21, 2016, Vancouver, British Columbia, Canada, PVP2016-63581.

27. Prueter, P.E. and Macejko, B., "Recommended Guidance for Local Post Weld Heat Treatment Con-figurations Based on Thermal-Mechanical Finite Element Analysis." In 20th Annual International Pres-sure Equipment Integrity Association (IPEIA) Conference. February 24-26, 2016, Banff, Alberta, Canada.

28. Prueter, P.E. and Macejko, B., "Local Post Weld Heat Treatment: Fundamentals, Case Studies, and General Recommendations." 7th Biennial API Inspection Summit, January 30 – February 2, 2017, Galveston Island, Texas.

29. NBIC, 2015. National Board Inspection Code: Part 3 - Repairs and Alterations. The National Board of Boiler and Pressure Vessel Inspectors, Columbus, OH.

Risk Reduction Projects Combating High Temperature Hydrogen Attack

This paper describes case study projects where Johnson Matthey (JM) has assisted operators in reducing or removing the risk of high temperature hydrogen attack (HTHA) by design of upgraded equipment or by changing the process operating conditions such that a required safety margin can be reinstated. The following plant case studies are reviewed with learnings gained from these projects over several years: a reformer upgrade project replacing non-PWHT carbon steel refractory lined equipment; the replacement of carbon-½molybdenum (C½Mo) equipment found to be suffering HTHA and; a change of catalyst allowing new process conditions reinstating the safety margin of derated C-½Mo equipment

John Brightling, Stephen Shapcott
Johnson Matthey Plc

Introduction

Managing the risk of damage due to HTHA presents a risk to the synthesis gas industries (ammonia, methanol, hydrogen/CO). The likelihood of HTHA occurring has proven difficult to predict, as witnessed by various updates to the relevant design codes over the years. Also, detection of HTHA by inspection is itself problematic.

It is well known that HTHA affects carbon and low alloy steel equipment and piping. A recent catastrophic release of hydrogen which was attributed to HTHA was presented by the Tesoro Anacortes incident in 2010. The consequences are illustrated by Figure 1, from the US Chemical Safety Board's final report, which shows the hydrotreater exchanger that failed at Tesoro and resulted in seven fatalities. The incident was investigated in detail by the US Chemical Safety Board (CSB) [1]. The incident also led to a revision to API 941[2]. The Tesoro incident serves as a powerful reminder of the damage potential from unchecked HTHA damage.

Figure 1: Failure of heat exchanger at Tesoro (Image from U.S. Chemical Safety Board Final Report to Tesoro Refinery fatal explosion and fire).

HTHA – What is it?

High temperature exposure of the carbon and low-alloy steels used for piping and pressure vessels in hydrogen service leads to a special form of degradation known as HTHA, or simply 'hydrogen attack'. HTHA causes degradation of the material at elevated temperatures and can result in sudden and catastrophic brittle fracture.

The process of HTHA requires the dissolving of atomic hydrogen into the steel. This is normal as all ferritic steels operating at elevated temperature and pressure will contain dissolved hydrogen. With the amount of hydrogen penetrating into the steel being linked to temperature, hydrogen partial pressure, and time, the higher these are the greater the amount of dissolved hydrogen.

Once in the steel, the hydrogen reacts with any free carbon and will reduce carbides (Fe_3C or M_3C in low alloy steels) in the steel to form methane (CH_4). Methane is not soluble and accumulates as a gas in small pockets at grain boundaries and inclusions causing fissuring and resulting in a reduction in mechanical properties.

Figure 2 from API 941 shows an example of micro fissuring caused internally by HTHA.

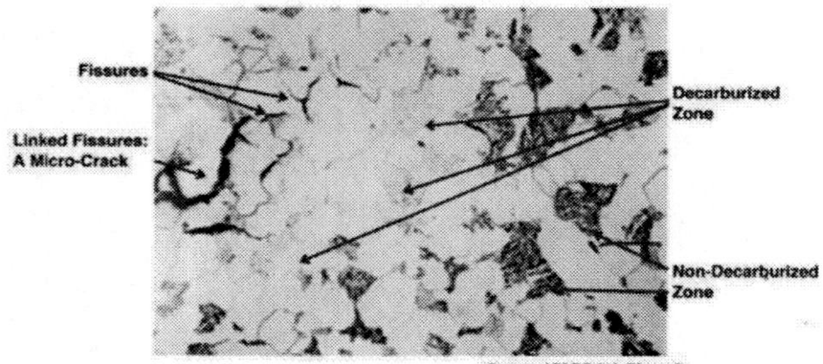

Figure 2: Image from API 941 showing fissures formed as a result of HTHA linked together to form a microcrack. Decarburized regions appear lighter in colour (because of an absence of carbon) than unaffected regions.

Inspection for HTHA has proved to be problematic. API 941 contains discussion on the various possible inspection techniques and concludes that no one method in isolation is ideally suited to detection of HTHA, especially during its incubation stage when micro fissures are just starting to develop. Ultrasonic inspection techniques have been found to have the best chance of detecting HTHA, although only once fissures have already started to develop. One of these techniques well-documented in API RP 941 is called Advanced Backscatter Ultrasonic Technique (ABUT), and is often used as an initial screening method before the use of other follow-up techniques. In an ABUT inspection, a pattern-based backscattering technique is used as the initial screening method. Depending on the backscatter pattern observed, one of several follow-up techniques, including frequency dependent backscatter, direction dependent backscatter, velocity ratio, spectral analysis and spatial averaging will be used to determine the cause of backscattering signal.

However as the 8th edition update to API RP 941 indicates, the most likely location of HTHA attack is at or near to welds. An issue being ABUT is unsuitable for use in these locations as this straight beam technique cannot interrogate the full volume of the weld region. Therefore angle-beam techniques based on backscatter and spectral analysis principles should be used for welds and heat affected zones adding to the complexity of inspection. A challenge is that this ultrasonic inspection requires a very high degree of skill in interpreting pulse-echo patterns on the oscilloscope interface.

Nelson Curves and API 941 Revisions

The history to the development of API 941 "Nelson Curves" is that operating limits were determined empirically by G A Nelson of Shell Development Company and presented to API in 1949. Throughout the 1950s and 1960s they continued to be revised by Nelson.

Although the curves can yield safe operating limits, they are actually curves of indicated industrial failure experience and have no safety margin. It is the responsibility of the designer/engineer/client to determine and include a safety margin. Including a safety margin is a safe practice that must be incorporated into a proper design.

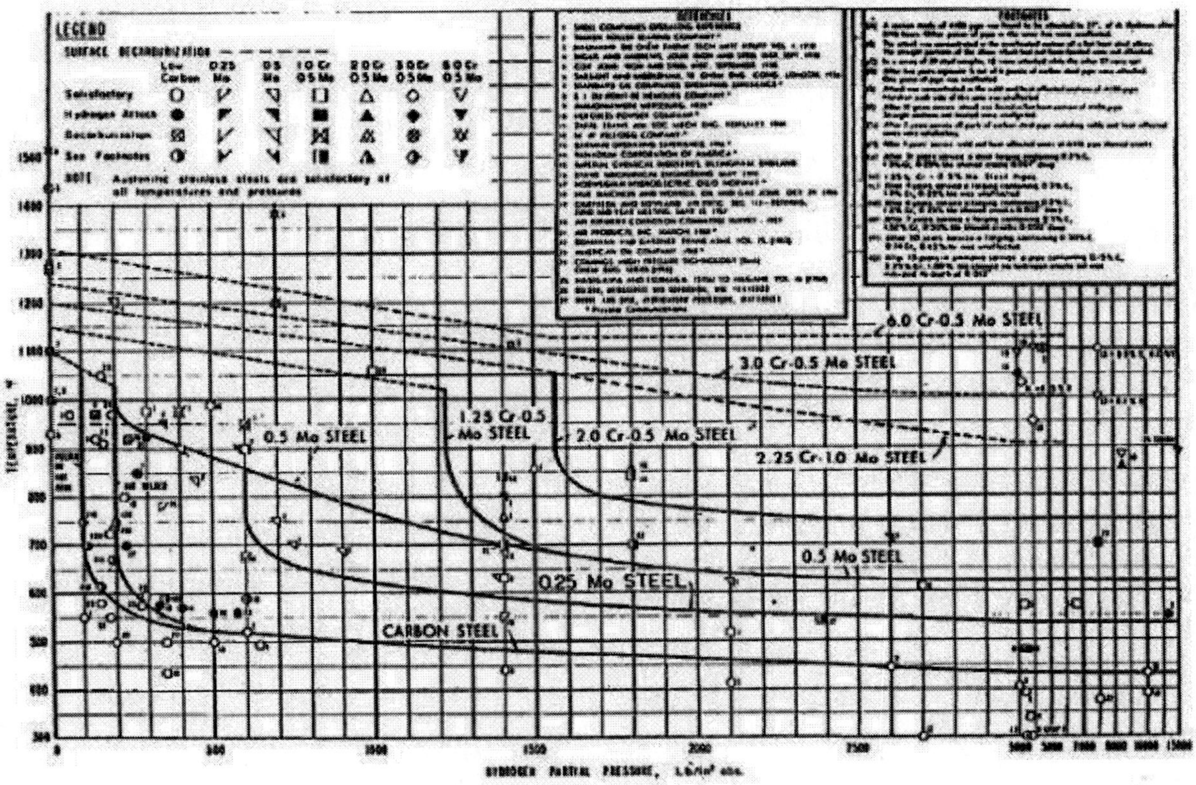

Figure 3: API 941 curve 1st Edition 1970 – including C½Mo curve (later lowered and finally withdrawn in 1990).

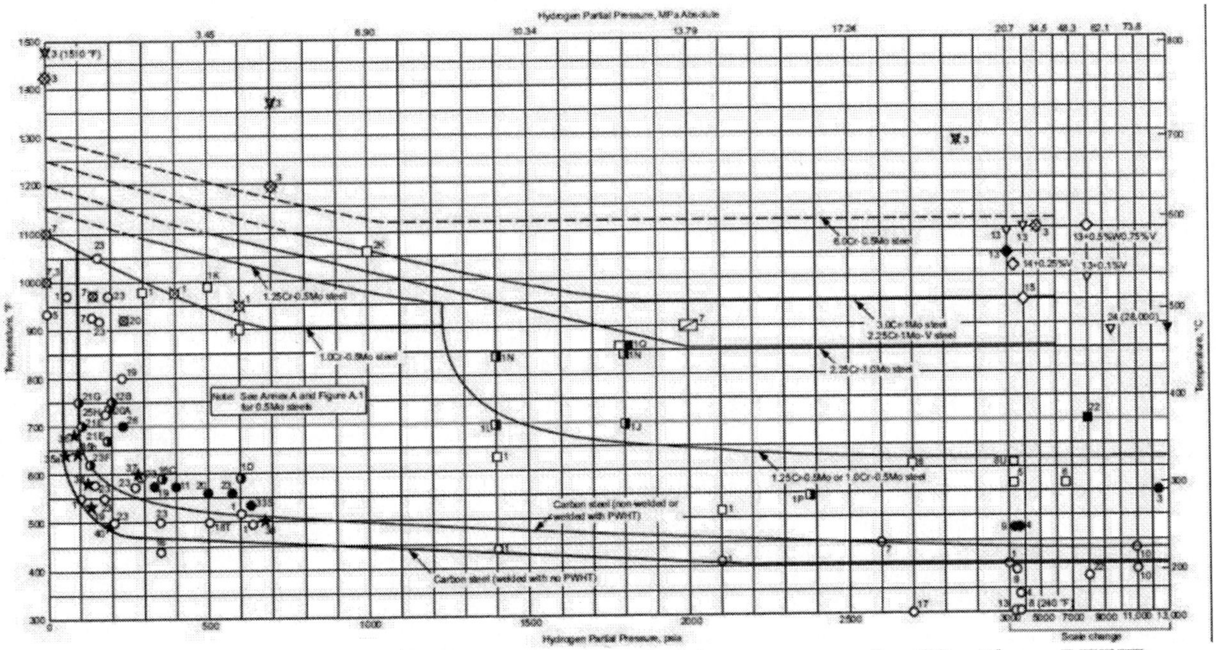

Figure 4: API 941 curve 8th Edition February 2016 – new curve for CS with no-PWHT.

The curves were first published as API 941 (Steels for Hydrogen Services at Elevated Temperatures and Pressures in Petroleum Refineries and Petrochemical Plants) in July 1970. They are still described in API 941, and are referred to as the Nelson Curves.

The curves are based on observed performance, and over the years they have been revised as new empirical data has come to light. The evolution can be seen by comparing Figure 3 which shows the 1st Edition 1970 curve and Figure 4 which shows the most recent 8th Edition 2016 curve.

In earlier editions of API RP 941, the recommended operational boundary of C½Mo steel in hydrogen service was described by a separate Nelson curve, well above that for carbon steel, see Figure 3. One of the first major decisions which impacted ammonia plants concerned HTHA in C½Mo steels. Due to new failures having been documented at conditions around the curve, in the 2nd Edition 1977 the curve for C½Mo was lowered. Also, in the 3rd Edition 1983 both the C½Mo and 2¼CrMo curves were lowered, with other changes having been made, all toward a more conservative position.

Through the 1980s there were more HTHA failures of C½Mo in the safe area of the Nelson curves, leading the API Committee to withdraw the C½Mo curve in the 4th edition API 941, 1990 revision, downgrading C½Mo to the same curve that applies to common carbon steel grades.

This meant a previous benefit in terms of hydrogen resistance, allowing a higher operational temperature for C½Mo steels versus carbon steels, was removed and that C½Mo steel should be treated the same as carbon steel with respect to operating temperature and HTHA resistance.

The second significant change to the Nelson curves has come following more recent incidents, including the Tesoro incident, where investigations concluded that the failure of equipment occurred due to HTHA in a carbon steel that was operating below the Nelson curve. The equipment had not been post weld heat treated and all the observed damage was in the heat affected zone adjacent to a weld. The latest 8th Edition of API 941 (2016) contains relevant information concerning the influence of stress on HTHA risk and contains a new lower curve for carbon steel welded without post-weld heat treatment.

Engineering Reviews

The API 941 changes mean that the original choice of metallurgy may not be appropriate. Plants should therefore consider performing an engineering review on equipment and piping operating under conditions covered by the most up to date Nelson curves to confirm that the metallurgy is considered suitable for continued operation. [4].

API RP 581, Risk-Based Inspection Technology, is a recommended practice developed and published by API to provide quantitative risk-based inspection (RBI). The 2nd Edition published in 2008 attempted to assist in quantifying risk factors. It used a parameter, P_v, which was a function of the hydrogen partial pressure (pH_2), temperature and time. This P_v parameter has since been removed from the latest version on API 581 3rd Edition 2016.

The 2016 revision adopts what is viewed as a much more conservative approach, especially for carbon steels and C½Mo steels for which the latest version of API 581 assigns high susceptibility to equipment operating above 177°C (350°F) and a hydrogen partial pressure exceeding 0.345Mpa (50psia). These parameters are considered very conservative with respect to the both the PWHT and non-PWHT carbon steels curves in the AP 941.

Typically, an engineering review includes an assessment of susceptibility to HTHA, risk assessment and management plan.

Susceptibility review

In API 581 3rd Edition 2016, an example guideline uses 27.7°C (50°F) increments to represent relative change in susceptibility ranking for Cr-Mo low alloy steels (Figure 5).

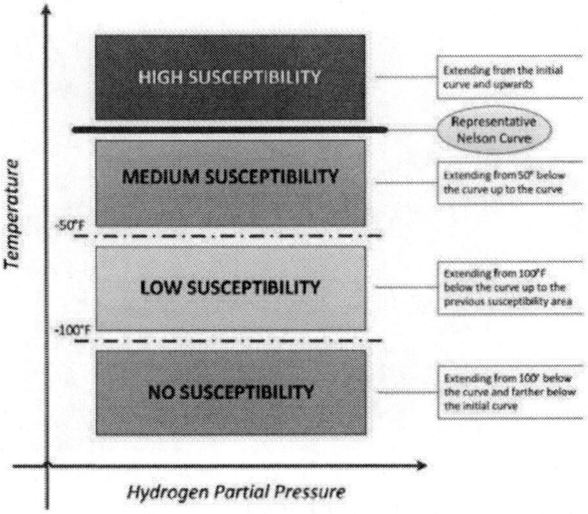

Figure 5: An example of HTHA Susceptibility Rankings for Cr-Mo Low Alloy Steels, API 581 3rd Edition 2016.

Whilst API RP581 provides this screening criteria the owner-user has the responsibility to determine the applicability for their assets by risk assessment and inspections.

Risk Assessment

For all equipment identified as being susceptible to HTHA, a risk assessment is undertaken using a standard risk methodology to develop a risk matrix considering likelihood of failure occurring versus loss consequence of it.

Management Plan

Based on the level of risk, equipment is targeted with risk management plans developed with the primary objective to reduce items of equipment deemed high risk in the risk assessment. Solutions may involve:

1. Increased inspection at outages, possibly including destructive testing where equipment is retired or by taking a 'boat-sample' from the equipment.
2. Replacement of equipment with more resistant metallurgy.
3. Reducing operating conditions, usually temperature, as HTHA is closely influenced by operating conditions.
4. Establishing Integrity Operating Window (IOW) to be fully understood in operations including the actions in the event of any excursions.

In many cases the safest course of action will be to consider upgrading to more resistant metallurgy. The first two case studies cover equipment upgrade projects to reduce risk, and the third case study is an example of changing operating conditions to reduce risk.

Refractory Lined Systems

With the latest amendment to API 941, an area of risk to plants that needs additional management attention is refractory lined equipment. A typical example of such a system is the transfer line from the primary reformer to the secondary reformer which historically was often designed as a refractory lined (non-PWHT) carbon steel system. As originally designed, the carbon steel part of the transfer line would operate with a shell temperature below 200°C (392°F). However, over time the refractory may degrade, crack and deteriorate meaning that shell temperatures have increased. Typically, such equipment is inspected by infrared thermal imaging inspection. Figure 6 shows an example of a transfer line operating with hot areas.

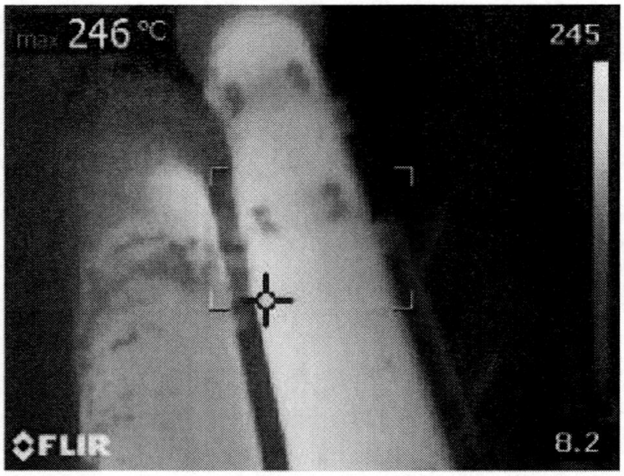

Figure 6: Thermal Image scan showing transfer line operating with hot areas.

In such circumstances, the steel casing is operating with areas exceeding 240°C (464°F) and outside the recommendation of the latest API 941 for non PWHT-carbon steel.

Case Study 1

For a European ammonia plant designed and built in the 1970s with a refractory lined, non-PWHT carbon steel transfer line system which had areas with refractory hotspots, a major refurbishment project for the transfer system was justified by the plant to improve plant reliability and integrity. The project being scheduled to coincide with a radiant section re-tube.

To support the project JM performed an engineering design for the following items of the reformer - inlet pigtails, replacement reformer tube supports, reformer tubes, outlet pigtails, hot collectors (outlet manifolds) and the replacement refractory lined transfer mains.

The new transfer mains were designed to have a larger internal diameter to enable easier inspection and future maintenance as well as a lower pressure drop.

With a new pressure shell design based on 1¼Cr½Mo material, temperature of concern with respect to HTHA would be circa 450°C (842°F). The associated refractory system was designed to limit the shell temperature to less than 180°C (356°F) in still air.

The extent of the equipment replacement in the refurbishment project is shown in Figure 7.

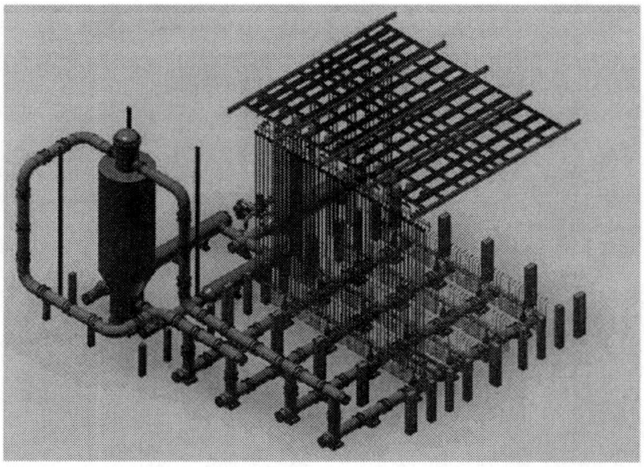

Figure 7: Case Study 1 - Equipment replacement in the refurbishment project.

Methanation Change in Vessel

A catalyst vessel in the ammonia plant which has proven vulnerable to HTHA is the methanator.

Case Study 2

A European ammonia plant [4] was commissioned in 1970 and is currently operating at 1050 MTPD (1160 STPD). As was typical for plants designed in the 1960s, the original material selected for the methanator in hydrogen service was C½Mo, based on the Nelson curves published in API 941 at the time.

The vessel had been subject to regular inspections during its working life. Because the lifetime of the methanator catalyst is long, internal inspections were infrequent.

Some cracking was initially detected during an internal inspection in 1992. At the time, metallography indicated the cracks could be original plate defects so it was unclear whether HTHA was occurring. Further inspection was called for at the next inspection opportunity in 2000.

The cracking found during the internal inspection in 2000 extended on either side of the 1992 cracking for over five times the length, exceeding half the circumference of the vessel (>2700 mm, >106 inches). The height of the band had extended from 105 to 600 mm (4 to 24 inches). Isolated cracks were ground out to determine their depth.

Following ultrasonic examination in February 2003, the cracking above the band was concluded to be present in excess of 10% of the wall thickness of the vessel. It was clear that the vessel was suffering from HTHA. A complete vessel repair would require complete vessel post weld heat treatment, which would not be possible within the timescale of a plant shutdown. The repair would also retain risk so replacement was planned for the 2005 shutdown.

A replacement vessel was designed by JM to modern design standards and several improvements in the design were made:

- Material change to 1¼Cr ½Mo to give greater HTHA resistance.
- Design temperature increase to 500°C (932°F) from 427°C (800°F) to improve ability to handle any high temperature excursions in the event of a process upset.
- Inlet and outlet connections and the connecting pipework were increased in diameter to give reduced pressure drop.
- Catalyst now supported directly on support media in the vessel base. Previously a grating supported on a welded ring inside the vessel was used. The welded ring promoted stresses and cracking.

Many of these improvements simplified the vessel design and construction. The improvements have a further benefit that, looking forward to the future, the reduction of internal features and attachments to the vessel shell mean that requirements for internal inspection of the vessel may be reduced.

The replacement vessel was commissioned in 2005 with the **KATALCO**™ 11-series methanation catalyst charge achieving <1ppm $CO+CO_2$ slip which it has maintained over its 12 years run.

In addition to the vessel and catalyst replacement, a safety integrity level (SIL) assessment was performed and the instrumentation was changed providing additional layers of protection in both the initiating and final elements of the instrumented protection system [5]. Part of the instrumentation upgrade included fast acting thermocouples to detect the very high rate of temperature rise in a methanator if a process upset results in CO_2 breakthrough from the upstream removal section [6].

Recently in 2017, a similar project has been completed for a North American ammonia plant. The plant which dated from the 1960s has an existing C½Mo methanation vessel. JM assisted with replacement by designing a modern vessel to ASME Section VIII Division 1:2017 using 1¼Cr ½Mo material (ASME SA387 Gr11 Cl2), see Figure 8 using **KATALCO** 11-series methanation catalyst and **STREAMLINE** low pressure drop media.

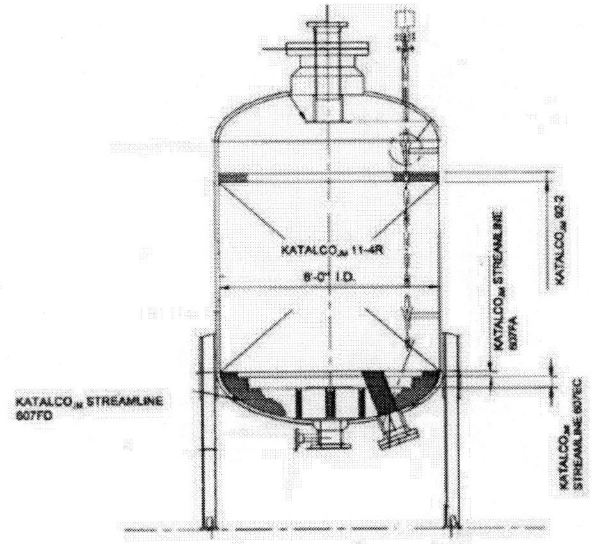

Figure 8: Case Study 2 – Replacement Methanator vessel.

Methanation Change in Process Conditions

The removal of the C½Mo curve from the API guideline placed a large number of C½Mo equipment items in an indeterminate state. Good practice is to maintain an operating safety margin of 30°C (50°F) below the Nelson curves at any time.

Case Study 3

For a 1250 MTPD (1380 STPD) European ammonia plant, the inlet/outlet heat exchanger for the methanator was made of C½Mo. The maximum recommended outlet temperature dictated by the revised Nelson curve is 290-300°C (550-570°F) as compared to the original design temperature of 427°C (800°F) from the withdrawn C½Mo curve [7].

However, the exchanger operated from 1970 to 2007 right on or slightly above the Nelson curve. Therefore, regular 100% ultrasonic testing (UT) was performed to check for cracks on the main welds of the pressurised shell (i.e. longitudinal, circumferential, butt welds and man hole welds). In addition, during each turnaround, a replica test was performed. However, no cracks or fissures were revealed during these investigations.

The change to a more active catalyst capable of operating at a much lower inlet temperature, reinstated a safe operating margin below the Nelson curve as shown in Figure 9.

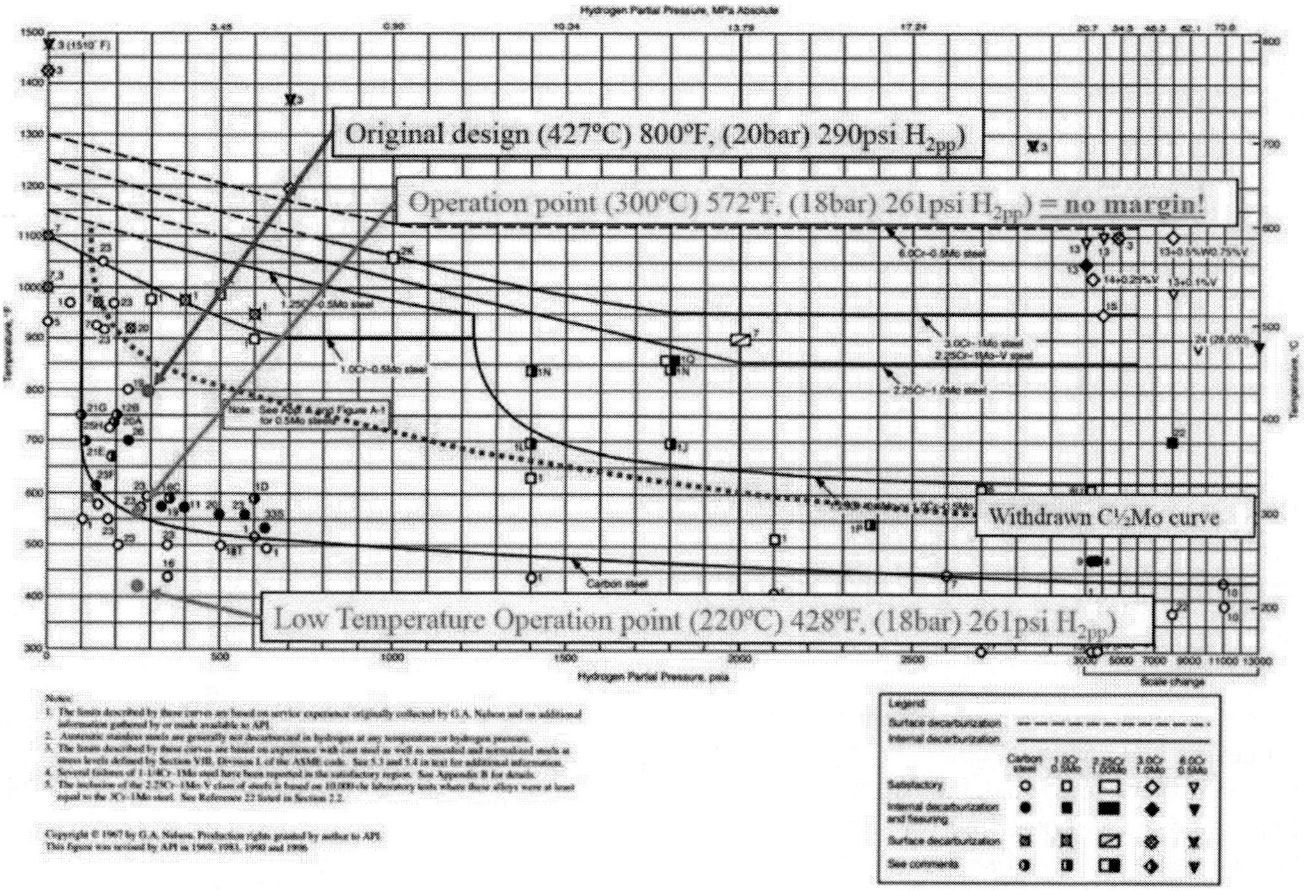

Figure 9: Case study 3 - Impact of lowering methanator operating temperature shown on Nelson curve from API 941 6th Edition, 2004.

By taking the process operations to a lower temperature it was possible to achieve a 30°C (54°F) safety margin below the Nelson curve, thus low susceptibility as per Figure 5

To achieve this required lowering the inlet and exit temperatures of the methanator by 65°C (117°F) by lowering the inlet temperature to 220°C (428°F), as shown in Figure 10.

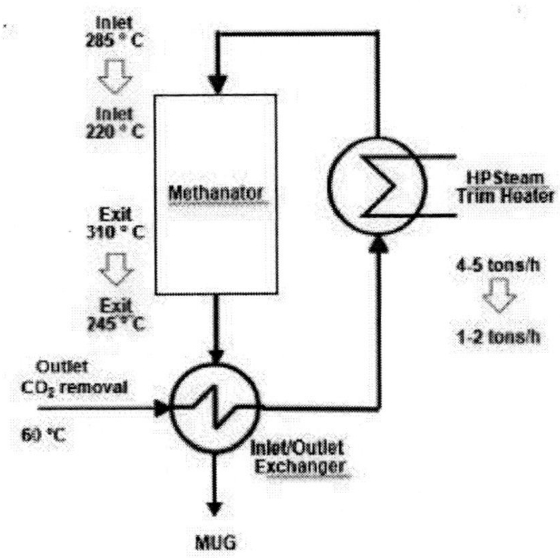

Figure 10: Process flow scheme around methanator.

The driver behind the choice of **KATALCO** 11-series as a low temperature methanation catalyst, was reinstatement of a safe operating margin in respect of Nelson curves. A second benefit was the improved plant efficiency due to HP steam savings on the trim heater.

By installing **KATALCO** 11-series low temperature methanator catalyst, less duty is required from the trim heater fed by HP steam (100 bar, 1450 psi), as shown in Figure 10. A gain in HP steam consumption of 2-3 ton/h is obtained, depending on the temperature rise over the methanator catalyst. This corresponds to an improvement of the energy efficiency of 0.15 GJ/ton (0.14 mmBTU/ton).

At start-up, it was proven that the CO/CO_2 slip was lower than 1 ppm and the inlet temperature was set at 220°C (428°F). After 10 years of continuous operation, the CO/CO_2 slip is still low and the inlet temperature remains at 220°C with a sharp and steady reaction profile, Figure 11.

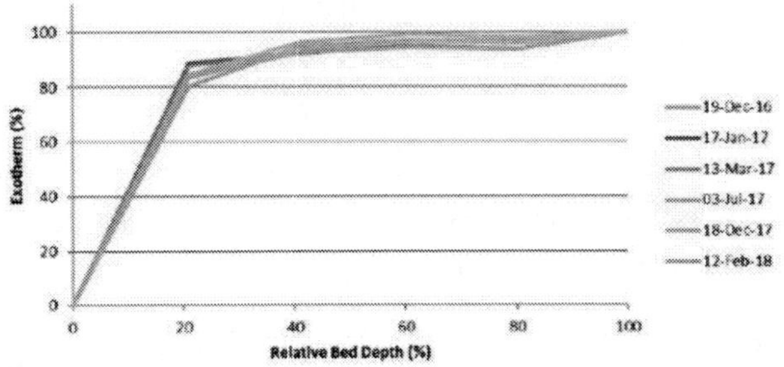

Figure 11: Methanator exotherm profile.

In this case study due to operation with the new catalyst charge at low temperatures, C½Mo steel equipment is now operating with a safer margin within the Nelson curve.

Conclusion

On-going inspection for HTHA can be a difficult task requiring significant expertise. Depending on the age of the plant many items of equipment can be affected. Appropriate asset management may require planned capital expenditures for timely renewal of vulnerable items of high risk equipment.

The case studies give some examples of how the risks inherent from HTHA have been managed as part of asset integrity and renewal programs at a number of sites, with JM providing support in the design of the new equipment constructed from inherently more resistant metallurgy.

An example was provided where selecting a catalyst with a lower operating temperature range allowed a safe operation window to be established by moving from a high susceptibility to low susceptibility risk and eliminating the need to purchase new equipment.

References:

[1] US Chemical Safety and Hazard Investigation Board Report 2010-01-IWA, May 2014 (www.csb.gov/tesoro-refinery-fatal-explosion-and-fire/)

[2] API Recommended Practice 941. Steels for Hydrogen Service at Elevated Temperatures and Pressures in Petroleum Refineries and Petrochemical Plants. Eighth Edition February 2016.

[3] API Recommended Practice 941. Steels for Hydrogen Service at Elevated Temperatures and Pressures in Petroleum Refineries and Petrochemical Plants. First Edition July 1977.

[4] D Keen, C Jones & C Thomas. Inspection for high Temperature Hydrogen Attack, Proc Nitrogen+Syngas 2018 conference, Gothenburg, February 2018.

[5] M Walton, T Southerton & P Sharp. Safety Improvements in a Methanation Reactor , AIChE Ammonia technical manual Vol 48, 2007

[6] A Janssen, N Siraa, J M Blanken, Temperature Runaway of a Methanator Ammonia technical manual Vol 23 , 1980

[7] S Van Den Broeck, M Fowles. Low temperature methanation: Operating experience and something unexpected!, AIChE Ammonia Technical manual Vol 49, 2008

Fleet-Wide HTHA Risk Assessment Using Latest Methodologies

Due to recent guidance issued by API 941 and the United States Chemical Safety Board (CSB), understanding the risk of assets with regard to High Temperature Hydrogen Attack (HTHA) has been a challenge for individual sites as well as major international owner-users. In this paper, a corporation-wide assessment at 10 different production sites on four different continents will be presented.

This assessment utilized the most recent HTHA damage progression modeling technology in order to create a methodology to rank risk. The outcome of the study was used by the client to supplement their previously established internal criteria. The improved criteria allowed for better discrimination of risks considering time in service and Post Weld Heat Treatment (PWHT), among other factors. The project allowed the owner-user to better prioritize future actions such as replacement and inspection. A high-level overview of this process will be discussed.

Giuseppe Franceschini
Yara International

Trace P. Silfies, Michael Nugent, Joseph Bednarz
The Equity Engineering Group, Inc.

Introduction

The purpose of this paper is to document one major international ammonia producer's response to the 8th edition of API Recommended Practice 941 (API RP 941)[1]. In the recently-issued 2016 version, there are now two carbon steel Nelson Curves, distinguished by whether the equipment has been post-weld heat treated (PWHT) or not (as-welded) (as shown in **Figure 1** at the end of this paper). This change requires High Temperature Hydrogen Attack (HTHA) risk re-assessment for carbon steel equipment in hydrogen service. In addition, CSB issued an alert recommending replacement of carbon steel and C-0.5Mo equipment that operates above 400°F and greater than 50psia hydrogen partial pressure.

In response to the API 941 change and the CSB alert, Yara International updated its risk assessment for HTHA using the most recent HTHA damage progression modeling technology.

CSB Safety Alert: Preventing High Temperature Hydrogen Attack (HTHA), 2016

The U.S. Chemical Safety and Hazard Investigation Board (CSB) published an Alert on the 22nd of August 2016 indicating:

"The CSB ultimately believes that the stronger option for industry to protect against HTHA is to focus on upgrading equipment susceptible to HTHA with inherently safer materials of construction rather than simply relying on administrative controls. Not only is HTHA very difficult to detect but equipment inspections and post-weld heat-treating rely on procedures and human implementation, which are low on the hierarchy of controls. These options are weaker safeguards to prevent HTHA failures than the use of materials that are less susceptible to HTHA damage...In the absence of a carbon steel Nelson Curve that adequately protects against HTHA and incorporates findings from the Tesoro Anacortes failure, the CSB provides the following guidance for industry:

1. *Identify all carbon steel equipment in hydrogen service that has the potential to harm workers or communities due to catastrophic failure;*
2. *Verify actual operating conditions (hydrogen partial pressure and temperature) for the identified carbon steel equipment;*
3. *Replace carbon steel process equipment that operates above 400°F and greater than 50psia hydrogen partial pressure; and*
4. *Use inherently safer materials, such as steels with higher chromium and molybdenum content."*[2]

Background

Driven by recent changes to API Recommended Practice 941 and the alert provided by the CSB, Yara International identified all equipment in hydrogen service operating in their fleet. The fleet consists of pressure vessels and piping operated in 10 plants on 4 continents. Yara defined hydrogen service as operating conditions above 200°C (392°F) and hydrogen partial pressures greater than 0.5 bar a (7.2 psia). Note that hydrogen partial pressure was determined on a "wet basis" and presents a more realistic approach than "any hydrogen".

After completing this exercise, Yara concluded they had 263 carbon steel and C-0.5Mo steel items (pressure vessels and piping) in hydrogen service. Operating conditions were simplified to typical operating conditions in regard to metal skin temperature, total pressure, and hydrogen partial pressure.

The first assessment used by Yara to risk-rank these carbon and C-0.5Mo equipment was based on the definition of subjective Probability of Failure (HTHA POF) categories. These HTHA POF categories were determined considering the equipment's typical operating temperature and proximity to its respective API 941 Nelson Curve (considering PWHT'ed Carbon Steel and non-PWHT'ed Carbon Steel). The strategy used by Yara is a typical strategy employed by many owner-users in ammonia, refinery, and petrochemical industries, which is to use a Factor of Safety of delta temperature range below the published API 941 Nelson Curve line at the determined hydrogen partial pressure. Note that C-0.5Mo steels are considered to be the same as PWHT'ed carbon steel in the scope of completing an API RP 941-based HTHA determination. The categories Yara used for determining HTHA POF are shown in **Table 1**, below.

HTHA POF	Assessment Point Margin against new API RP 941 curves for Carbon Steel
1, Very High	< 0°C (< 0°F)
2, High	≥ 0°C and ≤ 10°C (≥ 0°F and ≤ 18°F)
3, Medium	> 10°C and ≤ 30°C (≥ 18°F and ≤ 54°F)
4, Low	> 30°C and ≤ 60°C (≥ 54°F and ≤ 108°F)
5, Very Low	> 60°C (> 108°F)

Table 1. Assessment Point Margin (as a difference in temperature) against new API RP 941 Curves for Carbon Steel.

The outcome of Yara HTHA POF determination is shown in **Table 2** for both pressure vessels and piping. Based solely on the CSB Alert, replacement of 94 out of 109 pressure vessels in hydrogen service should have been planned. POF ranking based on proximity to respective Nelson

Curve position better defines the risk as a function of temperature from API RP 941 Curves.

POF Levels	Pressure Vessels	Piping
1, Very High	28	36
2, High	5	19
3, Medium	13	27
4, Low	14	32
5, Very Low	49	40
Total	**109**	**154**

Table 2. Yara HTHA POF determination.

To further determine a path forward for pressure vessels operating in hydrogen service, Yara developed the risk ranking by using the HTHA POF determination and meshing it with Consequence of Failure (COF) categories. These COF categories were based on safety losses already established with the Yara Risk-Based Inspection (RBI) program and documented below in **Table 3** as categories A-E. When the HTHA POF and COF are plotted on Yara's corporate risk matrix, hydrogen service pressure vessels are differentiated based on risk (as shown in **Figure 2** at the end of this paper).

Category	Safety Losses (sq ft affected)	Financial Loss
A	> 5,000,000	> $10 MM
B	500,000 – 5,000,000	$1MM – $10MM
C	50,000 – 500,000	$100K – $1MM
D	5,000 – 50,000	$10K – $100K
E	< 5,000	< $10K

Table 3. Consequence of Failure (COF) categories as determined by Yara RBI.

Advanced HTHA POF and Risk Determination

Based on the outcomes of the Yara Initial Risk Assessment, The Equity Engineering Group, Inc. (E^2G) was contracted to perform a detailed High Temperature Hydrogen Attack (HTHA) evaluation of carbon steel and C-0.5Mo steel pressure vessels operating in hydrogen service determined to be "High" and "Medium High" risk. In total, 48 unique pressure vessels were selected for HTHA POF determination based on advanced damage prediction models (note that the same COF categories from Yara's prior assessment were to be utilized). The key differentiator of the E^2G assessment compared to an API RP 941 Nelson Curve-like approach allows for consideration for time in service and applied stresses (residual stress from welds, applied stresses from total pressure, and predicted internal methane stresses).

Assessment conditions selected for the evaluation of the 48 pressure vessels were as follows:

- Fixed operating conditions determined as per Yara's process engineers (single point temperature, total pressure, and hydrogen partial pressure), which were the same as the ones used for the Yara Initial Risk Assessment.
- Time in service was considered and was defined primarily based on equipment's install date.
- Material condition was defined as either carbon steel or C-0.5Mo.
- Consideration was given to cladding in terms of reducing effective hydrogen partial pressures at the base metal.
- Differentiation for PWHT or Non-PWHT for both carbon steel and C-0.5Mo was provided. These details affected the assumptions on weld residual stress (WRS) considered in the modeling.
- Applied Stress State of the equipment was based on the nominal hoop stress (function of total pressure and vessel thickness).

The assessment methodology was based on an API 579-like Level 1 Fitness for Service (FFS) model approach recently developed by E^2G's HTHA Joint Industry Project (JIP), which had been ongoing for four years at the time with eight active major oil refinery sponsors.[3,4,5,6] While the model currently remains proprietary to the JIP members, the following details can be shared:

- Equipment was evaluated for both volumetric damage and crack-like flaw damage. Volumetric HTHA is defined as a layer-by-layer attack of the material where decarburization leads to methane generation and fissuring throughout the material as a more uniform attack, (this is further shown in **Figure 3**). Crack-like HTHA occurs when the growth of an existing crack is accelerated by the presence of hydrogen; the assessment assumes an initial flaw size at day zero and propagates the flaw with crack growth rates based on stress concentration, hydrogen partial pressure, and temperature. A number studies have helped provide the mechanistic details of HTHA damage progression and were considered in model development. [7-13]
- Based on the outcomes of the Level 1 FFS model, the following criteria were applied to determine discrete POF categories of 1-5 to match Yara's risk matrix, with Level 1 being the highest POF and Level 5 being the lowest POF:
    - Level 1: Equipment had no remaining life determined by the volumetric damage model (creep-like). Equipment was found to be very susceptible to rapid crack growth after initiation of a flaw based on outputs from the crack-like flaw model. In addition, all but two points were above their respective Nelson Curve.
    - Level 2: Equipment had minimal to no remaining life determined by the volumetric damage model (creep-like). Equipment was found to be very susceptible to rapid crack growth after initiation of a flaw based on outputs from the crack-like flaw model. In addition, all but four points (high stress) were above their respective Nelson Curve.
    - Level 3: Equipment had some predicted remaining life determined by the volumetric damage model (creep-like). In addition, all but two points were above their respective Nelson Curve.
    - Level 4: Equipment had reasonable predicted remaining life determined by the volumetric damage model (creep-like). In addition, all but one (C-0.5Mo ~10°C, 18°F above PWHT CS) of these points were below their respective Nelson Curve.
    - Level 5: Equipment had long predicted remaining life determined by the volumetric damage model (creep-like). In addition, all of these points were below their respective Nelson Curve (coincidentally).

All of the 48 pressure vessels were evaluated using the Level 1 FFS advanced damage progression model and ranked into a POF level. A summary of the determined HTHA POF compared to Yara's prior determined HTHA POF based on the API RP 941 Nelson Curves is shown in **Table 4**. Note that E²G evaluated 49 pieces because 1 pressure vessel was modeled as both hot and cold wall situations based on potential refractory condition.

	E²G HTHA JIP Methodology	Yara INITIAL RISK ASSESSMENT
1, Very High	14	22
2, High	11	3
3, Medium	4	11
4, Low	7	2
5, Very Low	13	10
Total	49	48

Table 4. Outcomes of HTHA POF level determination

Figure 4 at the end of this paper shows the E²G results for POF were combined with Yara's prior COF and plotted on the risk matrix along with the other 61 pressure vessels that were previously determined by Yara to be "Low" to "Medium" risk and were not analyzed by E^2G. A graphical representation of the 48 pressure vessels analyzed by E^2G is shown in **Figure 5** with the API RP 941 Nelson Curve (including the determined POF categories).

In addition to determining a POF category for each of the 48 pressure vessels, a recommendation was provided for each level to help guide the path forward in terms of replacement, inspection, monitoring, and re-assessment:

- For POF Level 1 (Highest likelihood of HTHA) vessels: Consider future replacement. Inspect at next T/A if replacement is not feasible.
- For POF Level 2 and 3 vessels: Consider future replacement. Inspect at next T/A if replacement is not feasible. Consider a Level 2 FFS model to further examine histogram operating conditions and custom material properties.
- For POF Level 4 and 5 (lowest likelihood of HTHA): Monitor the equipment to ensure operating conditions remain as modeled. If operating variation occurs, consider re-validating the POF study.

Following the above recommendations, Yara established the replacement, inspection, and monitoring plan for all of the assessed equipment. The number of pressure vessels to be replaced, according to the CSB Alert, the Yara Initial Risk Assessment, and the E^2G Advanced Assessment, is shown in **Figure 6** at the end of this paper.

As a note to the reader, HTHA inspection has been inconsistent and difficult using most methods indicated in API RP 941 8th edition. There are recent and significant developments regarding the recommended Non-Destructive Examination (NDE) methods for HTHA.[14-17] E^2G inspection recommendations were based on the application of these recently enhanced methods of Time of Flight Diffraction (ToFD) and Phased Array Ultra-Sonics (PAUT) utilizing trained and certified NDE technicians. A Level 2 FFS assessment, considering all operating trends and not fixed operating conditions, was also offered which can allow for more specific considerations on a per-vessel basis, which may remove some of the conservatism applied in a Level 1 screening.

Conclusions

The recent changes in API RP 941 and recommendations from CSB have placed heightened scrutiny on the classification of the operation of equipment and piping in high temperature hydrogen service. One methodology for further discrimination of this definition has been presented in this paper.

Recent NDE training using Performance Demonstrated Initiatives with ex-service HTHA components has greatly improved the reliability and repeatability of NDE for HTHA, and API will soon release significantly revised NDE guidance.

With these improved NDE tools, the ability to conduct Fitness for Service becomes a viable option, and API 579 has a set of draft procedures for committee review. The API Risk Based Community (API 580/581) can now revise the HTHA inspection effectiveness table.

Each owner-user should define their own internal HTHA risk assessment criteria. These can now be made on an internal basis (as presented in this paper) and then be validated by empirical analytical tools and models as discussed in this paper. For specific higher-risk components, detailed historical operation assessments may be applied to have more clarity in the once-nebulous arena of HTHA prediction.

Figures

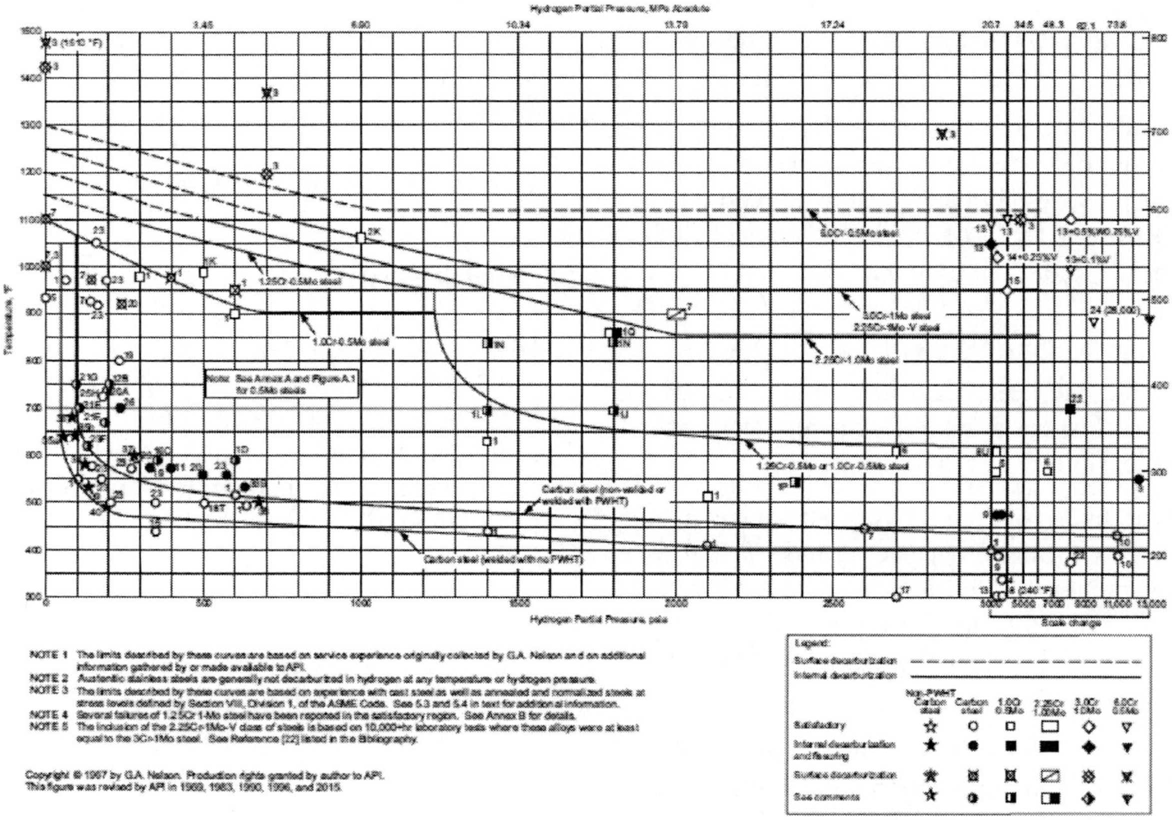

Figure 1. API 941 8th edition Nelson Curves. Note the new non-PWHT curve for carbon steel that is 50-70 °F below the prior-used carbon steel curve. The prior-used carbon steel curve is now referred to as the Carbon Steel with PWHT curve

Risk	No. of Components
Low	21
Medium	33
Medium High	23
High	32

	POF	E	D	C	B	A
	1	5		2	13	8
	2				1	4
POF	3	2			6	5
	4	3			9	2
	5	5		11	24	9
		E	D	C	B	A
		COF				

Figure 2. Yara's risk matrix with pressure vessels in hydrogen service plotted (right); Number of pressure vessels in each risk category (left)

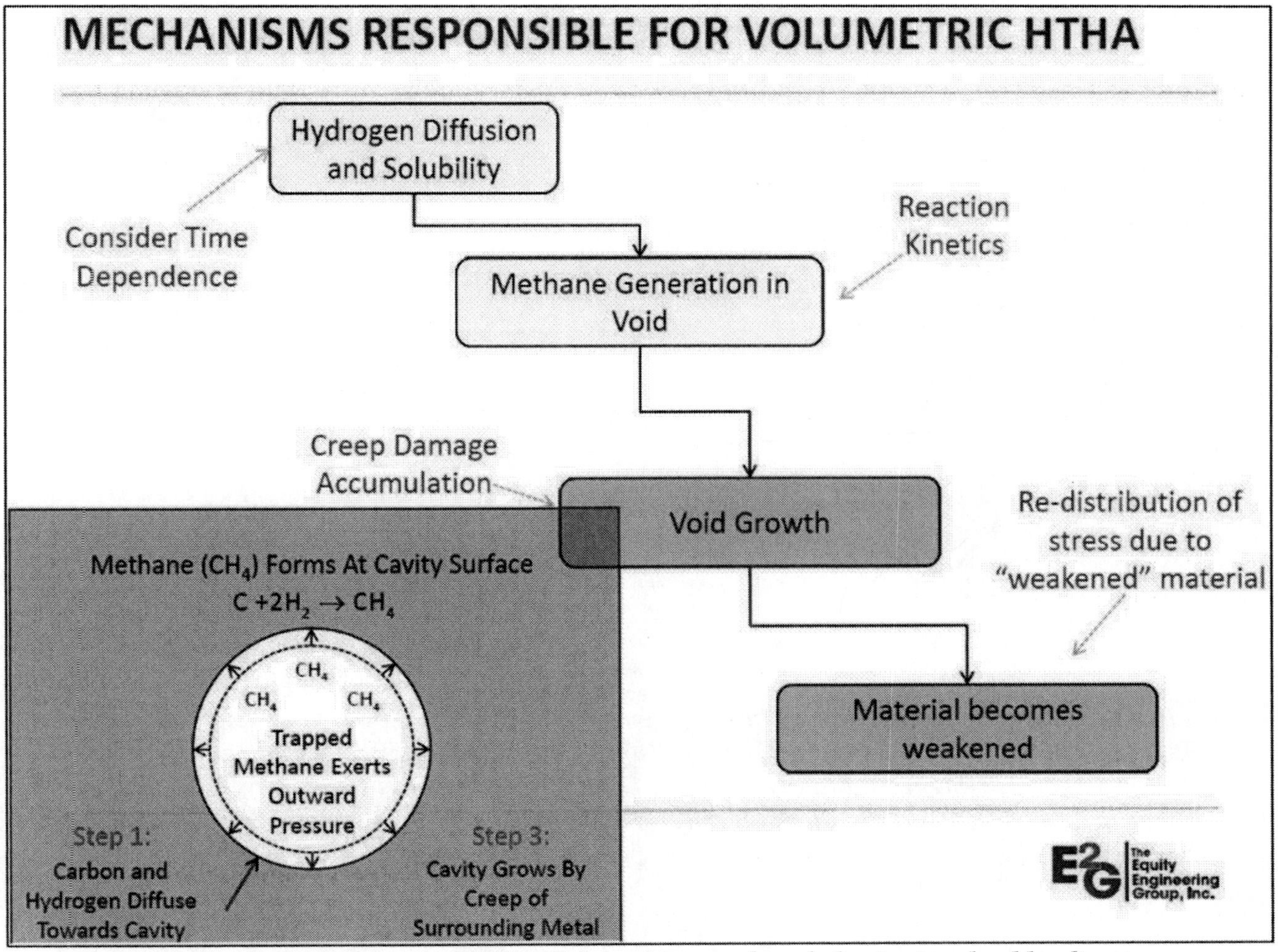

Figure 3. Demonstration of HTHA Volumetric attack and physics involved leading to formation and growth of voids

Risk	No. of Components
Low	22
Medium	41
Medium High	20
High	26

POF	1	3		2	9	5
	2	1			2	9
	3	3			2	1
	4	3			13	3
	5	5		11	27	10
		E	D	C	B	A
		COF				

Figure 4. Yara's updated risk matrix considering E^2G's HTHA POF determination for 48 pressure vessels combined with Yara's prior results for the rest of the fleet (right); Number of pressure vessels in each risk category (left)

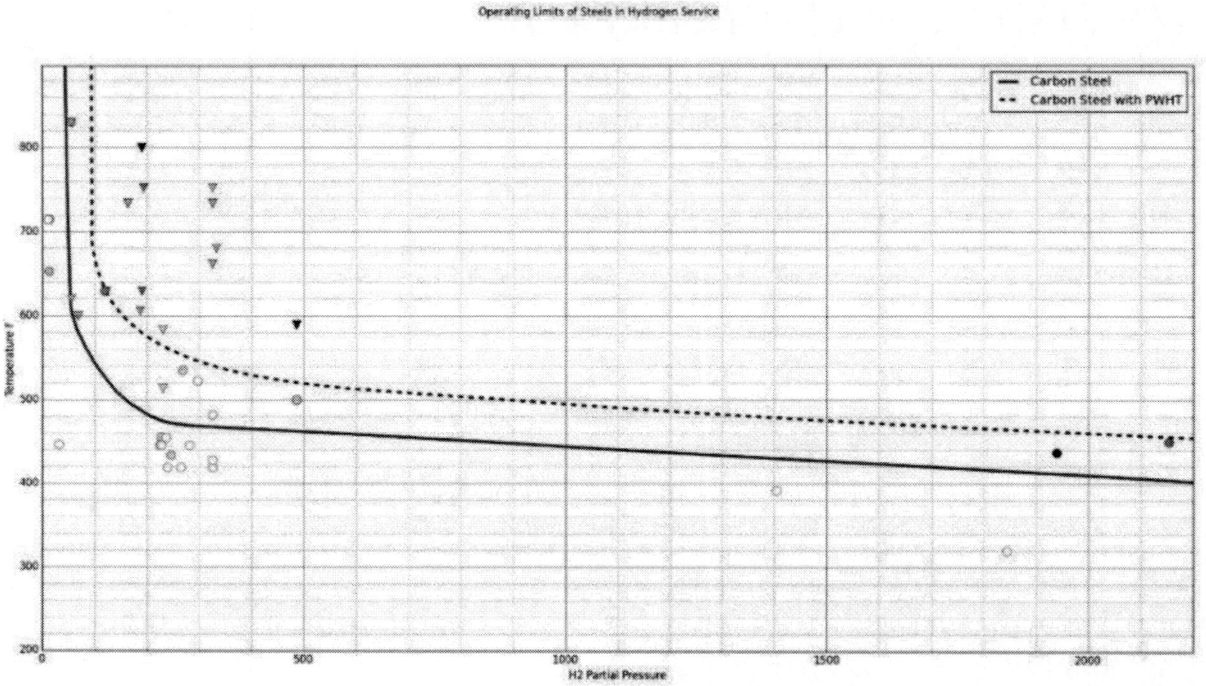

Figure 5. The 48 pressure vessels in hydrogen service, evaluated by E^2G to determine HTHA POF categories, shown on the API 941 Nelson Curve. Note that color indicates risk level: Yellow is Level 1 (highest), Black is Level 2, Dark Grey is Level 3, Light grey is Level 4, and White is Level 5 (lowest). Additionally, circles indicate items being within their respective Nelson Curve and triangles indicate items being over their respective Nelson Curve

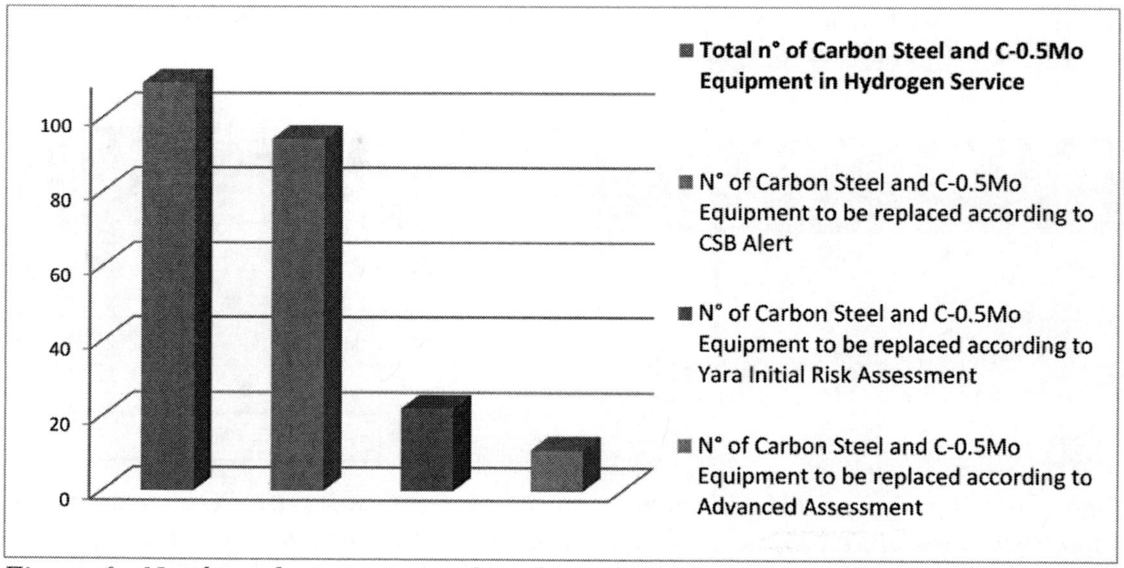

Figure 6. Number of pressure vessels to be replaced based on CSB Alert, Yara initial Risk Assessment, and Advanced Assessment

References

1. API, "API RP 941: Steels for Hydrogen Service at Elevated Temperatures and Pressures in Petroleum Refineries and Petrochemical Plants (8th Ed.)," American Petroleum Institute, Washington, D.C., February 2016.
2. "CSB Safety Alert: Preventing High Temperature Hydrogen Attack (HTHA)," The U.S. Chemical Safety and Hazard Investigation Board (CSB), Aug. 2016, www.csb.gov/assets/1/20/safety_alert_htha_final.pdf?15689.
3. M. Nugent, J. Dobis, T. Silfies and T. Armitt, "High Temperature Hydrogen Attack: Historic Review, Current Status and Future Perspective," in IPEIA, Banff, 2016.
4. M. Nugent, T. Armitt, J. Dobis and T. Silfies, "Advances in High Temperature Hydrogen Attack (HTHA) Non-Destructive Examination, Modeling, and Prediction, in Petrochemical and in Refinery Applications," in NACE CORROSION, New Orleans, 2017.
5. T. Silfies, J. D. Dobis and M. Nugent, "Looking behind the curtain of API RP 941 High Temperature Hydrogen Attack (HTHA) Data," in Corrosion 2016, Vancouver, BC, Canada, 2016.
6. M. Nugent, T. Silfies, P. Kowalski, and N. Sutton, "Recent Applications of Evaluations of Equipment in HTHA Service" in Corrosion 2018, Phoenix, AZ, USA, 2018
7. Materials Property Council, "MolyHi Joint Industry Project".
8. G. Manna, Factors Limiting the In-Service Lifetime of CrMoV Steels and Weldments of Pressure Equipment, Swansea: University of Wales, Swansea, 2004.
9. P. Shewmon, L. C. Chen and M. Prager, "Stress Assissted Hydrogen Attack Cracking in 2.25Cr-1Mo Steels At Elevated Temperatures," WRC Bulletin 527: Practical Aspects of Hydrogen Attack, pp. 79-110, 2009.
10. P. Shewmon and Y. H. Xue, "Effect of High-Pressure Hydrogen on Crack Grwoth in Carbon Steel," Metallurgical Transactions A, vol. 22A, pp. 2703-2707, November 1991.
11. L. C. Weiner, "Kinetics and Mechanism of Hydrogen Attack of Steel," Corrosion - NACE, vol. 17, pp. 110-115, 1961.
12. F. K. Naumann, "Influence of Alloy Elements in Steel Upon Resistance to Hydrogen Under High Pressures," Technieche Nittellungen Krupp, Zorechungeberichte, vol. 1, no. 12, pp. 223-234, 1938.
13. M. McKimpson and P. G. Shewmon, "Initial Hydrogen Attack Kinetics in a Carbon Steel," Metallurgical Transactions A, May 1981.
14. M. Nugent, T. Armitt, T. Silfies, J. Dobis, "State-of-the Art Prediction and NDE Techniques for High Temperature Hydrogen Attack in Piping and Equipment Paper 43" 7th Biennial Inspection Summit, Galveston Bay TX, 2017
15. M. Nugent, T Armitt "HTHA IN PETROCHEM EQUIPMENT:PREDICTION, PRIORITIZATION ANDNDE METHODS FOR DETECTION" International Chemical & Petroleum Industry Inspection Technology (ICPIIT), Galveston Bay TX, 2017
16. T. Armitt, M. Nugent, T. Silfies "HTHA Detection, Characterization and Verification" ASNT Fall meeting, 2017
17. T. Armitt, M. Nugent, T. Silfies "State-of-the Art Prediction and NDE Techniques for High Temperature Hydrogen Attack In Piping and Equipment Paper 407" 1st WCCM London UK, 2017.

A Rupture in the High Pressure Drain Line for HP Scrubber in Urea Plant

This paper highlights a critical leak incident that occurred in one of the high pressure pipes of the urea plant during an emergency shutdown, and how the situation was analyzed to prevent re-occurrence of similar incidents in urea plants.

Mohamed Mostafa Shams
Misr Fertilizers Production Company

Ashraf Mohamed Abdel-Hamid
Misr Fertilizers Production Company

Introduction

Misr Fertilizer Production Company (MOPCO) operates one of the largest urea ammonia plants in Egypt (2 million tons of urea per year). On 17th February 2016, an ammonia plant trip took place, and during the draining of the high pressure system of the urea plant a sudden rupture occurred on the high pressure drain line for the High Pressure Scrubber (HP Scrubber) resulting in major leakage of high pressure ammonium carbamate as illustrated in Figures 1 and 2.

Figure 1. Carbamate leak from drain line

Figure 2. Carbamate leak from drain

HP Drain Line Description

A drainage network is used to drain liquid from the high pressure (HP) synthesis equipment to the recycle section in the urea plant. This system comprises 7 points at different elevations to ensure complete drain:

1. Liquid line between HP Carbamate Condenser (HPCC) and Reactor
2. Gas line between HPCC and Reactor
3. Reactor down comer line to HP Stripper
4. HP Scrubber overflow line
5. Downstream Ejector HV222602 to HPCC

6. HP Scrubber heater recycle line
7. HP Stripper liquid outlet

During normal operation, the drain valves from the synthesis high-pressure lines are closed together with the drain valve in the common drain line. The block valve in the urea discharge line of the HP Stripper is open and the HP Stripper level control valve controls the level in the stripper. The pressure in the drain line is atmospheric, and the bleed in between the drain valves of the high-pressure synthesis lines and the drain valve in the common drain line are open. If one drain valve connected to one high-pressure synthesis line is open or corroded, this shall be detected by the vapor smell coming out of the bleed. During draining of the synthesis section, the bleed and the block valve in the urea solution line leaving the HP Stripper are closed. One by one, the drain valves from the HP synthesis lines are opened. The block valve in the common drain line is opened, and the stripper level control valve controls the flow to the recirculation section such that the pressure in that section is under control.

In the urea plant, the drain system was kept under pressure during normal operation by a continuous injection of condensate from the HP flush water pumps. The reason for keeping the drain system under pressure during normal operation was to avoid leakage of urea or carbamate into the drain/flush system in case of a leaking block valve. The flow of the HP condensate injection pump is about 20 liters per hour (5 gal/hr).

Function of the Common Drain Line in the Urea Plant

The HP synthesis section is provided with a drain circuit connecting all HP lines to the outlet of the HP Stripper, as illustrated in Figure 3. This drain line and other parts of the synthesis section are connected to the HP flush water circuit so as to enable them to be flushed. The HP flush water is delivered by the HP flush water pump 329P002A/B. The HP flush water line is provided with a pressure gauge PI329273, required for a special flush procedure. The injection pump 329P004A/B is connected to the common drain line, in order to keep it free of urea.

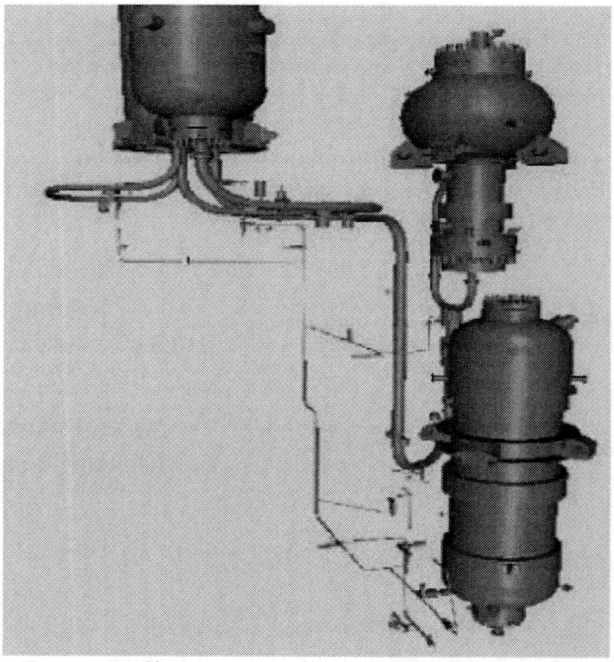

Figure3. Illustration of drain circuit (grey piping) for high pressure lines

The Incident

On 17th February 2016 8 years after the urea plant commissioning, a sudden rupture occurred on a 1-inch branch off the common drain line (actual thickness 3 mm, material AISI 310MoLN 1.4466) resulting in ammonium carbamate being relieved to atmosphere as illustrated in Figures 4. Figure 5 shows the ruptured drain line.

Figure 4. Ammonium carbamate being relieved to atmosphere

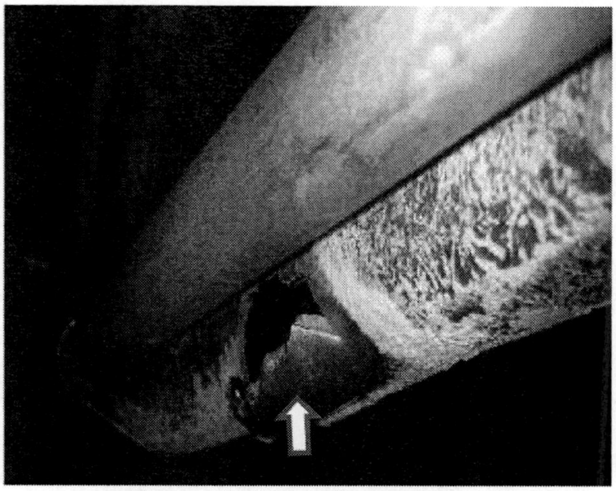

Figure 5. Drain line rupture.

Detailed Specification of the Pipe:

Line Number: 50-ACA1-322005-EK19L-D50
Pipe Material: 1.4466 X 2 CRNIMON 25 22
Process Fluid: Ammonium Carbamate
Line Size: 1" (OD 33.7 mm* thickness 2.6 mm)
Design Temp: 200°C (392°F)
Design Press: 167 barg (2420 psi)

Insulation design is D50 insulation with steam tracing to maintain a certain temperature requirement for process reason.

Investigation and Cause of Pipe Rupture

The insulation was removed from the ruptured pipe and an investigation was conducted. The cause of the failure is as follows:

1. Visual inspection of the pipe indicated some minor surface defects in the pipe spool and major cracks near the rupture location as illustrated in Figure 5.
2. Thickness measurement for the ruptured spool indicated no reduction of wall thickness.
3. Die penetrant test indicated branched shape surface cracks as shown in Figure 6.

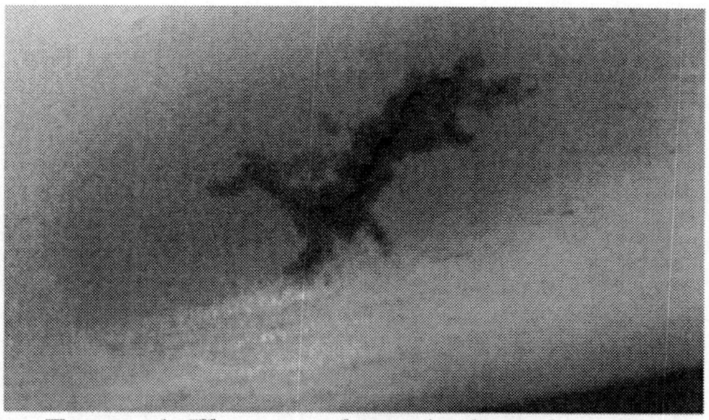

Figure 6. Illustrates branch-shaped surface cracks

4. Laboratory chemical analysis for insulation material indicated high chloride content in the insulation.

Conclusion of Pipeline Rupture

The inspection and laboratory results indicated that the cause of the line breakage was stress-corrosion cracking, which had originated on the outside of the pipe and was associated with the presence of chlorides. The austenitic stainless steel pipe ruptured as a result of stress-corrosion cracking. The presence of chloride ions, coming probably from the marine environment or other sources, caused chloride contamination of the pipe.

Repair and Rehabilitation

The corroded section of pipe was removed, and a new pipe was welded in as illustrated in Figure 7.

Figure 7. Pipeline after welding the new material.

Chloride Stress Corrosion Cracking (Cl-SCC)

Austenitic stainless steels suffer from stress corrosion cracking in hot solutions containing chloride. A high chloride concentration is required, although relatively small amounts of chloride are sufficient at heated surfaces, where chloride concentration can occur, or where chloride is concentrated by pitting or crevice corrosion. Surface initiated cracks caused by environmental damage of 300 Series SS and some nickel base alloys can be caused under the combined action of tensile stress, temperature and an aqueous chloride environment. The presence of dissolved oxygen increases propensity for cracking. The temperature usually needs to be above 70°C (158°F)

Affected Materials

a) All 300 Series SS are highly susceptible.
b) Duplex stainless steels are more resistant.
c) Nickel Base alloys are highly resistant, but not immune.

Critical Factors for Chloride Stress Corrosion Cracking (Cl-SCC)

The critical factors that result in Cl-SCC are as follows:

a) Chloride content, pH, temperature, stress, presence of oxygen and alloy composition are critical factors.
b) Increasing temperatures increase the susceptibility to cracking.
c) Increasing levels of chloride increase the likelihood of cracking.
d) No practical lower limit for chlorides exists because there is always a potential for chlorides to concentrate.
e) Heat transfer conditions significantly increase cracking susceptibility because they allow chlorides to concentrate. Alternate exposures to wet-dry conditions or steam and water are also conducive to cracking.
f) SCC usually occurs at pH values above 2. At lower pH values, uniform corrosion generally predominates. SCC tendency decreases toward the alkaline pH region.
g) Cracking usually occurs at metal temperatures above about 140°F (60°C), although exceptions can be found at lower temperatures.
h) Oxygen dissolved in the water normally accelerates SCC but it is not clear whether there is an oxygen concentration threshold below which chloride SCC is impossible.
i) Nickel content of the alloy has a major effect on resistance. The greatest susceptibility is at a nickel content of 8% to 12%. Alloys with nickel content above 35% are highly resistant and alloys above 45% are nearly immune.
j) Low-nickel stainless steels, such as the duplex (ferrite-austenite) stainless steels, have improved resistance over the 300 Series SS but are not immune.
k) External Cl–SCC has also been a problem on insulated surfaces when insulation gets wet.
l) Drains in liquid processing units are susceptible to cracking during startup/shutdown if not properly purged.

Appearance or Morphology of Damage

a) Surface breaking cracks can occur from the process side or externally under insulation (Figures 5, 8).

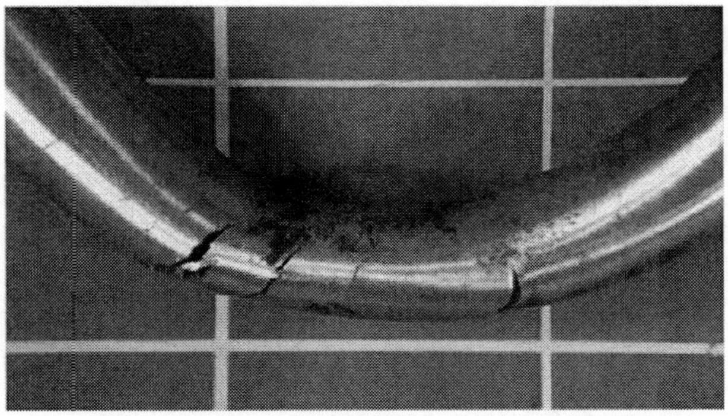

Figure 8. External cracking of Type 304SS instrument tubing under insulation

b) The material usually shows no visible signs of corrosion.
c) Characteristic stress corrosion cracks have many branches and may be visually detectable by an irregular cracked appearance of the surface (Figure 9, 10).

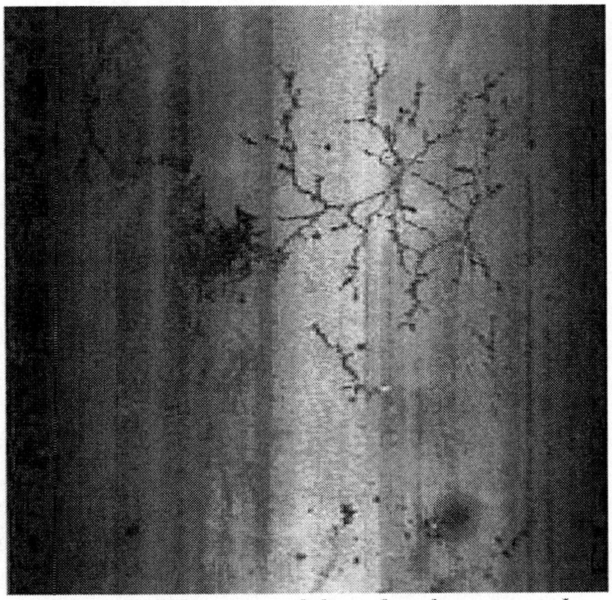

Figure 9. Close-up of the tube showing tight cracks with a spider web appearance.

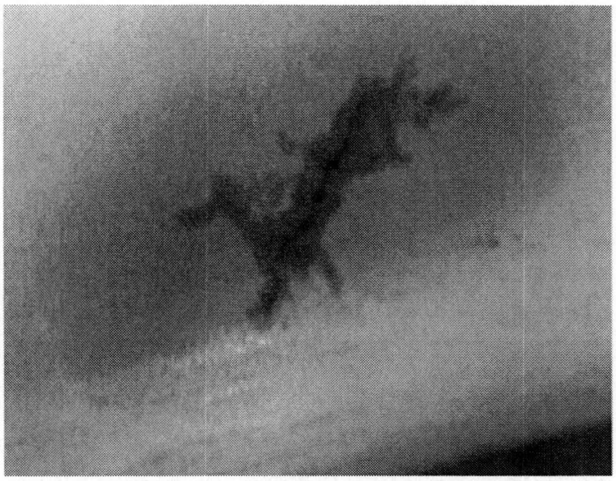

Figure 10. Close-up showing cracks in MOPCO HP pressure drain line.

d) Metallography of cracked samples typically shows branched transgranular cracks (Figure 11).

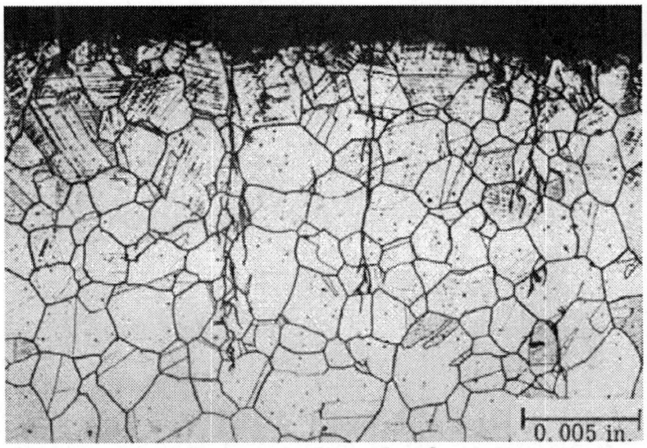

Figure 11. Photomicrograph of a cross-section of a cracked tube illustrating the transgranular mode of cracking initiating on the surface.

e) Sometimes inter granular cracking of sensitized 300 Series SS may also be seen.
f) Welds in 300 Series SS usually contain some ferrite, producing a duplex structure that is usually more resistant to Cl–SCC
g) Corrosion-resistant nickel-based alloys are also susceptible to cracking under severe conditions. Figure 12 shows severe cracking, including branching trans-granular cracks, of a finned Alloy C-276 tube in a reboiler after 8 years in service due to ammonium chloride carryover.

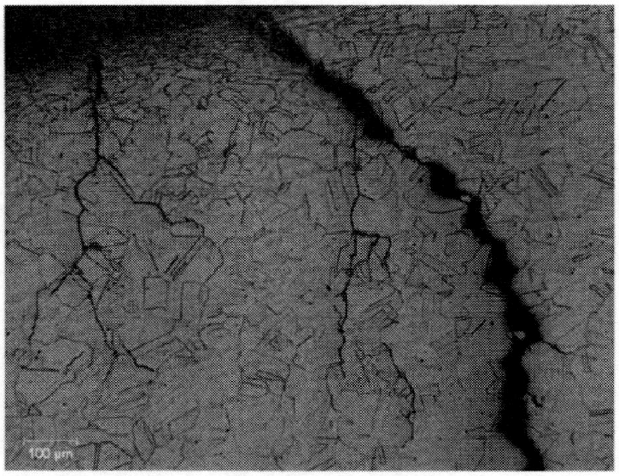

Figure 12. – Severe cracking of a finned Alloy C-276 tube in a reboiler after 8 years in service.

Prevention / Mitigation of SCC

a) Use resistant materials of construction.
b) When hydro testing, use low chloride content water and dry out thoroughly and quickly.
c) Properly applied coatings under insulation.
d) Avoid designs that allow stagnant regions where chlorides can concentrate or deposit.
e) A high temperature stress relief of 300 Series SS after fabrication may reduce residual stresses. However, consideration should be given to the possible effects of sensitization that may occur, increasing susceptibility to polythionic SCC, possible distortion problems and potential reheat cracking.

Inspection and Monitoring of SCC

a) Cracking is surface connected and may be detected visually in some cases.
b) Penetrant testing or phase analysis eddy current techniques are the preferred methods.
c) Eddy current inspection methods have also been used on condenser tubes, as well as piping and pressure vessels.
d) Extremely fine cracks may be difficult to find with die penetrant testing. Special surface preparation methods, including polishing or high-pressure water blast, may be required in some cases, especially in high pressure services.
e) Ultrasonic testing (UT) is an inspection method which uses high frequency sound waves, well above range of human hearing, to measure geometric and physical properties in material. Ultrasonic testing may be another alternative to find Cl-SCC.
f) Often, radioactive testing is not sufficiently sensitive to detect cracks except in advanced stages where a significant network of cracks has developed.

Lessons Learned

1. Inspecting and monitoring operating conditions of the high pressure lines periodically are critical to ensuring safe operation.
2. Periodically inspect insulation on high pressure pipes to ensure it is in good condition and has good sealing.
3. Avoid any source of chloride in urea high pressure lines to avoid stress corrosion cracking.
4. Time and money spent in inspection and maintaining the vessel in healthy condition will always be paid back and achieve the goal of preventing accidents and improving safety at the plant.
5. Emergency preparedness and Emergency Response Team (ERT) should be available.
   - Hazmat equipment should be found at the plant for containing small leakage until the unit can be quickly shutdown and depressured.
   - Emergency equipment should be distributed at all levels in the urea building and inside elevators (SCBA, escape cylinders, Full face masks, etc.) to protect employees' respiratory, eyes, and skin.
   - Be fully aware of the escape routes - escape lateral / upwind direction
   - Training and drills should be regularly performed both for hazardous materials response and rescue system.

References

1. Recommended Practice API 571 Second Edition, April 2011
2. UreaKnowhow.com

High Pressure Carbamate Condenser Liner Replacement

The High Pressure Carbamate Condenser (HPCC) liner was observed with bulges and cracks at the region where the pass partition plate attaches to the liner. The damage was discovered during the warranty inspection, which prevented a potential hazardous situation that could have resulted in a process leak and plant shut down. A level 3 fitness for service (FFS) assessment in accordance with the requirements of API 579 was performed. The assessment showed that protection against ratcheting was not met and recommended for rectification at the next scheduled opportunity. The liner segments were replaced as per specification via an innovative and delicate in situ replacement procedure. The paper explains the root cause of the damage and the challenges faced during development of the rectification procedure.

Sultan Al-Harthi, SAFCO, Saudi Arabia

Gys van Zyl, SABIC, Saudi Arabia

Ravinder Singh, Larsen & Toubro, India

Introduction

The High Pressure Carbamate Condenser (HPCC) is a critical heat exchanger in the high-pressure synthesis loop in any urea plant. During the first inspection after commissioning, the HPCC liner was observed with bulges and cracks at the region where the pass partition plate attaches to the liner. Figure 1 shows the location of liner damage.

Figure 1. Red arrows indicate liner bulges and cracks locations

This paper describes the root causes of the observed liner bulge, lesson learned and the fitness for service (FFS) assessment results. Furthermore, an innovative approach implemented to perform an in-situ replacement of the defective liner segments during a planned shutdown without any modification will be discussed.

General Technical Data

The HPCC is a horizontal U-tube heat exchanger with a thick mono wall carbon steel (CS) channel head (SA-266M-GR2). The channel head has buttered grooves of SS309 to which the internal loose liner of 6 mm thick 25Cr:22Ni:2Mo material has been welded. Figure 2 shows the liner welds with buttering layer. Design conditions for the HPCC are shown in Table 1.

Design Parameter	Tube side	Shell side
Pressure kg/cm^2 (psi)	165 (2347)	6 (85)
Temperature °C (°F)	210 (410)	180 (356)
Process	Carbamate solution	Steam condensate

Table 1. HPCC Design Conditions

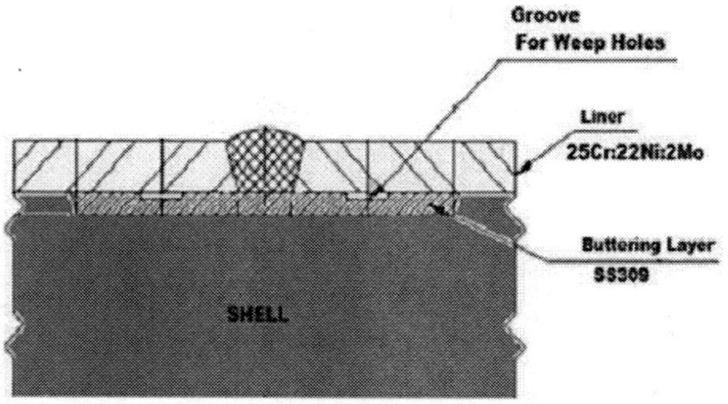

Figure 2. Cross sectional view for liner weld with buttering layer

First internal inspection

The first internal inspection one year after commissioning revealed bulges and cracks at both sides of the partition plate welds with the liner. Figure 3 shows the as found condition. An extensive inspection was conducted for all liner welds via dye penetrant test (DPT). In addition, the gap measurement at the bulge locations revealed a 10 mm air gap between the liner and the carbon steel, while the maximum acceptable limit is 3 mm. The bulge was only observed near the partition plate welds to the liner.

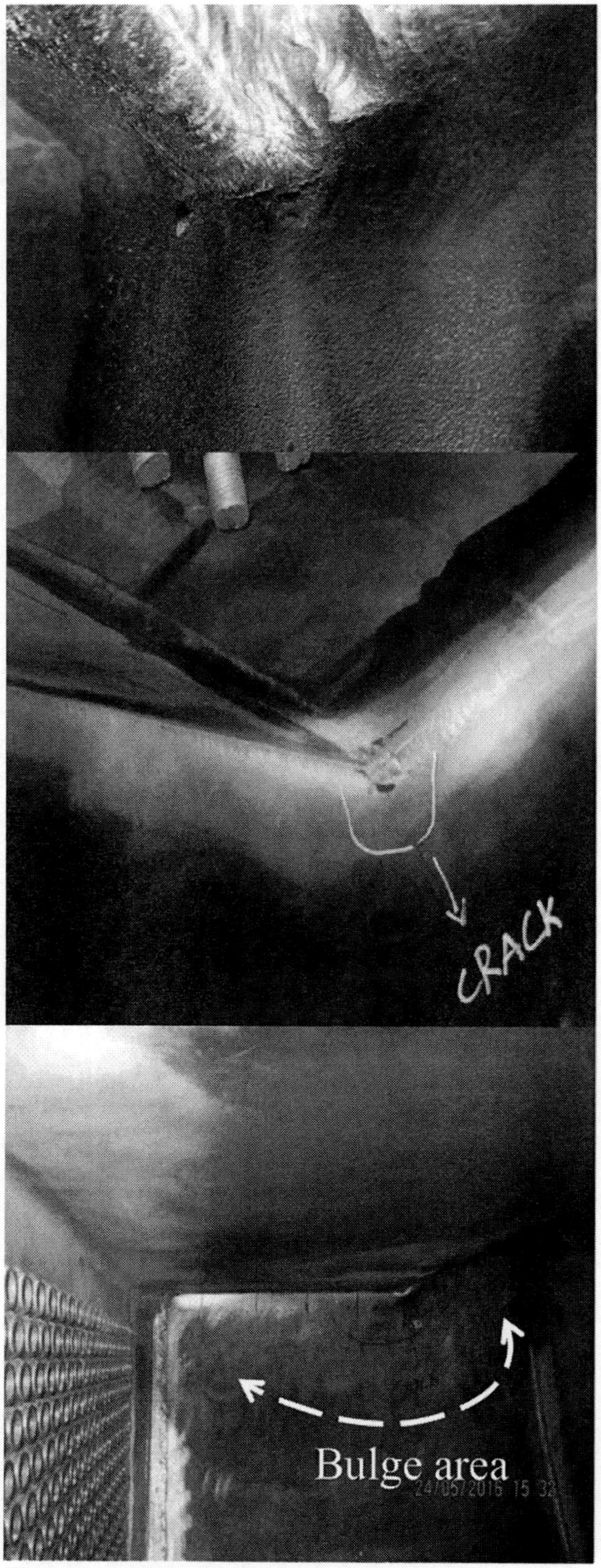

Figure 3. Photos show cracking and bulges

Fitness for service assessment

In order to evaluate the impact that the observed damage would have on the structural integrity of the equipment, a level 3 fitness for service assessment of the bulge was performed in accordance with the requirements of API 579 (1).

One challenging aspect of the assessment was obtaining an accurate representation of the deformed geometry. To accomplish this, a geometric scan of the bulged region was performed using a hand-held 3D laser scanner. The measured geometry and extracted bulge profiles are illustrated in Figure 4

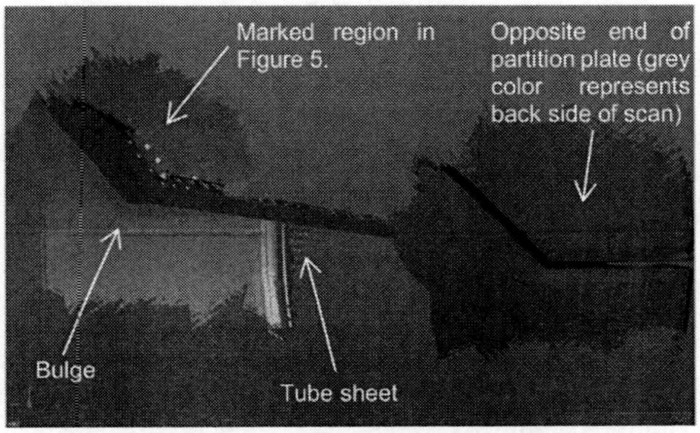

Figure 4. 3D scanned geometry of the bulged region

The finite element model constructed consisted of the channel, liner, partition plate and tube sheet, including the bulge in the liner modeled as per the laser scanned measurements (See Figure 5).

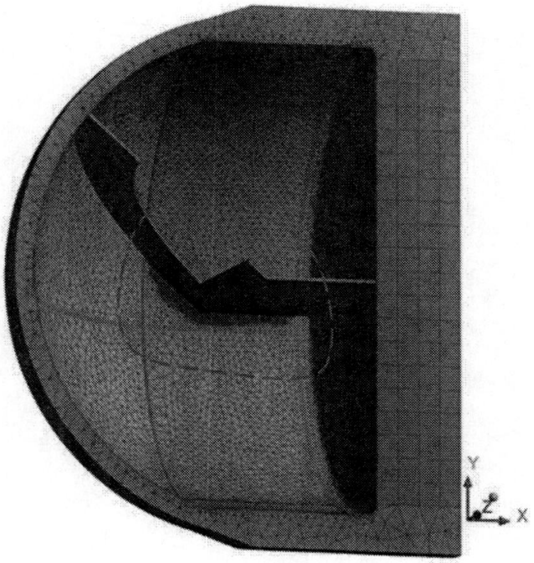

Figure 5. Finite element model

In accordance with the procedures of API 579 Annex B1, the following modes of failure were considered:

- **Plastic collapse**

An elastic-plastic analysis was performed in accordance with B1.2.4.

- **Local failure**

An elastic-plastic analysis was performed in accordance with B1.3.3.

- **Collapse from buckling**

There is no potential for failure of the liner through buckling. Therefore, a detailed assessment was not performed.

- **Cyclic loading**

The equipment is not in cyclic service. Therefore a fatigue assessment was not performed. An elastic-plastic ratcheting assessment in accordance with B1.5.7 was performed.

The assessment found that the damaged component has adequate protection against plastic collapse. As illustrated in Figure 6, convergence was achieved well beyond the required factored load. The damaged component also has adequate protection against local failure, with the maximum calculated plastic strain being lower than the limiting tri-axial strain by a factor of 4.7 (see Figure 7).

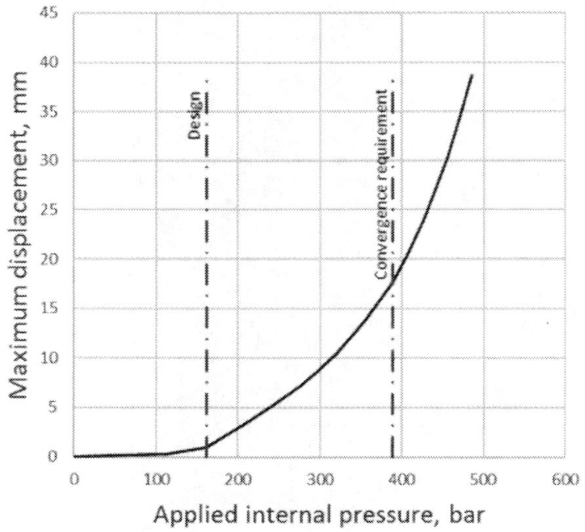

Figure 6. The FEA achieved convergence was well beyond the required factored load, illustrating adequate protection against plastic collapse

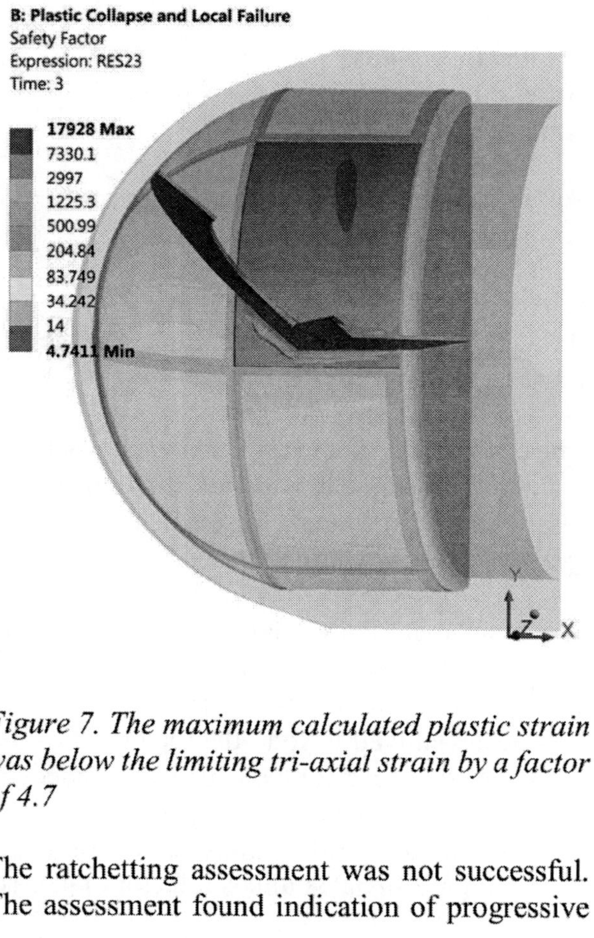

Figure 7. The maximum calculated plastic strain was below the limiting tri-axial strain by a factor of 4.7

The ratchetting assessment was not successful. The assessment found indication of progressive plastic deformation with successive applications of the design loads, as illustrated in Figure 8.

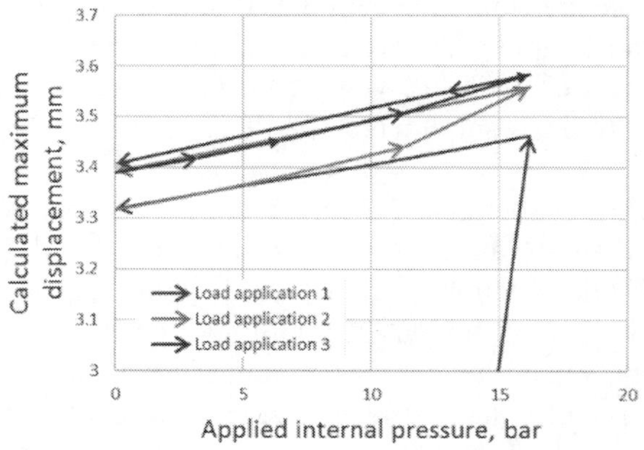

Figure 8. Progressive plastic deformation with successive applications of the design loads.

The fitness for service assessment concluded that the equipment could be returned to service for a limited time, as the damage mode that was identified as a concern would require a number of applications and removals of the design loads (startup/shutdown cycles) to initiate failure. As a result, this gave the opportunity for detailed development and qualification of a comprehensive repair procedure.

Root causes

Deep analysis and discussions were carried out to identify the causes of the liner bulges. Equipment engineering design, operations history and manufacturing records were reviewed during the investigation. Accordingly, the root cause was determined to be due to the welding sequence followed during manufacturing for the partition plate welding to the liner. In addition, hydrostatic testing was performed after completing the partition plate welding to the liner, which constrained the liner expansion and resulted in more gap around that area. Furthermore, the gap measurement tool used during manufacturing did not allow taking a reading very close to partition plate.

Lessons learned

1) Liner welding should be carried out using moon plate fixture to restrict lifting of liner during welding and creating excessive air gap.

2) An additional hydrostatic test was included in the inspection and testing plan to hydro-form the welded liner. This ensures the gap beneath liner is ≤ 1.5 mm.

3) The optimum weld sequence should be developed to reduce the liner shrinkage during installation to an acceptable limit. The same was considered during the liner replacement by keeping last weld of the partition plate to the liner side as marked in Figure 9 welds 11 and 12.

4) The weld process parameters and filler wire selection should be optimized to restrict the heat input during pass partition plate welds.

5) The gap measurement locations shall include the area near the partition plate and should be ≤ 10 mm prior to start of welding and hydrostatic test, then same locations to be taken again after completing both task.

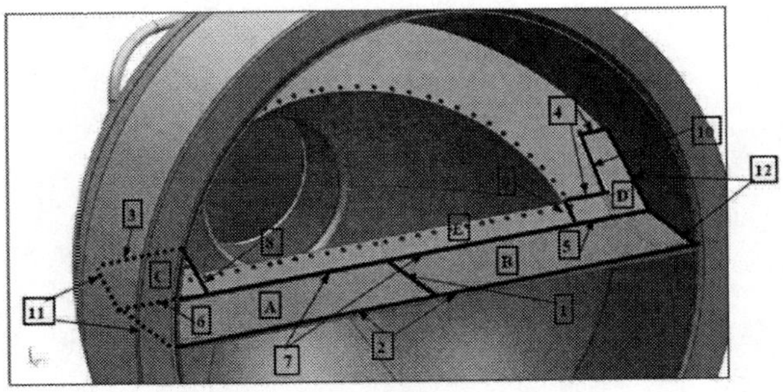

Figure 9. Sequence of welding partition plate implemented during liner replacement

Repair Procedure

The repair scope was to carry out rectification of liner segments 964 x 661 mm (38 x 26 inches) considering limited access through equipment manway with a 500 mm (20 inch) opening. To carry out such a repair, every step needed to be carefully evaluated to ensure safe, quality and timely execution.

Three different proposed options to rectify the observed damage were evaluated as follows:

Option-1: By jacking existing liner

This rectification is to apply mechanical force to the bulge area to depress it back to the shell side. The jack would be inserted in a cut slot between partition plate and the liner. Finite element analysis (FEA) was carried out to evaluate this proposal. The FEA result showed that 100 ton would not be enough load to reduce the gap to the acceptable limit of 1.5 mm. Afterward, when the applied force is removed, the gap "springs back" to 8.5 mm. More-over, the liner cut edges become wavy and pose challenges during liner set-up and welding. See the FEA results in Figures 10, 11 and 12. Hence, this proposal was not considered viable.

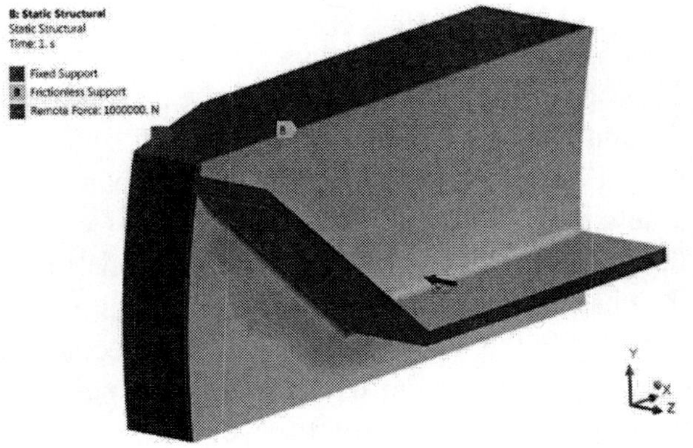

Figure 10. FEA Model used to evaluate application of external mechanical force to depress the liner bulge

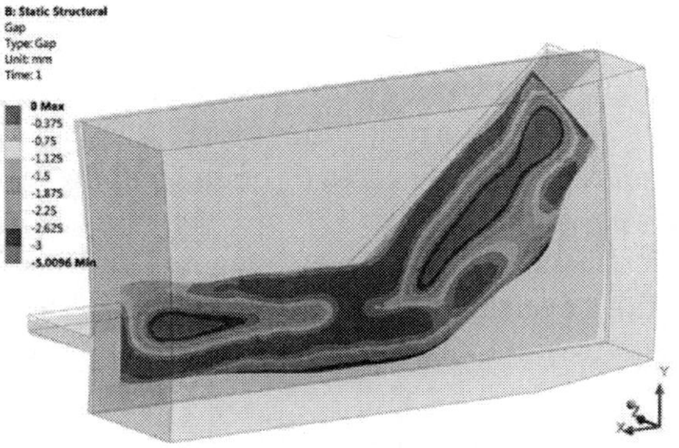

Figure 11. Liner gap with 100t applied load

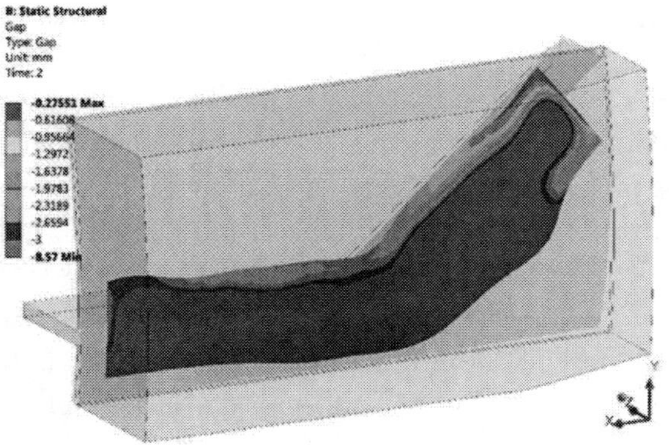

Figure 12. Predicted liner gap after load removal and "spring back" effect

Option-2: Replacement of the liner in 2 pieces to allow access through man way

This option considered inserting the liner through the manway as two segments. These segments would be welded together inside the channel by a new butt weld joint (see Figure13). However, this would result in two new seam welds inside the channel. In addition, this would require modification to the weep holes to ensure detectability of a leak in the event of a crack in the new butt weld joint. Additionally, the weep tubes require buttering on the shell to facilitate the attachment weld. As a result this option was not considered favorable and discarded.

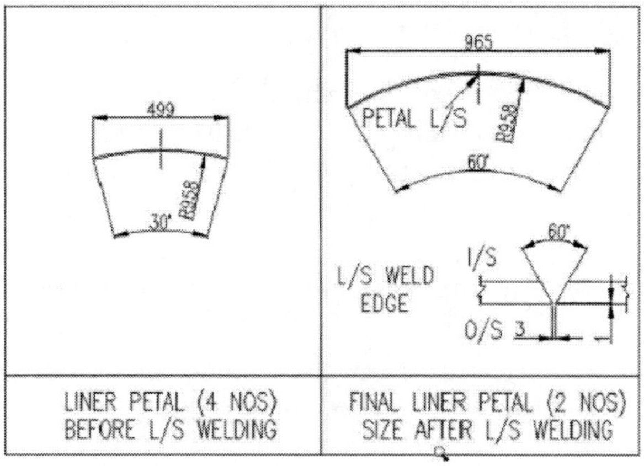

Figure 13. Liner segment cut two halves

Option-3: Replacement of the liner without modification by rolling the liner

This method was developed subsequent to the review of the previous two options, in order to overcome accessibility of manway, see Figure 14. In addition, this option avoids any change to the existing liner weld configurations and their weep holes. Rolling the liner to a smaller diameter to facilitate insertion through the manway and unrolling the liner to the exact shell curvature geometry inside the channel head required a special rolling machine. This idea was a challenge to develop and successful implementation was ensured through qualification of the procedure by using a mockup. Figure 15 shows the full scale mockup that proved the viability of this option, after which it was selected for liner rectification.

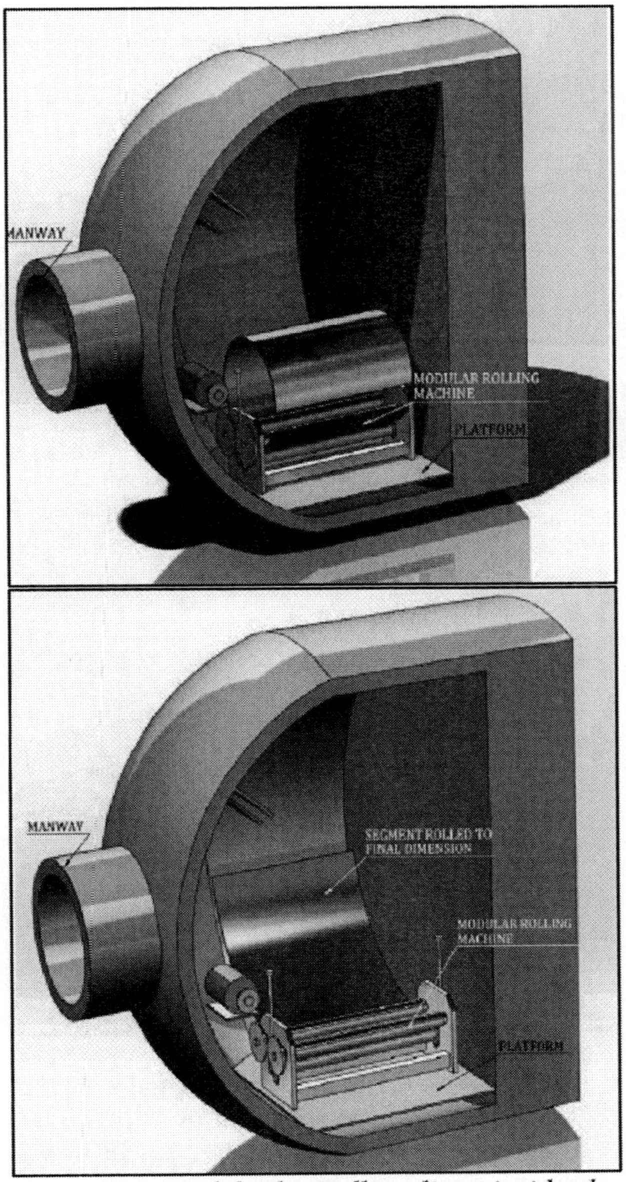

Figure 14. Model of unrolling liner inside the channel head

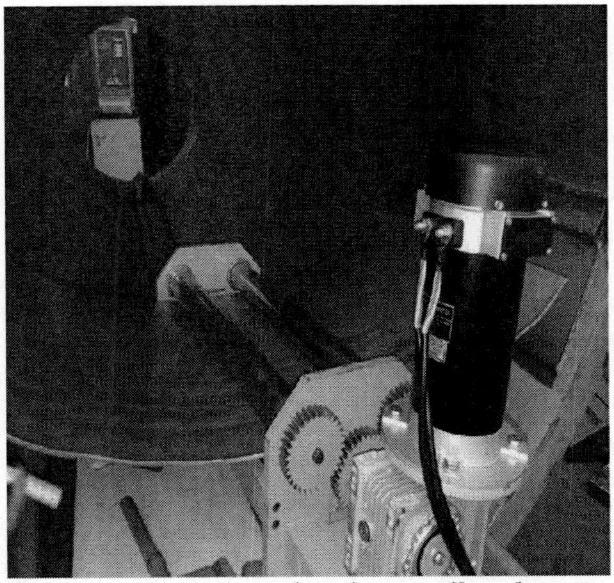

Figure 15. Actual mockup for unrolling liner inside the channel head

Bulged liner removal

Removing the bulged liners is a sensitive and major step because a significant amount of grinding is needed. Therefore, a detailed procedure was developed to ensure that the grinding would not damage the tube sheet area, adjacent liners, and SS309 buttering. The liner in the channel head consists of six petals marked in Figure 16 below. Petals intended for removal are petal no.1 and no.4.

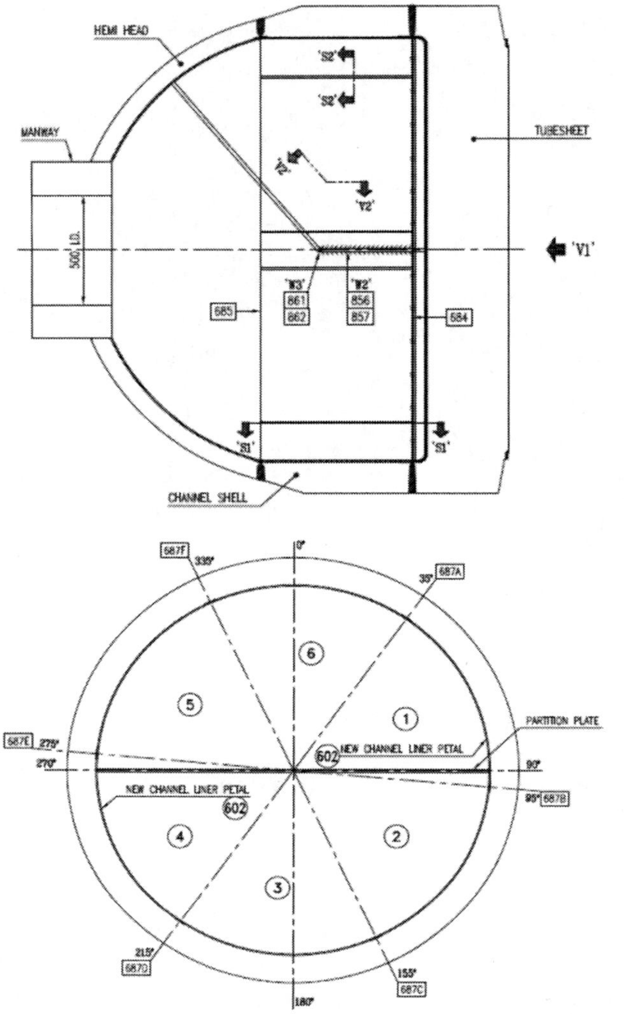

Figure 16. Liner petals arrangement and Partition plate in the channel head

The first step after marking the cut lines is cutting the partition plate, which is welded to the tube sheet and liners. The cut was 45 mm (1.77 inch) away from the tube sheet overlay weld. The small remaining portion will be used to reattach the new partition plate. The depth of cut is an important parameter and needs to be restricted to the liner thickness thus safeguarding against grinding into the buttering and exposing the carbon steel channel shell. Thereafter, liner welds were cut from the weld center while maintaining the root pass weld of adjacent petal, see Figure 17.

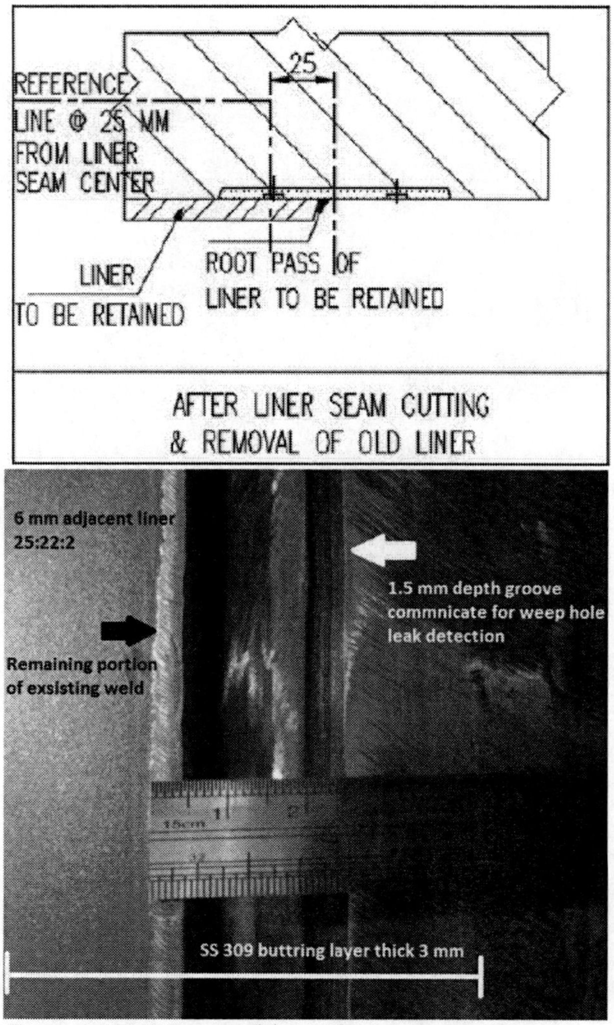

Figure 17. Liner weld cut from center

An important step after liner removal is protection of the weep holes from grinding debris. Visual inspection of carbon steel part and 100 % die penetrant testing for the remaining portion of the root pass weld was performed. In this stage, the weep holes were checked for any blockage by charging air from outside to the particular path.

New liner installation

The first step is inserting the rolling machine and preparing the rolled liner plates inside the channel head. Then, unrolling the liner plates, and trimming the edges to fit them in place using tag welds, see Figure 18.

Figure 18. Photos show prepared rolled plate, unrolling process and fit up

After completing the liner fit up came the step of the jacking arrangement installation to minimize shrinkage during welding, see Figure 19. A similar jacking arrangement was carried out during partition plate welding.

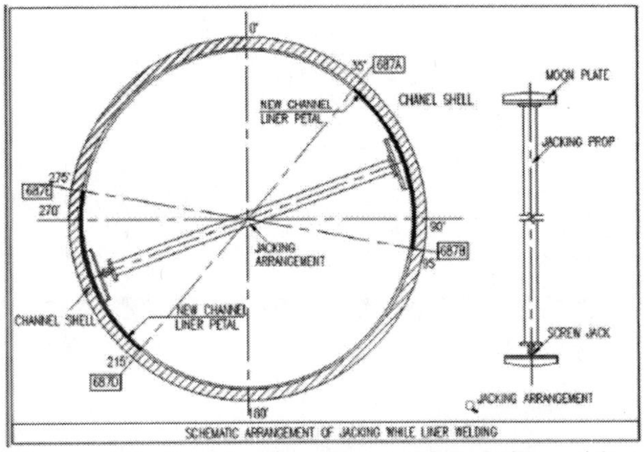

Figure 19. Jacking arrangement while welding liner plates

Welding the liner plates directly to the original SS309 buttering layer is the advantage in this case, since the unrolling process allowed having the same liner plate size. Subsequently, a gap check was done in an as welded condition before the intermediate hydrostatic test, which was in the acceptable limit ≤ 5 mm. Then, an intermediate hydrostatic test was performed for expansion of the liner plates towards the carbon steel part to have the gap ≤ 1.5 mm.

The partition plate welding was carried out as per the welding sequence illustrated in Figure 9. To keep the partition plate in place, a new jacking arrangement was installed prior to welding, see Figure 20. The welding was carried out with the leftover 45 mm piece of the original partition plate with a double V groove joint. The final weld is between the liner and the partition plate balancing both sides. Consequently, a final hydrostatic test was carried out followed by an ammonia leak test, and the equipment was handed over back to operations within planned mechanical window of 18 days. In Figure 21, several photos for all above mentioned steps are provided.

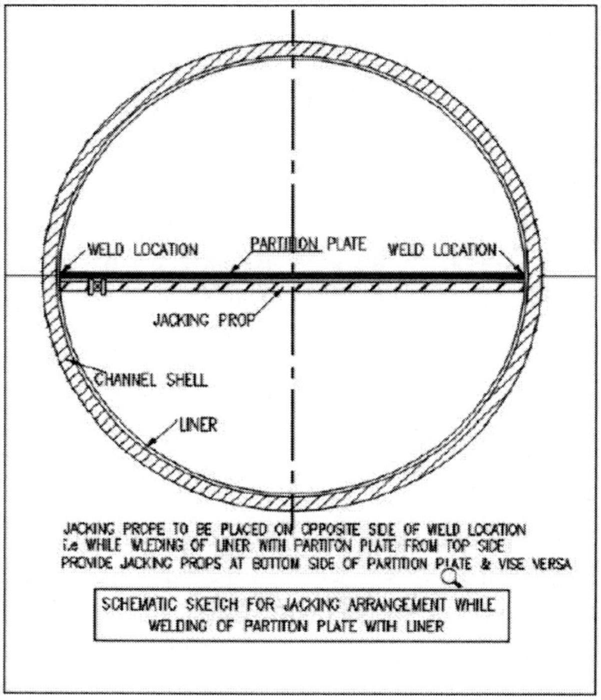

Figure 20. Jacking arrangement while welding partition plate to the liner

Quality control plan

A quality control procedure with a detailed inspection and test plan was developed for each step including a site work initiation meeting, procedures review, welding qualification and consumables inspection. In addition, material testing and certification was performed at a third party laboratory. Inspection started from the as found condition and included protection of the flange face seating area before cutting until job completion by cleaning and tools removal for manway box up.

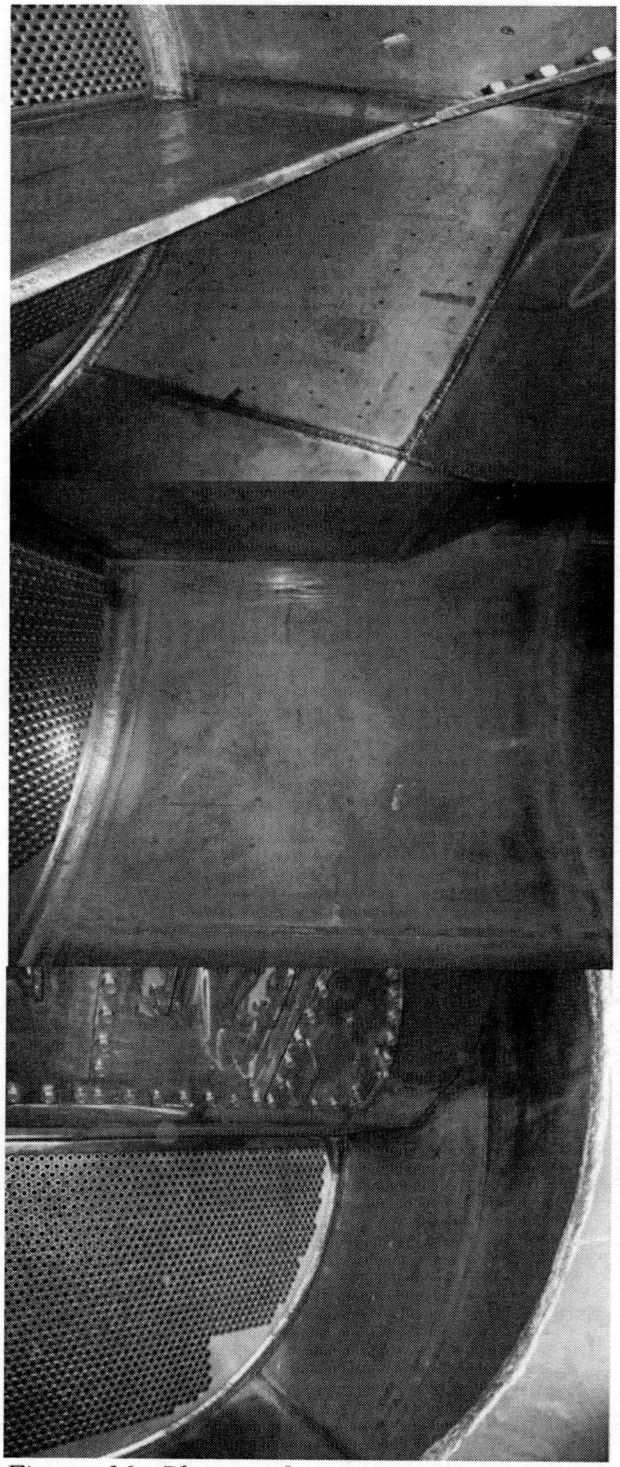

Figure 21. Photos of partition plate welding, NH3 leak test and final inspection before box up

Conclusions

- Welding sequence for the liner segments is a very crucial step to minimize the liner gap.
- Intermediate hydrostatic testing is essential after liner welding and before partition plate welding.
- Initial inspection after commissioning is also essential for reference and rectifying any manufacturing defect timely.
- FFS is a useful tool for assessment of the effect that damage has on structural integrity.
- Replacement of the liner with complex geometry is possible even though the liner plate is larger than the manway, by rolling the liner plate inside the vessel.
- A detailed repair plan including a mock up test is mandatory for efficient liner replacement.

References

1. API 579-1 / ASME FFS-1, Fitness-For-Service, Third Edition

Secondary Urea Reactor Top Head and Diaphragm Failure

During the process of urea production, multiple compounds will form that are highly corrosive to standard metals under the existing process conditions. Thus, it is imperative to use the correct materials when constructing piping and vessels. When a leak developed on the head of a high pressure urea reactor in a urea plant, the cause was the failure of the diaphragm that protected the pressure containing reactor bolted head cover. The diaphragm was found to have been constructed with 316 stainless steel, rather than the correct 317L. The diaphragm corroded and developed a pinhole leak after extended exposure to active corrosion from the top of the reactor. The failure of the diaphragm allowed the carbon steel bolted head cover to be exposed to carbamate resulting in significant corrosion to the cover.

This paper presents the process from the detection of the leak to the replacement of the diaphragm and repair of the bolted head cover. Also discussed will be the machining and weld repair of the corroded area of the cover, and the use of updated ASME standards to remove material from the cover without the need to change the pressure rating of the urea reactor.

Ryan Schlotfelt
Coffeyville Resources Nitrogen Fertilizers, LLC

Brandon Daniels
Coffeyville Resources Nitrogen Fertilizers, LLC

Introduction

Coffeyville Resources Nitrogen Fertilizers, LLC operates a Weatherly urea plant that was designed in 1998. The urea plant includes a Secondary Urea Reactor which provides retention time for the reaction of ammonia (NH_3) and carbon dioxide (CO_2) to generate ammonium carbamate followed by dehydration of ammonium carbamate to urea and water. The reactor operates at 3600 psi (24,821 kPa) and 390°F (199°C) to drive the reaction to equilibrium and maximize conversion.

In October 2017, a leak was found on the head of the Secondary Urea Reactor; a unit that had already experienced two prior failures. The resulting investigation found that the reactor's top diaphragm plate had failed and the bolted head cover had been damaged. An extensive repair process for the cover and re-evaluation of the original design calculations followed. In this paper we will discuss the previous incidents involving the Secondary Urea Reactor, as well as the discovery, investigation, and repair process involving the most recent incident.

Past Incidents

March 2009 Incident

The first prior incident with the Secondary Urea Reactor occurred on March 6th, 2009 at 8:50 p.m. The incident consisted of the formation of a flammable mixture of oxygen (O_2), ammonia, and hydrogen (H_2) that was not fully combusted through the oxidizing catalyst in the CO_2 Purification process. This mixture ignited in the headspace at the top of the Secondary Urea Reactor. The possible sources of ignition considered were auto-temperature ignition, static electrical discharge present from the flow of material through the system, or impact energy created by the inert gas purge line rubbing against the crown of the downcomer. The combustion caused significant damage to the internals of the reactor and was even great enough to damage the bolts and the foundation.

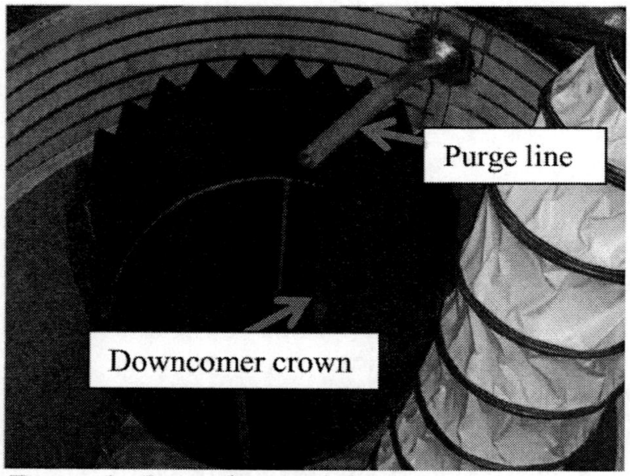

Figure 1. Purge line and downcomer crown

Several actions were taken to decrease the likelihood of this event occurring in the future. These included:

- The purge line was removed, causing the inert gases to exit through the downcomer.
- The crown atop the downcomer (see Figure 1 above) was removed.
- The downcomer height was raised to reduce the reactor headspace (see Figure 2 right).
- Hydrogen content to the reactor is regularly tested.
- Oxygen concentration to the reactor was re-evaluated.

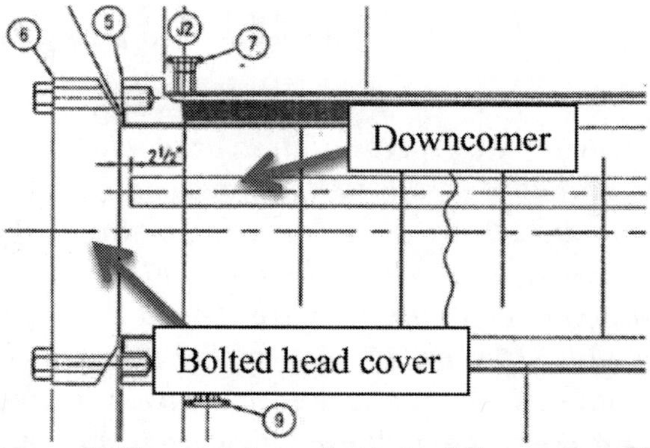

Figure 2. Top portion of original Secondary Reactor drawing

September 2010 Incident

The second incident occurred on September 30th, 2010 at 6:36 p.m. The Secondary Urea Reactor experienced a rapid depressurizing event when the reactor head separated from the vessel at normal operating pressure [1]. The event resulted in the complete loss of process contents consisting of ammonia, water, carbon dioxide, ammonium carbamate, and urea. The internal mechanical equipment including the reactor trays were destroyed. This event also caused damage to surrounding equipment and buildings.

The cause of this incident was found to be a weld that had failed between the forged head bottom flange, shown in Figure 3, and laminated cylinder.

Figure 3. Secondary Urea Reactor Head

Areas of cracking were found around the heat affected zone (HAZ) of the weld, and the hardness in the HAZ was found to be high. This indicated that the welds were not stress relieved. The high hardness and residual weld stresses made the welds susceptible to environmental stresses. It has been shown that environmental stress cracking can occur on welded carbon steel exposed to boiler feed water (BFW), anhydrous ammonia, nitrates, or caustic [2]. The design of the vessel included a boiler feed water jacket to maintain constant wall temperature. This water jacket exposed the weld to a BFW environment and prevented the weld between the head and cylinder to be inspected [1]. This resulted in multiple action items being created, including developing a new weld procedure, and not reinstalling the water jacket to ensure the weld could be inspected in the future.

The vessel was repaired by the original equipment manufacturer (OEM) following this incident. This included the installation of a new diaphragm plate. Documentation for this repair indicated that the new diaphragm plate was the correct 317L stainless steel (SS). It was found after the current incident took place that this metallurgy was incorrect, as discussed later.

Current Incident

While making a normal area walkthrough on the night of October 8, 2017, two urea plant operators found suspicious debris at the base of the Secondary Urea Reactor. The operators reported the debris to their supervisor and the debris was found to be ammonium carbamate. With the knowledge gained from the previous unit incidents, it was quickly discovered that the source of the carbamate was a leak on the Secondary Urea Reactor between the bolted head cover and the vessel. The urea plant was shut down so that the bolted head cover could be removed for inspection.

Investigation

Upon investigation, it was found that the diaphragm plate, which separated the reactor head face from the process, had developed a small hole. This was revealed by running a dye penetrant test (DPI) on the plate, shown in Figure 4.

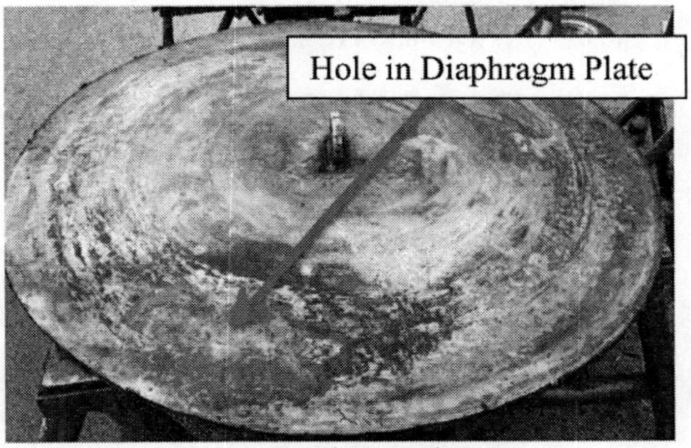

Figure 4. Diaphragm plate with DPI

The diaphragm plate was also inspected with a Positive Material Identification (PMI) tester, and it was found that the plate was made of 316 SS rather than the required 317L SS. In the presence of ammonium carbamate at 3600 psi (24,821 kPa) and 390°F (199°C), the 316 SS diaphragm plate failed. This allowed the carbon steel bolted head cover to be exposed to ammonium carbamate. Ammonium carbamate can corrode carbon steel at rates as high as 35 in (90 cm) per year.

The surface of the bolted head cover was found to have areas of low thickness and pitting, as shown in Figure 5. The area of the cover (Figure 6) that experienced the most corrosion and pitting correlated with the location of the hole on the diaphragm where the process carbamate was able to penetrate to the surface of the head. A decision was made to replace the diaphragm plate and to resurface the bolted head cover to remove the corrosion and create a flat surface.

Figure 5. Bolted head cover after removal

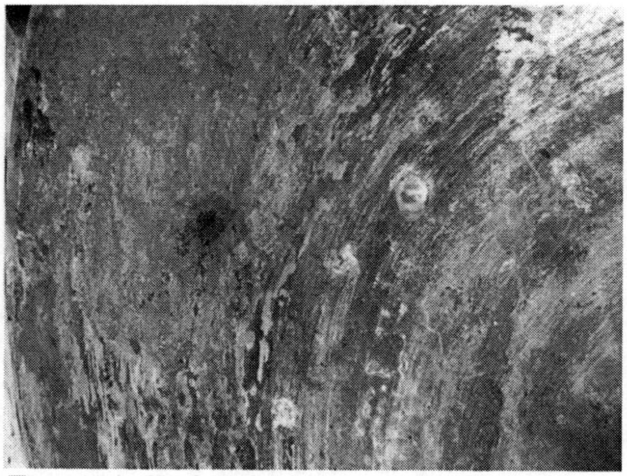

Figure 6. High corrosion area of bolted head cover

Repair

The bolted head cover was brought to a third party shop to be resurfaced on a lathe. The initial thickness of the cover was measured to be 15.25 in (38.735 cm). The vessel calculations from the original fabrication of the reactor in 1998 showed a minimum allowable thickness of the cover to be 15.130 in (38.430 cm). It was originally decided to take the cover down to minimum thickness to remove the corrosion and make the face of the cover true. After making several full passes on the lathe, it was discovered that the high corrosion area of the face was a deeply penetrating rust spot. The deepest extent of the rust penetration was not yet revealed when the minimum thickness was reached. It was then decided to resume machining until there was no rust remaining and re-rate the maximum allowable working pressure of the (MAWP) vessel to accommodate the new cover thickness.

Many more passes were made on the lathe and the cover face was turned down to a thickness of 14.94 in (37.85 cm) with the rust area still remaining, as shown in Figure 7. A decision was then made to return the cover to the plant and reevaluate the options. These options included replacing the cover entirely with a new one of original thickness, or resume machining on the surface of the cover and re-rate the vessel as mentioned previously.

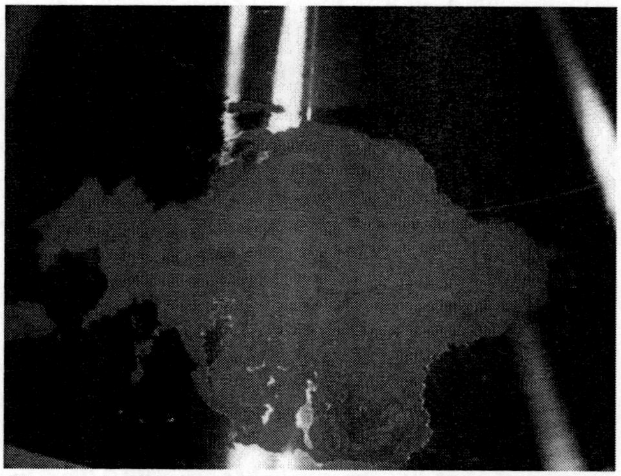

Figure 7. High corrosion area after machining

With the cover back at the plant, the maintenance group utilized an air powered chisel on the rust spot to remove as much of the corrosion as possible and find the minimum thickness that the cover would need to be machined to. This thickness was found to be 13.2 in (33.5 cm) which is much lower than originally estimated (see Figure 8).

Figure 8. Cover face after rust removal

After consultation with two engineering firms, it was determined that the calculated minimum thickness was based on an allowable stress for SA-508 Gr. 4 Class 3 Forgings from 1995 ASME Section VIII Division I. This allowable stress value is different in the 2015 ASME standard, and after updated calculations were completed, the new minimum thickness was determined to be 14.14 in (35.91 cm). These calculations will be discussed in further detail in the next section.

Along with the finding of this new minimum thickness value, it was also determined that the void in the bolted head cover face could be weld filled by a code shop with a qualified weld procedure that does not require post weld heat treatment. Post weld heat treatment would have added substantial downtime resulting from the amount of time required to get the bolted head cover to temperature. This weld was done only to build back material on the face of the cover, so local stresses on the welds would be much lower than a joint weld. This type of repair would prevent further loss of material from the cover and would give the repaired cover a thickness that was well above the minimum determined from the updated calculations.

The bolted head cover was taken to the code repair shop, where the void was cleaned out and filled by welding (see Figure 9). After welding was completed, the face was once again machined on a lathe (see figure 10) where it was turned to a final thickness of 14.94 in (37.94 cm). An "R" (repair) stamp was issued by the shop, and Authorized Inspector (AI) approval and state confirmation were obtained. The bolted head cover was reinstalled with a new 317L SS diaphragm on the Secondary Urea Reactor and the unit was put back into operation at its previously rated pressure.

Figure 9. Bolted head cover after weld fill

Figure 10. Bolted head cover on lathe for machining

Updated Calculations

As discussed earlier, based on consultation with two engineering firms it was determined the vessel calculations could be updated to the 2015 ASME Section VIII Division I standards without having to re-rate the vessel. The calculations were based on UG-34 in the standards, which are as follows:

$$t = d * Sqr\left(C * \frac{P}{S * E} + 1.9 * W * \frac{\text{hG}}{S * E * d^3}\right) + Corrosion$$

Where,

C = a factor depending upon the method of attachment of head, shell dimensions, and other items, dimensionless

d = diameter, or short span, in

E = joint efficiency

h_G = gasket moment arm, equal to the radial distance from the center line of the bolts to the line of the gasket reaction, in

P = internal design pressure, psi

S = maximum allowable stress value in tension, psi

W = total bolt load, lbs.

As stated earlier, the main difference between the 1995 and 2015 standards was "S", maximum allowable stress value in tension. The value changed from 22,300 psi to 25,500 psi, which resulted in a smaller minimum design thickness for the reactor bolted head cover. The calculations can be viewed below.

1995 Calculation

$$t = 52.2192 * Sqr\left(0.3 * \frac{4400.653}{22300 * 1} + 1.9 * 9419920 * \frac{4.3904}{22300 * 1 * 52.2192^3}\right) + 0$$

$$t = 15.1299\ in\ (38.43\ cm)$$

2015 Calculation

$$t = 53.25 * Sqr\left(0.3 * \frac{4400}{25500 * 1} + 1.9 * 9794032.88 * \frac{3.875}{25500 * 1 * 53.25^3}\right) + 0$$

$$t = 14.1381\ in\ (35.91\ cm)$$

Conclusion

Through the process of the discovery of the issue to the repair of the bolted head cover and finally its reinstallation and putting the unit back into service, there were many steps where improvements could be made to significantly decrease the downtime should a similar future event occur.

First, all metallurgies should be verified at the facility before putting a unit back into service. The cause of the current incident was that the diaphragm plate was not the correct grade of stainless steel. The MTR (material test requirements) from the OEM shop that had repaired the head after the 2010 incident indicated that the diaphragm was the correct grade of stainless steel. Checking the diaphragm with a PMI tester before installation could have prevented the most recent incident.

Second, it would have been beneficial to do a thorough inspection of the cover at the facility before sending it to be machined off site. The cover could have been cleaned and the rust removed with hand tools so that the depth of the damage was found. A plan would have then been based on the known extent of the damage.

Third, checking the original calculations with current standards should have taken place before work was started. Fortunately, this resulted in a more favorable minimum thickness of which the final thickness of the head was sufficiently greater than.

Finally, knowledge of available repair procedures would have led to the weld repair taking place first. This would have saved many hours of downtime and minimized thickness loss on the cover.

This facility believes in a policy of "You see it, you own it." This led to operators seeing something that was not normal for the plant, and taking action to assure that something was done to fix the issue. Damage to the diaphragm and bolted head cover could have caused catastrophic failure of the Secondary Urea Reactor. This could have resulted in loss of property, extended downtime, or most importantly injuries or fatalities. Without the diligence of these operators this incident would have been much more severe.

References

[1] *Multi-layered Secondary Urea Reactor Failure*, M. Gilbertson and D. J. Benac, AIChE Vol 53, 2012.

[2] *Environmental Stress Cracking of a Shell and Tube Heat Exchanger*, J. Moen and G. Roener, International Pressure Equipment Integrity Association (IPEIA) Conference February 1-3, 2006.

Alarm Management
Prevent Your Operator From Drowning

Due to the complexity of the ammonia manufacturing process, several stakeholders are involved when it comes to engineering activities in an ammonia plant. Each discipline and each equipment vendor recommends a huge number of alarms and messages to trigger human intervention. Most of this information is sent to the DCS screens of the control room operators. Depending on the status of the production plant, the operator is regularly overloaded with alarms during plant shut downs and start ups. After reviewing of process safety incidents, inadequate alarm management was identified as one of the common contributing factors.

The aim of the this paper is to describe the group initiative concerning the management of alarms in Borealis Group companies and the resulting initiative of the ammonia plant in Grand Quevilly, France. The result will be discussed by review of process safety relevant cases, which have initiated the implementation.

Laurent Barthalay
Borealis Group France

Introduction

The automation of industrial processes is based on systems of increasing size and complexity that present operators with tailored and structured information to guide their decisions. On the other hand, an excess of complexity could also be a risk of misunderstanding by the operators resulting in a counter performance of the overall process. Humans still play a decisive role in the safe and efficient operation of complex systems. However, people responsible for operating these systems must have relevant, accurate, and especially easy to access information to make the right decision quickly.

We will see in this paper some examples of incidents first, and how industry reacted. We will then understand the different sources of progress we own, the technical one and the Lesson's Learned. The key being the awareness of the importance of how efficient the system is.

Lessons' Learned - International Awareness

Three Mile Island

In 1979, a number of events caused the partial melt down of the reactor core[1]. Initially, a failure of the cooling system resulted in opening of a pilot operated relief valve. The pilot operated relief valve failed in the open position. The operators in charge were not aware that limited cooling of the reactor caused it to warm up. The water level in the reactor was not measured directly, but was deducted from the level of water in the pressurizer. As the pressurizer level was high, operators assumed that the heart of the reactor was properly cooled. Although flashing lights and buzzers in the control room indicated a number of alarms, operators did not realize the seriousness of the situation. On the contrary, a decision to reduce the flow of water into the pressurizer only exacerbated the problem. In the investigation of the incident, it was concluded that control operators did not have all the information to understand the actual course of events. Due to their distorted perception of the situation, all their interventions were based on a wrong model. Even when specific actions did not result in consequences compliant with their understanding of the situation, the operators in charge did not question their judgment of the situation although several alarms could not be acknowledged.

The accident served as a catalyst for the re-definition of systems design. Many technical aspects had to be reviewed, as well as the role of human beings in some processes. As a consequence of all the investigations performed, emergency management, training of operators and, above all, the weight of human factors in power plants were deeply reviewed. A large part of these adaptations were subsequently taken over by other industrial sectors.

Milford Haven

On Sunday July 24th 1994, a severe electrical storm caused plant disturbances in the Pembroke Cracking Company plant (PCC), at the Texaco Refinery[2]. This led to several explosions in the unit, which were caused by flammable liquid hydrocarbons continuously pumped into a vessel with the outlet valves closed.

Consequentially, 20 tonnes were released and exploded. Among the many contributing factors the investigation team found, one factor was the operators had difficulty identifying the cause of the upset due the configuration of the display-screen. In addition, too many poorly prioritized alarms were overwhelming the operator in the control room during the process upset. Investigation of the data recorded by the distributed control system (DCS) indicated that alarms were presented to operators at the rate of one every two to three seconds. Consequentially, alarms were acknowledged without necessarily recognising their meaning. In the last eleven minutes before the explosion, the two operators had to acknowledge and review 275 alarms. There were 1365 measuring points connected to instrumentation, of which 55% had one alarm and 31% had two alarms configured. 190 additional other alarms were configured in the system resulting in more than 2000 alarms configured, of which 87% were high priority and 13% were low priority alarms. As there was no philosophy to determine the priority of an alarm, all were presented in the summary without the possibility for the operators to figure out the critical ones.

Texas City

On March 23rd 2005, an explosion and fires occurred at the BP refinery in Texas City after a raffinate splitter section of the Isomerization unit (ISOM) was restarted following a maintenance turnaround[3]. Due to overfilling of the raffinate splitter tower, flammable liquid was released, vaporized, and ignited resulting in an explosion with several consequential fires.

After the startup, the unit had to be stopped again due to delayed turnaround works in other units. Consequentially, the raffinate section was shut down and had to be re-started during the next shift. Interrupting the startup of the unit was unusual and was not covered in the procedures.

The splitter tower was equipped with a level transmitter that measured the tower's liquid level in a 5-foot (1.5 m) span within the bottom 9 feet (2.7 m) of the 170-foot (52-m) tall tower. Two separate alarms were configured. One was programmed with acoustic sound when the transmitter reading reached a height of 7.6 feet (2.3 m) in the tower corresponding to 72%, and the other was a redundant high level alarm that was designed to indicate when tower level reached 7.9 feet (2.4 m) corresponding to 78%.

However, when the raffinate splitter tower was filled beyond the set points of both alarms to a level reading of 99% in the early morning on March 23, 2005, only one alarm was activated. The high-level alarm was triggered at 3:09 a.m.

The redundant hardwired high-level alarm never indicated. Material balance calculations showed that the tower had actually filled to 13 feet (4 m), four feet over the top tap of the level transmitter. Because the failure of the high-level switch was unnoticed, it was not reported by the Night Lead Operator or other operations personnel. Consequently, no work order was written and the malfunction was not noted in the logbook. The high-level alarm associated with the level transmitter remained in the alarmed state throughout the incident. Filling the bottom of the tower until the level transmitter read 99% was not unusual, even though the startup procedure called for the level in the tower to be established at a 50% transmitter reading. Operations personnel explained that this was experienced in previous start ups, and that additional liquid level was needed to avoid losing the liquid contents of the tower and potentially damaging equipment. DCS operators typically maintained the tower level well above 50%. Once the raffinate section equipment was filled, the startup was stopped and the splitter tower feed and bottoms pumps shut off. The circulation was shut down and the tower level control valve remained in the "closed" position for the next shift to resume the startup.

What operations did not identify during those hours was that the tower was overflowing into the overhead pipe and down the side of the tower to pressure relief valves 148 feet (45 m) below. Pressure increased at the relief valves, and flammable liquid was discharged out of the relief valves overfilling the blowdown drum and discharged out of the vent stack, which was producing a cloud of flammable material in the plant area causing a violent explosion.

The ISOM operators stated they had insufficient time to sound the emergency alarm before the vapor cloud explosion occurred. Approximately 15 seconds after hearing from the field operators via radio message that the blowdown drum was overflowing, both the Board Operator and the Lead Operator said they started the process of shutting off the fuel to the furnace using the computerized control system. Their testimony is substantiated by the computerized control system data, which showed that the fuel gas flow control valve was shut five seconds before the explosion. Hundreds of alarms registered in the computerized control system at 1:20:04 p.m., including the high-level alarm on the blowdown drum; the flood of alarms appeared when the explosion occurred.

The DCS operator's belief that the tower level was accurate was supported by the fact that the redundant high-level alarm had failed to activate. This alarm should have provided a redundant high-level indication in case of a malfunction of the main level transmitter. However, this alarm's set-point was not known to operations personnel or provided in the procedure, control data, or training materials. Due to the fact that the set-point was not known by operational personnel, the probability of realizing the malfunction was significantly reduced.

It was stated in the investigation report that the control system display was found to be of poor design.

Birth of a Standard

Several essential factors come into account during the design of a system where the operator is part stakeholder:

- Size and complexity of the process
- Variability of the process
- Variability of production programs
- Impact of the process on the environment
- Economic consequences of judgments of production
- Security and protection of persons and facilities

Operators properly trained are, at first hand, able to handle a large number of these factors including identification of abnormal trends and events from a large amount of data; adaption of the procedures to a new situation; memorizing of a huge mass of knowledge over long periods; ability of logic as well as empirical reasoning and intuitive judgment.

To act effectively, the operator should have a global and dynamic vision of the process. Also, the "right" information has to be available at the "right" moment enabling an understanding of the situation and enhancing the ability to make the right decision. The design of the DCS synoptic should provide a high density of information presented in such a way that all-important factors "jump immediately to the eyes" allowing even in critical situations efficient mitigation actions.

From the good old days

The old-generation industrial control rooms were equipped with an ergonomic system where alarms were presented with bulb indicators together with a process flow diagram on a huge panel. The incident at Three Mile Island significantly influenced the improvement of the ergonomics of control rooms including optimizing the layout of the controls and information transfer. These improvement measures included colored panels clearly showing the functional units. Nevertheless, the amount of information displayed was dependent on the space available inside the control rooms. Consequently, a prioritization process was performed to select what information needed to be displayed for efficient mitigation of all operational disturbances.

Advances over time in human-machine interface technology has allowed more information to be displayed, including alert or alarm messages. In the end, technical capabilities made it possible to remove all constraints forcing the prioritization of alarms, which finally resulted in overloading the operator with an unrealistic amount of information.

Emergence of Industry Standards

In 1994, the Abnormal Situation Management Consortium (ASM) was formed by some petroleum companies and a DCS provider company to develop standards ensuring efficient application of new visualisation and control technologies. ASM provided data and helped the Engineering Equipment Material Users Association to create EEMUA 191[4]. This document was first released in 1999 and has since been referred to as the "main" guideline for alarm management. EEMUA is written in a text book format and provides some excellent examples in the appendixes. The release from 2014 is still valid today.

International Society of Automation (ISA) developed ISA 18.2 which is written more like a standard and has many similarities to the safety instrumented standards[5].

EEMUA 191 and ISA 18.2 complement each other. In summary, EEMUA 191 describes in detail the tools and techniques for various aspects of alarm management (rationalization, risk assesments), whilst ISA 18.2 clearly defines the required key performance indicators (KPIs) and the overall lifecycle approach to alarm management.

Figure 1 below shows the history and therefore the emergence of Industry standards from 1994 to nowadays. In parallel, the quality of alarm management has improved from ineffective alarm management to the establishment of good engineering practice.

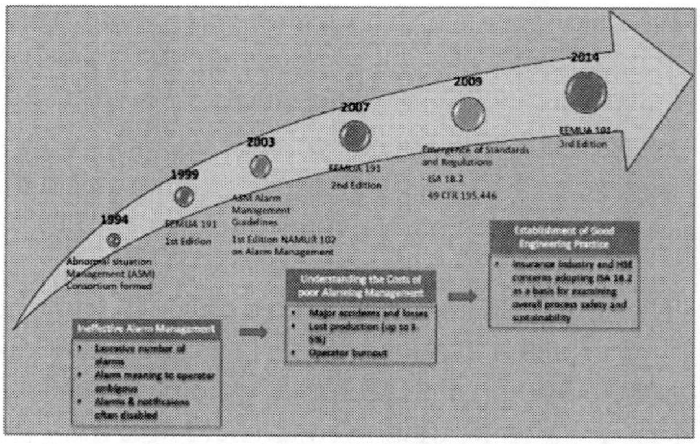

Figure 1. Emergence of Industry standards

Comparison of EEMUA 191 and ISA 18.2

EEMUA 191 suggests three main KPIs on a per operator basis for 10-minute time periods (see Tables 1 and 2):

- Average alarm rate
- Maximum alarm rate
- % of time alarm rates are outside of acceptability target

Long term average alarm rate	**Acceptability**
> 1 per minute	Very likely to be unacceptable
1 per 2 minutes	Likely to be over-demanding
1 per 5 minutes	Manageable
< 1 per 10 minutes	Very likely to be acceptable

Table 1. EEUMA 191 Alarm rates for steady operation

Number of alarms displayed in 10 minutes	**Acceptability**
> 100	Definitely excessive and very likely to lead to the operator abandoning use of the system
20 to 100	Hard to cope with
Under 10	Should be manageable – but may be difficult if several of the alarms require a complex operator response

Table 2. EEUMA 191 Alarm rates following a major plant upset

EEMUA also states that a site which has a DCS with greater than 1000 configured alarms should be targeting fewer than 10 standing alarms and fewer than 30 shelved alarms (excluding maintenance shelved alarms).

ISA 18.2 has, for its side, very clear alarm system performance KPIs (see Table 3). On the other hand, ISA 18.2 focuses on an alarm system lifecycle. This is an excellent method of representing the overall process of alarm system management.

KPI	Target Value	
Displayed alarms per 10 min	~1 (average) very likely to be acceptable	Maximum manageable ~2 (average)
Percentage of hours containing > 30 alarms	<1%	
Percentage of 10-min periods containing > 5 alarms	<1%	
Maximum number of alarms in a 10-min period	10 or less	
Percentage of time the alarm system is in a flood condition	<1%	

Table 3. Examples of ISA 18.2 KPIs

Good Practices

The published standards EEMUA 191 and ISA 18.2 are the references regarding the alarm management. However, before companies adopt these standards and implement them into their organisations, they need some understanding of best practices. These are some of the questions to help reflect if the implementation of such a process should be considered:

- Can operating companies really judge the effectiveness of alarm systems?
- Are operators acting on alarms, or just acknowledging or shelving without acting?
- Do we register incidents or near-miss incidents where operators have missed alarms?
- Do we measure the number of alarms per hour, disabled or suppressed?
- Do you have a documented alarm philosophy in order to describe roles and responsibilities?

Alarm Management Lifecycle

The alarm management lifecycle (Figure 2) is an ongoing process that is suitable for new or existing systems. It has been designed to represent sequential stages, which run simultaneously mostly with other stages that are linked horizontally. In some cases, a sequential stage may be skipped. For example, during a rationalization phase the outcome may be a change of alarm settings which may not require detailed design to implement. It is possible to develop a lifecycle model from Philosophy, Monitoring and Assessment or Auditing stages. For an existing site, the ideal entry point is usually the Monitoring and / or Assessment stage, which will provide a benchmark for the site alarm system.

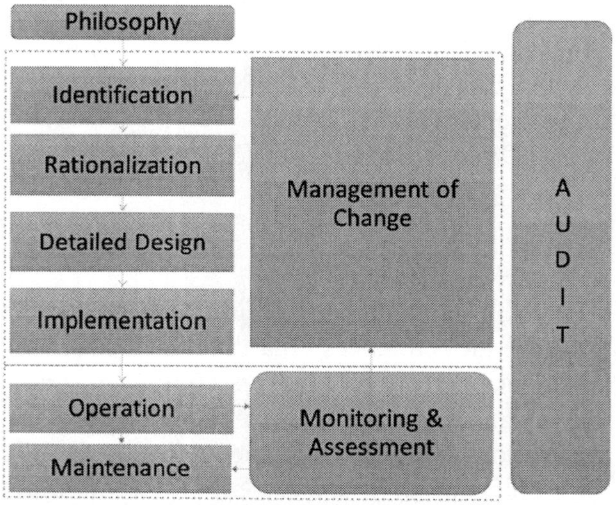

Figure 2. Alarm Management Lifecycle

There are three loops in the alarm management lifecycle model that are represented by the dotted lines in Figure 2.

If an issue concerning alarms is detected in Monitoring and Assessment stages, it would be resolved via the Monitoring and Maintenance loop (e.g. chattering alarm) or the Monitoring and Management of Change loop (e.g. more complex issue). The Monitoring and Maintenance loop typically is applied for simple maintenance related issues whereas the Monitoring and Management of Change loop is required for any design changes. The Auditing and Philosophy loop basically entails the whole lifecycle process. Periodic audits of your alarm system are a requirement of the standard.

Alarm philosophy

An alarm philosophy documents the approach of a production site towards alarm management and shall be seen as a mandatory requirement. It includes the definitions (what is an alarm, alarm class definitions, alarm priorities, etc.), principles (roles and requirements), and details of the practices and procedures (alarm shelving and suppression rules) for each of the remaining life cycle stages. It provides a lasting reference to sustain an effective alarm system.

To achieve this, it is advised to form a committee of stakeholders and involve an alarm management expert if possible. You will have to draft, review, edit, review, and repeat as necessary. Once approved, review the alarm philosophy regularly for any necessary changes.

Alarm rationalization process

Alarm rationalization is about reconciling each individual alarm against the principles and requirements of the alarm philosophy. All alarms require an operator response and a consequence. Alarms must be presented at a rate that the operator can respond.

For every event suspected to require an alarm, it is necessary to determine if the event requires a corrective action to be performed by the operator. What is the immediate consequence if an action is not taken? Events that are not alarms must be prioritized accordingly with a lower priority. These can be stored in databases and used, for example, as prioritisation of maintenance activities, without interfering with the operator's activities.

After documenting consequences and desired response time after an alarm, the priority can be derived from a matrix based on the consequence severity and response time. The matrix shall be defined by the alarm philosophy.

For every alarm, the alarm type, class priority limit or condition have to be documented. Also, the consequences of not carrying out the operator action in timely manner should be included. Nowadays, most of the DCS suppliers offer documentation modules directly linked with the DCS itself.

Detailed design

This step includes the basic DCS/PLC configuration of the alarm, Human Machine Interface (HMI) for the alarm, and any advanced methods of alarm management. The implementation is based on the understanding of the capabilities and limitations of the process control system. This step often involves more complex configuration steps than operational units have considered yet. In order to reduce complexity, common standard treatment per advanced alarming scenario can be developed.

Among the different techniques used in managing alarm configurations, group-based suppression, time limited suppression (shelving), or counter-based suppression should be considered.

HMI configuration is also part of this step. Additional considerations should be taken into account such as color conventions, iconic representations, rules for acknowledgement, representation in the alarm banner summary, as well as, area overviews, unit displays, detail pages, faceplates, etc.

Monitoring and assessment

As mentioned in the different standards, setup and control of pragmatic alarm system KPIs is absolutely crucial. Without those it is virtually impossible to maintain an effective alarm system. Therefore, a frequent assessment should be performed to determine problems such as nuisance alarms, state alarms and alarm floods.

Management of change and audit

Management of change is a critical stage that helps ensure the ongoing integrity of the alarm system. It needs to be a structured process of approval and authorization for any additions, modifications, and deletions of alarms from the system.

A periodic audit of the alarm system and the processes detailed in the alarm philosophy may determine the need to modify processes, the philosophy, or the design.

Borealis Standard: A Group Initiative

Amongst the Borealis Group process related to Production Excellence, one initiative was dedicated to improving and standardizing the Alarm Management over the plants and sites. A procedure was then written based on the good practices and international standards.

Managing Alarms

The aim of this procedure is to describe a uniform process across all functions within Borealis to ensure that operators are not over flooded by alarms. This procedure is based on the main standards EEMUA 191 and ISA 18.2, and proposes different measures for alarm prioritization. The Alarm management process is described in five steps:

1. Define

The first step is to configure alarm system performance indicators that will define key parameters and consequences ranging from unacceptable to excellence.

The measurements of the alarming performance shall be averaged over a period of six months. It is important to distinguish between steady operation and the aftermath of a major plant upset.

Moreover, alarm management organizational performance indicators should also be set up to measure the organizational performance such as completion of the documentation (philosophy, etc.), implementation of an alarm monitoring system, or use of an alarm reference database.

2. Analyse

During annual self-assessment, plant management evaluates the overall performance of the system. This period can be extended when the target level in all KPIs are reached.

3. Solve

This step describes the process of prioritizing alarms from low to emergency. During this phase, it is absolutely necessary to describe what is an alarm event, what does not constitute an alarm (event), and how the safety alarms have to be treated.

4. Implement

In this step, the predefined measures are implemented. For changes of existing situation, the management of change (MoC) procedure needs to be applied.

During this step all the training packages must be delivered. It is mandatory that those who design, install, operate, and maintain the systems are properly trained.

5. Review

Thanks to the monitoring tools, the effectiveness of implemented improvement activities are required to be checked.

Management has the responsibility to ensure that a continuous improvement process is in place and is updated at least once every year.

Implementation at GQ Ammonia Plant

In the Grand-Quevilly ammonia plant - after around thirty years of operation on the old-generation industrial control room - a DCS retrofit was performed in 2008. After implementation, operators were suffering from alarm floods. In 2011, the local management decided to implement an improvement initiative utilizing a company with alarm management expertise so as to rationalize the alarm system.

As a first step in this new initiative, the assessment stage was performed. In parallel, the local DCS expert put in place a routine to extract DCS data for all events to be used in later analysis. The team involved consisted of a DCS expert, a production engineer, and two panel operators. The first results of the assessment were not surprising and confirmed the bad health of the system.

After receiving specific training, the team began to establish a philosophy for the plant and started the rationalization phase using that philosophy. The more the rationalization was performed, the better the philosophy was tuned; always going back and forth between rationalization and philosophy. "Top alarm classification" was used in prioritizing the work during this stage. Considering that only a few alarms generated most alarm events, only alarms appearing in an established timeframe were collected. Consequentially, these alarms were also classified by the number of appearances. Almost 80% of the analog alarms were treated according to the philosophy rules over approximately one year.

In January 2015, a severe process safety incident occurred involving the failure of the flash tank in the Pressure Swing Absorption Unit (discussed in more detail in the next section of this paper). Poor alarm management was considered as one of the major root causes.

During the same period, a group wide "Production Excellence" program was launched at all locations including the element of alarm management. Since the local initiative of Grand-Quevilly ammonia unit had been dropped a couple of years before, this opportunity was used to reinstall implementation teams.

Again, an internal assessment of the overall process was performed according to the group procedure (alarm management organizational performance). Not surprising, this initiative revealed that alarm management still required significant attention. So a new team was created consisting of an operator, shift team leader, an engineer, a DCS expert and a process engineer, who was a part time member. In addition, there was strong support from the management. It was crucial to put in place a strong organisation in order to support the process. As an element of empowerment, a representative regarding alarms was named "alarm champion". Moreover a group wide network has been supporting the location allowing an exchange with other ammonia plants in Borealis.

After assessment of deviations from the Borealis Standard, several details had to be analysed to determine meaningful KPIs to stir the improvement initiative. Furthermore, automatic software was installed for data extraction and compilation to allow follow up of KPIs. Finally, a new Borealis standard tool was implemented in Grand Quevilly ammonia plant, as well as, in one plant in Linz.

This chosen approach resulted in reviewing all the alarms configured one after one during the rationalization stage and checking about their relevance and effectiveness. All the information regarding the alarm and events configured were analysed during this stage. The priority of the alarm was revised according to the matrix (requiring time for the operator action versus consequence), and the threshold validated. This resulted in reducing the number of alarms. For the more complex changes, a management of change process was initiated. In parallel, it was also prioritized to work on the steady state alarms. The steady state alarms are alarms that are almost permanent in that they stay in this status for days or months.

The conditional masking of alarms on equipment shutdowns was considered afterwards. Conditional masking consists of reconciliation of alarms that appear following the shutdown of an equipment and analysis of the conditions for masking and un-masking alarms. The organization put in place included a project manager who organized a follow-up meeting every two or three months (depending on the availability of each) with the Alarm Champions and the DCS specialist. The purpose of the meeting was to provide an update of the situation and define the conditions for masking to put in place. The goal was to reduce the number of permanent alarms related to units or stationary equipment.

Regarding the performance evaluation, the KPIs are calculated from the information available in the DynAmo Metrics and Reporting software and this report is sent to the Excel KPIs Measurement file. Depending on the period evaluated, it is possible to calculate the KPIs on:

- the overall period
- a "stable" period (without triggers of an alarm flood)
- a "stopped" period (during a unit shutdown or following a trigger of an alarm flood, for example)
- another global period

A performance evaluation is done quarterly and annually to:

- see the evolution of KPIs by quarter and by year
- define areas for improvement
- follow the impact of the actions carried out by the team

KPIs are reported quarterly from the "Alarm Management KPIs Reporting" template (see Figure 3). The template provides:

- feedback on the evolution of the performances of the system
- highlights the results of the actions carried out by the team
- reminds the team of the next improvement actions identified by the team

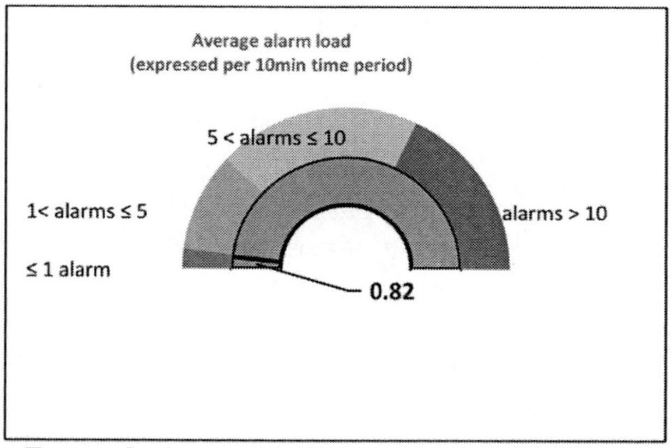

Figure 3. Ammonia Plant Grand-Quevilly KPI achievement-Example

January 2015 Pressure Swing Adsorption Incident

In January 2015, an incident occurred on the Pressure Swing Adsorption (PSA) unit during the restart of the plant just after a short partial stop.

The unit consists of three pressure vessels, which are containing absorption material. These vessels are operated in a sequence, allowing the separation of the purge gas into a hydrogen rich fraction, which is sent back to synthesis compressor and an inert rich fraction, which is depressurized into a flash vessel. From there the gas is evacuated with an ejector. This device is creating a vacuum by sending the heating gas to the reformer through a nozzle into an expansion chamber. The under pressure created there is used to pull the gas from the flash vessel and mix it with the natural gas stream to the burners.

During the incident, fluctuation of heating gas flow was already creating an under pressure although the flow from the PSA unit was not yet started. This sudden flow of motive fluid through the ejector was not yet compensated with gas from the outlet pipe of the flash vessel. Consequentially a large vacuum in the system developed.

The outlet pipe of this vessel has a pressure control valve and an automated shut off safety valve to isolate the system. The designed protection against vacuum was the safety shutoff valve and corresponding interlock, which was bypassed in the DCS. In addition, the pressure control valve was not closed by the panel operator, even though the alarm of very low pressure in the pipe was acknowledged.

The result was a collapse of the flash gas tank caused by vacuum, and, as a consequence, a loss of energy and production. On top of that, downstream units had to be stopped because of the vicinity of ammonia pipes that were threatened by the damaged tank. No people were injured during this incident.

One contributing factor to the incident is that many changes were made to this part of the unit through the years since 1988 when the PSA section was built.

A second contributing factor was that the project management of the "DCS project" between 2004 and 2008 was substandard of Borealis. The needed risk analyses were not done properly. Also, the risk of deviating out of the operating window of the PSA section was not well understood. Therefore, compliance with the operating window was also substandard of Borealis.

A third contributing factor was the lack of administrative controls. This was an area not adequately addressed by the alarm management project in 2011. The following administrative controls were added following the flash tank failure:

1. Implementation of the systematic check of DCS interlock status in prestart-up review checklists
2. Review of DCS interlocks (management, control) including LOPA analyses and SIF identification
3. Continue with the alarm management rationalization. Development of action plan with OPEX managers support
4. Training of OPS (production, maintenance and process engineers) on Process Safety risks

The incident makes us think about the Swiss Cheese Risk Model (Figure 4) where an organisation's defences against failure are modelled as a series of barriers, represented as slices of cheese. The holes in the barriers represent the weaknesses of the system. In our case, all these holes were crossed leading to the incident. This model includes active and latent failures that may lie dormant for days, weeks, or months until they contribute to the accident such as an inadequate alarm system.

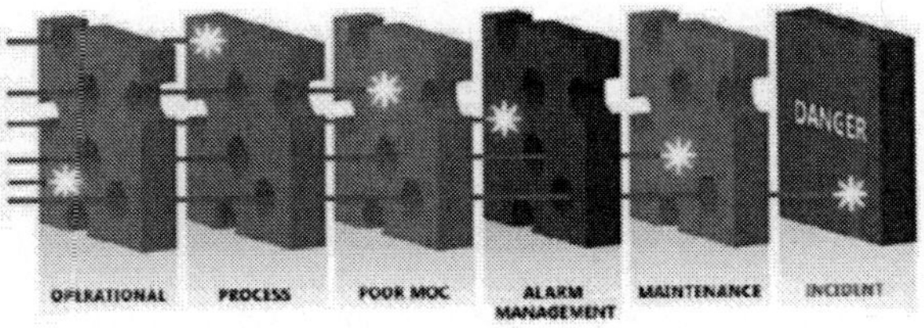

Figure 4. Swiss Cheese Risk Model

Conclusion

The awareness of the management is crucial to avoid these types of major incidents. An employer cannot nowadays ignore these kind of standards due to not having a sufficient number of personnel to develop such an alarm management program.

Secondly, depending on the result of the assessment, resources will be needed to launch and anchor the process. The amount of resources for this effort should not be under-estimated.

Time after time, the lesson learned is to dedicate people in charge of implementing a continuous process and follow up. That should be done through a leader assigned to this effort as his normal job function.

If alarm management is absolutely essential for our industry, then it must be a part of the excellence in production in which the MoC and bypassing of interlock processes belong too.

While we are talking about this topic, why not think about the next step: what about Artificial Intelligence? In any case, if we agree to think about new systems that can help operators in the understanding of the situation they are facing, it is today essential to understand that people are a key link in the automation chain.

References

1) U.S.Nuclear Regulation Commission – NRC Library: Backrounder on Three Mile Island Accident (2014)
2) HSE Books: The explosion and fires at Texaco Refinery, Milford Haven, 24July 1994; ISBN 077614131 (1997)
3) U.S. Chemical- Safety and Hazard Investigation Board: BP Refinery Explosion and Fire – Final Report (2007)
4) EEMUA Publication 191 Alarm systems
5) ANSI/ISA-18. 2-2009

Increase Safety by Using an Automated Protection System to Avoid Over-Firing of Primary Reformer

In early 2015, when Profertil was performing a hot restart of their ammonia plant, an uncontrolled and rapid increase in flue gas temperature was noticed in the primary reformer. After natural gas feed was introduced to the primary reformer, it was observed that a rupture of several catalyst tubes had occurred and the plant was tripped immediately. Profertil requested Haldor Topsoe A/S to take the lead technical advisor role in a thorough root cause analysis, performed to identify the reason(s) for the tube rupture incident.

The results of this thorough root cause investigation will be described in detail, and the solution to assure such incident will not happen again will be introduced. The incident with ruptured tubes not only impaired plant capacity utilization until the re-tubing of the furnace, but more importantly, the leaked gas from the ruptured tube could have resulted in unsafe conditions. Therefore, the possibility of avoiding this type of incident is of paramount significance to all plants operating with any type of tubular reformers. Following this incident, an automated over-firing protection (OFP) management system that provides four elements of protection against over-firing of primary reformer tubes was developed.

Peter Bruun Jensen
Haldor Topsoe A/S, Denmark

Sudip De Sarkar
Haldor Topsoe A/S, Denmark

Gaston Schulz
Profertil, Argentina

Introduction

Reforming section is the heart of the synthesis gas generation unit in an ammonia plant. Any operational problem in the reforming section impacts production of ammonia and thus, it is extremely important to ensure continuous and reliable operation of the reforming section.

Over the years, it has been observed that over-firing and subsequent rupture of primary reformer catalyst tubes during operation, especially during start-up, has been one of the common problems in the reforming section in ammonia plants [1, 2]. Apart from its economic impact, such incident might also create potentially unsafe condition. Therefore, Haldor Topsoe A/S and Profertil delved deeper into this incident by going through a thorough root cause analysis. A few action points were suggested to avoid similar incidents in the future. Later, Haldor Topsoe A/S, with valuable assistance from Profertil, has further worked on the recommendations to develop an automated system to provide better protection against over-firing of primary reformer catalyst tubes, especially during start-up.

About the Plant

The Profertil ammonia plant (in Bahia Blanca, Argentina) was designed for 2,050 MTPD (2,260 STPD) of ammonia based on licensed technology from Haldor Topsoe A/S. The plant was commissioned in 2001.

Figure 1. Profertil's ammonia plant

The reforming section consists of a primary reformer followed by a secondary reformer. The primary reformer is a side-fired Haldor Topsoe design with 264 catalyst tubes installed in a duplex furnace box. The necessary heat for the endothermic reforming reaction is supplied uniformly to the catalyst tubes by burning fuel through the radiant wall burners, arranged on both sides of the catalyst tubes in seven rows on the furnace walls. The process feed gas to the primary reformer is evenly distributed to the catalyst tubes through inlet hairpins at the top of furnace. The reformed process gas, coming out of the catalyst tubes at a normal operating temperature of 800°C (1472°F), passes through the outlet hairpins, below the bottom of the furnace box to six hot collectors. The reformed process gas further flows down to the refractory lined cold collector and transfer line, going to the secondary reformer. The hot flue gas with a normal operating bridge wall temperature (BWT) of 1045°C (1913°F) leaves the top of the furnace box and flows in the downward direction through vertical part of the convection section.

Figure 2. Primary reformer in Profertil

Some modifications have been made over the years in the ammonia plant to boost the production capacity by approximately 15% to 2,360 MTPD (2,600 STPD) of ammonia.

The catalyst tubes in the primary reformer were replaced with the advanced HP-MA-R (Centralloy® G 4852 Micro R supplied by Schmidt + Clemens) tubes and was in operation for five years before the tube rupture incident occurred.

Incident

The tube rupture incident occurred in the primary reformer in 2015. The sequence of events, which finally culminated in several tube ruptures, is described in the following section. This information is based on retrieved data from the distributed control system (DCS).

Sequence of Events / Observations

<u>11:00 hrs., February 9, 2015</u>
The ammonia plant was running stable at 112% of nameplate capacity when the plant tripped due to power failure.

<u>04:00 hrs., February 11, 2015</u>
The plant was restarted with nitrogen circulation through the reformer, prior to ignition of burners according to the normal start-up procedure.

21:00 hrs., February 11, 2015
The plant tripped again when operating at 24% load due to trip of the auxiliary boiler.

00:48 hrs., February 12, 2015
A hot restart (i.e. reintroduction of process steam while reigniting burners) was commenced but with a low process steam flow of approximately 3 t/h (6600 lb/h). Due to declining temperatures in the cold and hot collectors, the natural gas fuel flow was stepwise increased to approximately 7500 Nm³/h (280 MSCFH).

01:14 hrs., February 12, 2015
The increasing fuel gas flow and low process steam flow resulted in steep increase in flue gas temperature, but only modest increase in the hot collector temperatures.

01:57 hrs., February 12, 2015
When the BWT exceeded 1000°C (1832°F), one of the six hot collector temperatures suddenly increased by 115°C (207°F) above the other hot collector temperatures (shown in red in *Figure 3*) and the process vent valve to the flare system closed completely.

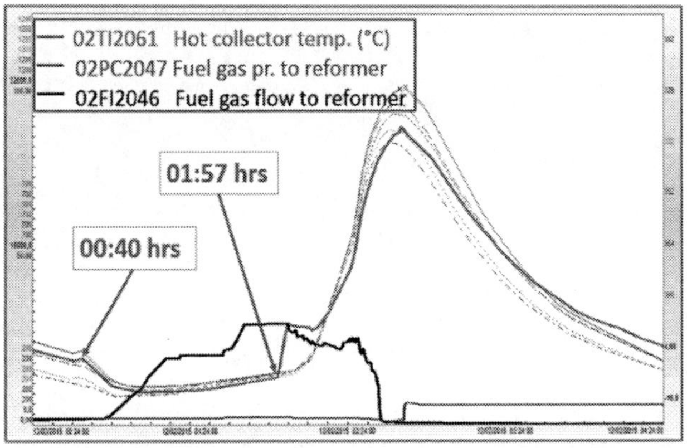

Figure 3. DCS plot of hot collector temperatures

02:10 hrs., February 12, 2015
Gradually, about 40t/h (88,000 lb/h) of process steam was introduced to the primary reformer over 30 minutes, which initiated a sharp temperature increase in the hot collectors followed by a significant temperature increase in the BWT. The fuel flow to the burner was then reduced in steps to approximately 6,500 Nm³/h (243 MSCFH), and natural gas feed was introduced. The feed flow was further increased to 4,000 Nm³/h (149 MSCFH) over 15 minutes to absorb the excessive firing in the furnace box by the endothermic steam methane reforming reaction inside the catalyst tubes.

02:46 hrs., February 12, 2015
Instead of cooling down the furnace, the introduction of natural gas feed to the primary reformer resulted in a dramatic temperature increase of BWT and subsequently in the hot and cold collectors. The fuel gas flow was stopped when the BWT exceeded 1200°C (2192°F), and the plant was tripped when rupture of several tubes was observed in the furnace.

February 27, 2015
The plant was restarted after inspection of the furnace box, transfer duct to the convection section, convection section, hairpins, and hot collectors. All of the ruptured tubes, except one, were replaced due to shortage of spare tubes. All damaged burners were replaced as well. The damage of the remaining tubes was assessed by metallographic replica tests and ultrasonic and eddy current inspection.

All significant process parameters were carefully monitored during re-start of the plant. Heating was done slowly and a burner light up sequence was followed to minimize firing in the expected damaged areas where potential over-firing has taken place.

13:03 hrs., March 4, 2015
The plant was in stable operating condition when BWT suddenly increased by 60°C (108°F) and oxygen content in the flue gas reduced from 2.1 to 1.6 mole%.

Visual observation from the end peepholes into the furnace box confirmed a tube rupture in the area next to the replaced catalyst tubes (as shown in *Figure 4*) and the plant was tripped immediately.

Figure 4. Tube rupture in March, 2015

Subsequently, three tubes ruptured (in several different events) before the plant was shut down in June, 2015 for replacement of the ruptured and remaining catalyst tubes in the furnace. Profertil managed to keep the plant in continuous operation during this period by successfully 'nipping' [3] (i.e. making the catalyst tube inlet and outlet hairpins flat by compressing between two steel blocks and thus preventing feed gas flow) of the ruptured catalyst tubes.

Furnace Inspection

Rupture of 19 catalyst tubes in the first section of chamber AB was observed during inspection. Ten of them were completely ruptured with circumferential cracks (as shown in *Figure 5*) and the rest had longitudinal cracks, facing the reformer furnace wall (as shown in ***Figure 6***).

Figure 5. Tubes with circumferential cracks

In chamber CD, 18 and 3 catalyst tubes were ruptured in first and second section respectively. All of these tubes had longitudinal cracks on the side facing the outer furnace wall.

Figure 6. Tubes with longitudinal cracks

Many of the burners were exposed to extremely high temperature causing melting of the burner tip (as shown in *Figure 7*).

Figure 7. Damage of the radiant wall burners

Root-cause Analysis

The tube rupture incident was followed by a thorough root cause analysis (RCA). The RCA consisted of metallurgical examination of both ruptured and remaining catalyst tubes in the furnaces; and analyses of the collected operating data retrieved from the DCS.

Metallurgical Analysis

Over-heating of the reformer tubes lasted for approximately 1.5 hours and resulted in rupture of 40 catalyst tubes. For the remaining tubes in the furnaces, the significant over-heating resulted in thermal damage of the tube material by changing the microstructure, i.e. coarsening of the primary carbides in the grain boundaries and dissolution of the secondary carbides in the grains. Dissolution of secondary carbides and coarsening of primary carbides decrease the creep strength significantly and therefore, accelerate the creep damage when the tubes are put back in operation.

Metallographic replica of the tube surface were made on the remaining tubes to assess the surface microstructure. It was found that the secondary carbides were absent in most of the catalyst tubes, which confirmed that the catalyst tubes had been exposed to a metal temperature above 1150°C (2102°F) and had experienced severe thermal damage. It was likely that the thermal damage had propagated all the way through the tube wall due to the relatively long exposure time, low flow through the tubes and high heat conductivity of the tube material. Based on this information, it was decided to replace the remaining 224 tubes in June 2015.

Formation of creep defects, like cavities, and alignment of cavities and formation of microcracks are normally not expected to occur in connection with short duration of overheating. Normally, it takes longer time for cavities to form, align and develop into cracks. However, the remaining tubes in the furnace were subjected to ultrasonic and eddy current inspection to ensure that the tubes were not at the brink of rupture due to severe creep defects when the plant would be restarted.

DCS Data Analysis

The operating data from DCS were analyzed to get a better overview of what went wrong during the hot re-start in February, 2015. The actual operating data was compared with the maximum allowable parameters derived from computer simulation of the reformer operation. The firing intensity during the start-up in February 2015 exceeded the recommended practice by Haldor Topsoe A/S.

As shown in *Figure 8*, during start-up in February, 2015 the actual heat input (Q_{act}) was more than the maximum allowable duty (Q_{max}) in the reformer for approximately 1.5 hours. At a certain point during this over-firing, the actual heat input (Q_{act}) was approximately 1.5 times of maximum allowable duty (Q_{max}) with no feed flow and steam flow of approximately 3 t/h (6600 lb/h). Moreover, the actual BWT was way above the maximum allowable BWT for an extended period of time. These factors together led to the rupture of 40 catalyst tubes and severe damage of the remaining tubes in the primary reformer.

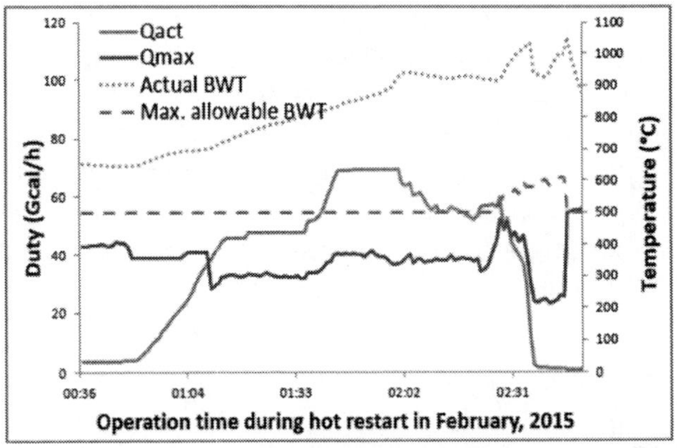

Figure 8. Data analysis from Feb,2015 start-up

The tube rupture started on February 12, 2015 at 01:57 hrs, when one of the six hot collector temperatures suddenly increased above the other hot collector temperatures and the process vent valve to the flare system closed completely. The reason for the sudden temperature increase in one of the collectors was due to the back flow of steam from the other hot collectors to the ruptured tubes. When this steam flow passed the refractory lined cold collector, having a temperature of approximately 450°C (842°F), it was heated up from 325°C (617°F) to 440°C (824°F). The vent valve gradually closed, as the process steam escaped through the ruptured tubes.

In addition to that, some data from previous start-ups was also analyzed. An excerpt of DCS data analysis from similar start-ups is provided in *Figure 9*. During start-up in February, 2014, both BWT and actual duty in the primary reformer went above the maximum limit (highlighted by a red circle).

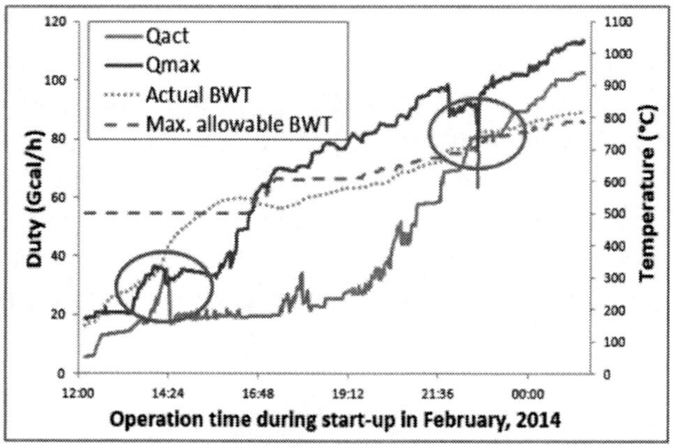

Figure 9. Data analysis Feb, 2014 start-up

Although the maximum allowable BWT and maximum allowable duty (Q_{max}) were marginally exceeded, lifetime of some of the catalyst tubes might have been reduced as a consequence of this upset.

Four Protection Elements

The analysis of the DCS data revealed more than required heat input to the primary reformer during start-ups. Dissolution of secondary carbides in metallurgical examinations further corroborated the fact that the primary reformer catalyst tubes were over-heated for an extended period of time during start-up.

As a follow-up action, a number of suggestions were provided to Profertil in order to avoid overheating of catalyst tubes. These recommendations have further been developed into the current over-firing protection management system (OFP). This system enforces a safe way of starting up a primary reformer by preventing the possibility of both 'local' and 'global' over-firing. The 'local' over-firing may happen due to higher fuel pressure, especially during start-up. Asymmetrical heating of catalyst tubes during start-up can also result in 'local' over-firing. 'Global' over-firing considers the possibility of higher heat input than what is actually required. Thus, four elements of protection are:

- Fuel header pressure limitation
- Symmetric burner ignition sequence
- Duty limitation
- Bridge wall temperature (BWT) limitation

Fuel Header Pressure Limitation

To ensure efficient heat distribution without over-firing during start-up, it is essential to keep fuel pressure as low as possible by igniting as many burners as possible. This is achieved by forcing fuel header pressure to a value that corresponds to stable flame with minimum heat release of the burner, until a sufficient number of burners are ignited. The number of burners used for this limitation should be defined during commissioning.

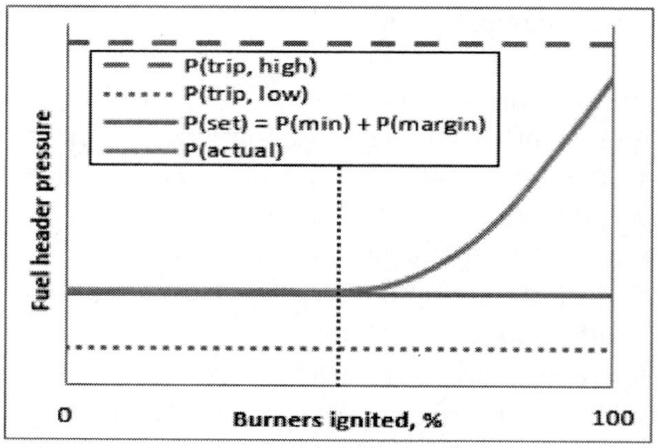

Figure 10. Fuel header pressure limitation

Figure 10 is a schematic of the basic concept of the fuel header pressure limitation. As it can be seen, OFP forces the actual fuel header pressure to a pre-set value until a certain number of burners are ignited. This pre-set value is marginally higher than the lower fuel pressure trip value to ensure stable firing without trip due to low fuel pressure.

Symmetric Burner Ignition Sequence

During start-up, it is important to ignite burners in a particular pattern to ensure that the heat distribution is as even as possible and there is no concentrated local heating. This can be achieved by following a particular burner ignition sequence. OFP does it proactively by enforcing the operation to follow symmetric burner ignition sequence. Moreover, it supervises the compliance with the suggested burner ignition sequence and, reactively, limits the fuel flow to the burners, if the sequence is not followed.

OFP system ensures the following:

- similar burner ignition pattern on opposite walls in the same furnace chamber
- similar burner ignition pattern on walls inside different furnace chambers
- minimum difference between number of ignited burners in each column
- avoidance of adjacent burner ignition unless decent heat distribution is obtained
- no ignition of top row burners during initial start-up

If the abovementioned suggestions are not followed, then an 'asymmetric firing' signal is sent to the DCS. The asymmetric signal does not allow any increase in heat input to the reformer. However, reduction in heat input to the primary reformer, if needed, is possible in this situation. Increase of heat input is possible only when the problem(s) for asymmetric signal is resolved.

Furthermore, OFP also estimates the number of ignited burners based on the required heat input to the reformer and burner heat release curve. Actual numbers of ignited burners in operation must be close to the estimated numbers of ignited burners, otherwise heat input to the primary reformer is restricted by sending 'asymmetric firing' signal to the DCS.

If the reformer is provided with hot collectors, the deviation in temperatures of the collectors is monitored, and in case the deviation exceeds a pre-defined value, heat input to the primary reformer is limited and no increase in heat input is allowed unless the temperature deviation among the hot collectors is acceptable.

So far, operator action ("call-do-respond") is required in Profertil ammonia plant to register which burners are ignited. The DCS operator informs the field operator which burners shall be ignited in accordance with the symmetric "Burner Ignition Sequence" module. Once the field operator ignites the burners, he / she reports back to the DCS operator which burners are ignited, and the DCS operator subsequently registers the ignited burners in the "Burner Ignition Sequence". Furthermore, the burner ignition is independently monitored by temperature deviations in hot collectors and fuel flow versus number of burners ignited. However, it is recommended to install the Topsoe Furnace Management (TFM) system [4] for automatic monitoring of burner ignition sequence. Alternatively, the fuel valves could be provided with limit switches.

Duty Limitation

Mismatch between heat input to primary reformer and heat uptake as reaction heat and sensible heat can result in over-heating of the primary reformer catalyst tubes. Therefore, it is important not to provide more heat than required. The heat required for heating up the furnace box, catalyst tubes, and catalyst has also been taken into consideration for heat balance. A simplified heat balance is shown in *Figure 11*.

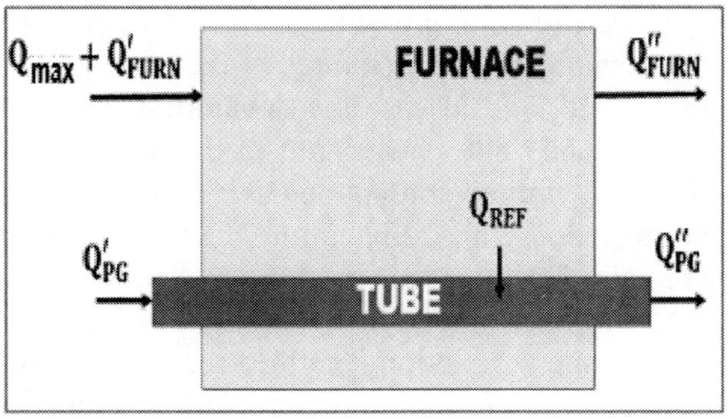

Figure 11. Energy balance in primary reformer

where,

Q_{max} maximum allowable duty to reformer
Q_{REF} required transferred duty to process gas
Q_{FURN} required transferred duty to flue gas side and heating of the furnace box
Q'_{PG} process gas energy flow in to reformer
Q''_{PG} process gas energy flow out of reformer
Q'_{FURN} flue gas energy flow in to reformer
Q''_{FURN} flue gas energy flow out of reformer

The duty input to the reformer is partially taken by the process gas via catalyst tubes and the rest goes to flue gas side. Thus, maximum required duty (Q_{max}) can be expressed in terms of the duty transferred to the process gas (Q_{REF}) and to the flue gas in the furnace (Q_{FURN}) as follows:

$$Q_{max} = (Q_{FURN} + Q_{REF}) \cdot FAC$$

$$Q_{REF} = (Q''_{PG} - Q'_{PG}) = (Q_{reac} + Q_{heat})$$

$$Q_{FURN} = (Q''_{FURN} - Q'_{FURN}) = M_{FL} \cdot Cp_{FL} \cdot \Delta T_F$$

where,

Q_{reac} required duty for endothermic steam reforming reaction
Q_{heat} required duty to heat up reformed gas at reformer outlet temperature
FAC duty adjusting parameter
M_{FL} flue gas mass flow
Cp_{FL} specific heat capacity of flue gas
ΔT_F temperature difference between set point for BWT limitation and ambient

The reformer duty (Q_{REF}) consists of required duty for endothermic steam reforming reaction (Q_{reac}) and sensible heat transferred to process gas (Q_{heat}). Both Q_{reac} and Q_{heat} can be estimated, based on measured feed flow, process steam flow, reformer inlet and outlet temperature, set point of steam to carbon ratio, and manual input for carbon number in the feed. FAC is a dimensionless parameter which can be adjusted, if Q_{max} is not appropriate due to deviation in heat loss or other uncertainties. The default value is 1.0 and it can only be adjusted with approval from the shift supervisor.

The duty transferred to the flue gas side is estimated based on flue gas flow (M_{FL}) and measured increase in flue gas temperature (ΔT_F) in the reformer. Flow of flue gas (M_{FL}), used for the duty estimation, is the lowest value among three different measurements / calculations. One of them is the direct flow measurement at the stack. Apart from this, flue gas flows are also estimated by measuring excess oxygen at the bridge wall and pressure at the suction of flue gas fan. The impact of false air damper opening is considered while using the last method. Specific heat of flue gas (Cp_{FL}) remains constant during both start-up and normal operation.

Bridge Wall Temperature Limitation

The impact of the extent of firing is reflected in two temperature measurements – bridge wall temperature (BWT) and reformer outlet process gas temperature. Between these two, the second has a lag and therefore, the fastest response, in case of over-firing, can be obtained by monitoring the BWT. This concept is used to provide more protection against over-firing due to mismatch in heat input and utilized heat in the reformer.

The BWT is limited by a function that increases proportionally with the plant capacity. This function, to limit BWT during operation, was obtained by performing simulations at various capacities. The constants, used in the function, must be defined after commissioning. BWT is limited to 500°C (932°F) during initial start-up. Protection against over-firing by BWT limitation remains active at all plant capacities.

Conclusions

The over-firing protection management system (OFP) addresses both 'local' and 'global' causes of primary reformer catalyst tube over-heating.

The current system does not allow fuel header pressure increase during start-up and enforces a symmetric burner ignition pattern. During start-up, these two protection elements not only help avoiding catalyst tube over-heating but also ensures better heat distribution inside the furnace.

Moreover, double protection elements of duty and BWT limitation ensure parity in duty input and duty uptake by adjusting fuel flow to the burners based on maximum estimated duty for the primary reformer at a particular capacity.

The over-firing protection management system (OFP) is not only applicable to Haldor Topsoe designed side-fired reformers, but can also be implemented in any type of tubular reformers.

Future Plan

The over-firing protection management system (OFP) has already been implemented in the distributed control system (DCS) in Profertil's ammonia plant and is running as an alarm system. It is being tested in operator training simulator (OTS) and is expected to be re-commissioned as an over-firing protection management system (OFP) in June - July, 2018.

The over-firing protection management system (OFP) is an integral part of all new Haldor Topsoe designed plants utilizing tubular reformer technology.

References

[1] D.H. Timbres, M. McConnell, "Primary Reformer Failure", AIChE Ammonia Technical Manual (2002), p 268 - 278,

[2] B. Rani, "Catastrophic Failure of Reformer Tubes at Courtright Ammonia Plant", AIChE Ammonia Technical Manual (2006), p 187 - 196

[3] R. Stevens, "Nipping of Reformer Tubes. Precaution and Mitigation of incidents.", AIChE Ammonia Technical Manual (2009), p 71 – 82

[4] S.W. Sexton, "Topsoe Furnace Manager Technology Benefits for Ammonia Producers", AIChE Ammonia Technical Manual (2016), p 31 – 40

Safety and Efficiency Improvement of Reformer Mitigation of Post Combustion Problem

After a plant revamp, poor combustion in the radiant section of the Primary Reformer was observed largely as indicated by measured carbon monoxide (CO). Flame patterns were irregular, including flickering with visible flame impingement on the reformer catalyst tubes. Apart from low methane conversion, poor combustion in the radiant section was also causing post combustion in the convection section where the oxygen was in excess. The post combustion was raising the convection coils' temperature too high, close to their operating limits, and thus limiting plant load. This paper illustrates the findings leading to poor combustion in the Primary Reformer furnace, actions taken to fix the problem, and key learnings to avoid such problems during plant changes.

Shahbaz Ali
Unit Manger-Process Engineering, Fatima Fertilizer Company Limited

Zubair Talha
Process Engineer- Ammonia Plant, Fatima Fertilizer Company Limited

Introduction

Fatima's fertilizer ammonia plant was designed by CF Braun & Co. and originally built in 1967 with a nameplate capacity of 1360 MTPD. The plant was originally located in Rozenburg, Netherlands and operated by Exxon and later by Kemira. In 2007, Fatima moved this plant from Netherlands to Pakistan. It started successful production on 15 March 2010 with a design capacity of 1500 MTPD. The plant was successfully revamped to 1650 MTPD in November 2015.

Background

During the 2015 revamp, the Primary Reformer burners were modified for higher furnace heat load with support of the burner manufacturer.

The Primary Reformer design is a Foster-Wheeler terrace wall furnace, in which the two radiant cells (A and B) are connected by a common convection section. The connection between the common convection section and the two radiant cells is called the "hip section" (HIP) with each radiant cell being a "leg" to the convection section.

In the summer of 2016, jets of flame were observed in the HIP section (flue gas entering from radiant to convection section) of furnace. Initially, the cause was suspected to be due to a leak in Feed Preheat Coil (E-169). Later on, as it diminished with decreases in plant load, it was established to be caused by after burning of fuel in the convection section. Excess oxygen measurements (2.1% to 3.5%) in the radiant and convection sections of the furnace indicated sufficient combustion air, but the visual inspection showed signs of oxygen deficiency and improper combustion.

Plant load had to be reduced by 1 to 2% during this observation, and more air had to be introduced through operation of the Auxiliary Combustion Air Fan (B-201).

In January, 2017, a Primary Reformer assessment survey was carried out with the help of Haldor Topsoe (HTAS). Burner performance was checked by measuring carbon monoxide and excess oxygen at the burners.

Primary Reformer Background

The reformer was supplied by Foster Wheeler with two levels of burners in each wall (terrace wall type). It has two radiant cells containing the catalyst tubes. The flue gases flow from the radiant section to the single, centrally mounted convection section through horizontal ducts which contain the HIP coils. The flue gases are extracted from the convection section by two Induced Draft Fans mounted on the top. The combustion flow scheme is shown in Figure 1.

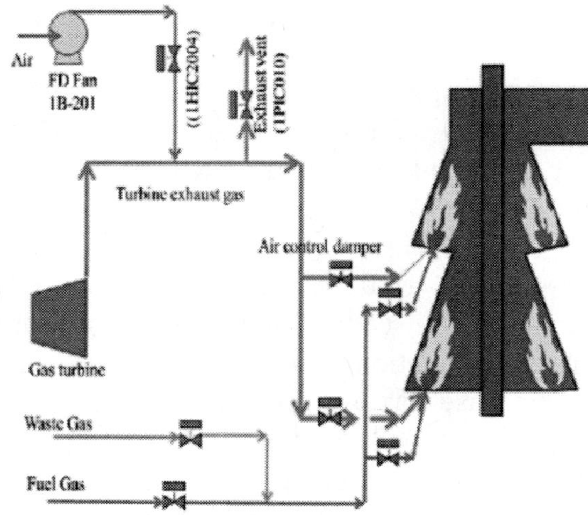

Figure 1. Basic combustion flow scheme

The furnace is arranged to use gas turbine exhaust from the process air compressor driver as the oxidant. Any excess exhaust which is not needed in the radiant section for combustion is injected into the end of the convection section to recover the heat. A new Forced Draft Air fan (B-201) was also installed to meet the extra combustion air requirement that arises in summer operation.

There are 160 turbine gas exhaust type burners (TGEW) installed in the Primary Reformer, split on two levels. Burner operation is on mixed gas fuel (natural gas and recovered waste gas from the cold box). The gas gun of each burner is a long tube, with defined gas drillings arranged along the same to provide a stable, flat flame. During revamp of the plant, these burners were replaced with a new design for higher heat loads while no change was made in burner tiles and burner air ducts.

Key Activities and Observations:

A number of activities were performed to check impact on combustion:

1. Visual inspection of the reformer furnace showed clear signs of severe oxygen deficiency.
   - Visible flames appearing in the furnace away from the burners (ghost flames), especially around the reformer tubes at level 2.
   - Flames from the burners were yellow and flickering instead of blue/invisible and stable.
   - Visible flame impingement, i.e. flames in direct contact with the reformer tubes.

2. Sodium Bicarbonate was injected in a burner port to make the entire flame visible so that further judgment of the actual status of combustion could be made. As per design, flame height should be less than three meters, but actual height observed was much higher. Level 1 flames were mixing with level 2 flames and even touching the roof (shown in Figure 2).

Figure 2. Left) picture of cell A, side A, level 2. Right) picture of cell A, side A, level 2, just after addition of Sodium Bicarbonate at a burner port at level 1

3. Oxygen analyzers (radiant section and exhaust) were showing sufficient excess oxygen (2.2% to 3.7%), but it was continuously varying without any particular change in combustion fuel. There were no calibration errors in oxygen analyzer readings. Refer to Figure 3. Trend for Excess Oxygen-2017, at the end of this paper

4. A jet flame was present from Feed Preheat Coil in the HIP section (refer to Figure 4). It was suspected to be from a pin-hole leak from the coil. It disappeared with the decrease in plant load. The Forced Draft Fan (B-201) was started to provide additional combustion air. After this, the jet flame almost completely disappeared and was only briefly observed every 20-30 seconds. However, no major improvement was observed in the furnace internal condition.

Figure 4. Projection of visible flame jet from Feed coil in HIP section

5. After the FD fan (B-201) startup and minimization of the jet flame in the furnace, CO and O_2 measurements were taken. It was attempted to measure CO and O_2 concentrations at all burners in cell A where the burner ports were not blocked. Due to the unstable flames and false air ingress from the burner port, it was not possible to get reliable readings for O_2. The CO readings were also fluctuating, but it was possible to get an approximate measurement. These CO measurements are shown in Figure 5a and 5b below. 5000 ppm was the maximum reading of the analyzer used. So for most measurements of 5000 ppm, the true value was likely much higher.

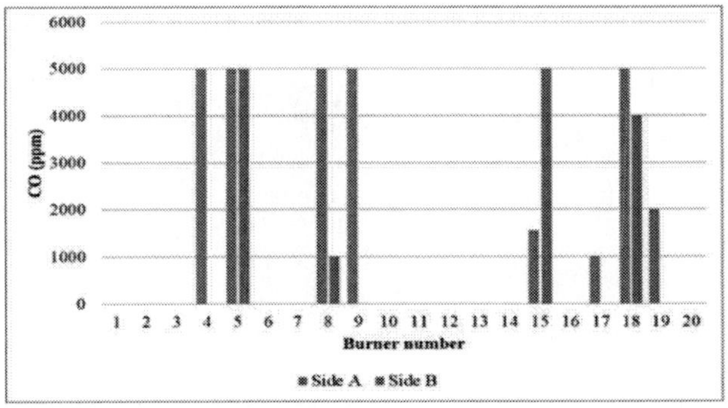

Figure 5a. Measured CO concentration (ppm) at level 1 burner in cell A. At the following burners, no measurements were made due to blocked burner ports: (1-07A, 1-20A, 1-01B, 1-02B, 1-03B, 1-04B, 1-11B, 1-16B, 1-17B, and 1-20B)

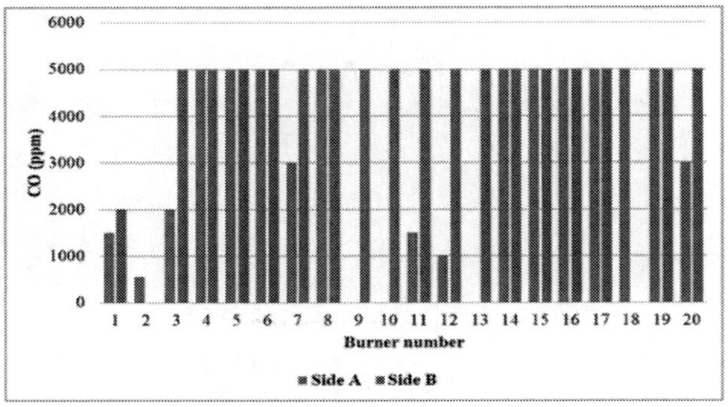

Figure 5b. Measured CO concentration (ppm) at level 2 burners in cell A. At the following burners, no measurements were made due to blocked burner ports: (2-09A, 2-10A, 2-13A, 2-02B, and 1-12B)

As expected from the visual inspection, it was found that a majority of burners showed high CO levels. For efficient combustion with sufficient oxygen available, the CO level should be nil. For measurements taken at the level 2 burner locations, all measurements showed high levels of CO. However, it was later found that this is due to the flames from the level 1 burners as it is practically impossible to measure at level 2 without measuring directly in a flame (see Figure 6). Inside a flame, the combustion is not yet complete, and as such, there will always be a high CO level.

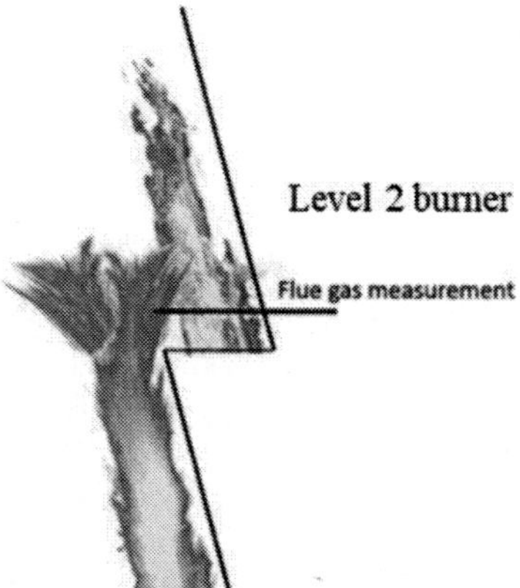

Figure 6. Illustration of flames from level 1 burners influencing measurements at level 2 burners

6. The FD Fan (B-201) discharge flow was increased to 12,000 Nm^3/h (447.9 MSCFH). Visual inspection of the furnace still showed clear signs of oxygen deficiency. CO concentration was measured again at a number of burners on cell A, side A, bottom level, and no significant change was found. The FD Fan was kept running at maximum allowable discharge valve opening after this.

7. The draft was decreased from -10 mmWC to -7 mmWC (-0.4 to -0.3 inches water), by reducing the opening of the flue gas stack dampers. Thereafter, combustion air flow was increased by increasing the opening of the discharge valve from the FD fan. Visual inspection showed significant improvement in the combustion though there were still clear signs of post combustion and flame impingement. CO levels were measured at some burners, and as can be seen in Figure 7, CO levels had significantly decreased but were still high in some places. This confirms the conclusion from the visual inspection.

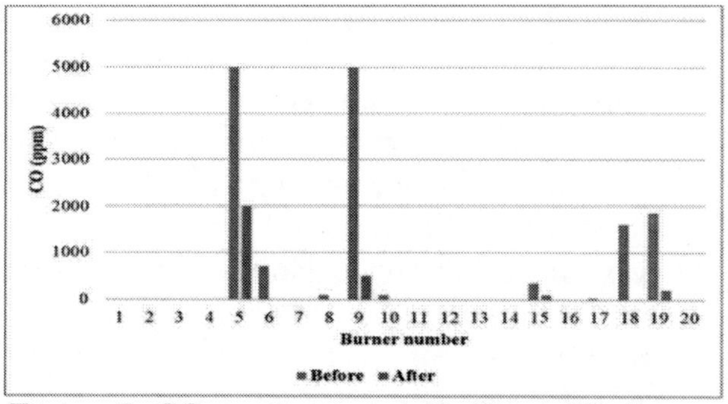

Figure 7. CO concentrations before and after reducing draft.

8. The natural gas fuel flow to furnace was reduced by 2.5%. As a result, the reformer outlet temperature decreased by 3 to 4°C (5 to 7°F), giving increased methane slip. In order to compensate for the increased methane slip, the air flow to the secondary reformer was increased. These changes gave additional waste-gas flow from the cold-box. So the

overall effect was that the fuel flow was not changed significantly, but the heating value was decreased (see Figure 8).

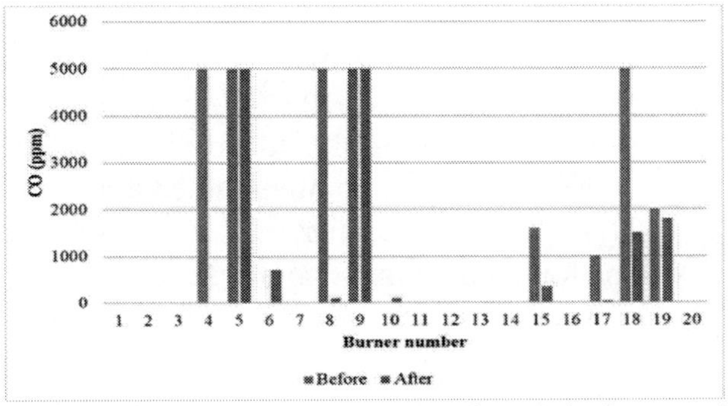

Figure 8. CO concentrations (ppm) before and after reducing fuel to the Primary Reformer. At the following burners, no measurements were made due to blocked burner ports: (1-07A, 1-20A)

9. After the draft was reduced, the amount of waste gas from the cold box being sent to the reformer as fuel was reduced by 5,000 Nm^3/hr (187 MSCFH) (from 30% to 15% waste gas mixture with natural gas). This waste gas was instead sent to the flare. The waste gas was replaced with 900 Nm^3/hr (33.6 MSCFH) of natural gas in order to maintain the reformer outlet temperature. The affect of this is that the amount of firing is kept constant, but the higher heating value of the fuel was giving a lower gas velocity at the burner tip. The combination of reduced draft and increased heating value of the fuel significantly reduced the fuel header pressures.

10. It was found that there were large variations in pressure of the different combustion air headers, as high as a factor two difference between headers at the same level, and with similar valve positions at the individual burners. For this reason, the position of the control valves for the individual combustion air headers was inspected further. The air header control valves had not been operated since the plant was initially started up. It was attempted to adjust the valve openings, but it was not possible due to condition of gear box.

As per above activities, significant improvements were gained by turning on the FD fan with maximum allowable opening of the discharge valve (1HIC2004) and with the draft reduced to -7 mmWC (0.3 inch water). Further improvement was possible by decreasing the amount of waste gas being used as fuel; however, since the excess waste gas was sent to flare, this was not a viable solution. However, in spite of the improvements, there were in all cases still clear signs of poor combustion in the reformer furnace. Though some burner's combustion condition was improved by frequent ON/OFF operation after online adjustments, the desired results were still not achieved. It was then decided to remove one burner from the furnace for inspection and cleaning. Combustion fuels (natural gas and waste gas) are clean and burner cleanliness was not checked after installation of burners during the revamp.

Burner inspection

One burner was removed and inspected to look for any areas of improvement. The following were checked:

- Blockage of air path
- Blockage of gas path
- Measurement of hole sizes (ignition, firing and air baffle)
- Measurement of burner hole angles (ignition, firing and air baffle)
- Installation dimensional check

The air and the gas path was observed clear. However, there was a clear mismatch between the installed and actual design of burners. Below were deviations:

- Burner hole angles (ignition, firing and air baffle) should be 20°/20°/40° while actual measurements were 28°/12°/38° (see Figure 9).

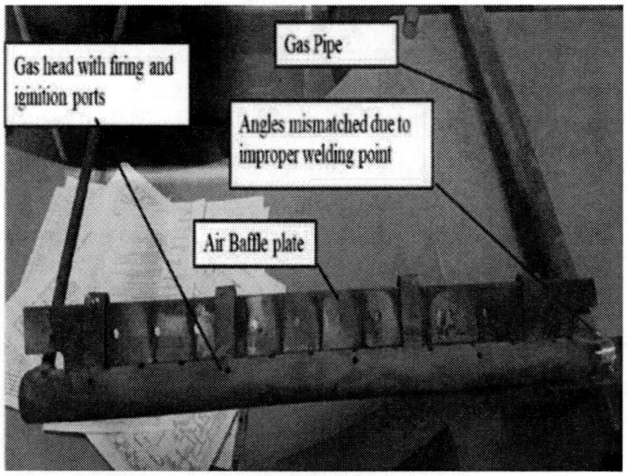

Figure 9. Typical TEGW gas head

- Orifice (hole) sizes for ignition and firing ports were as per drawing. While for air baffles these were 6.5 mm instead of 9.6 mm (0.25 inch instead of 0.40 inch).
- Gas head installation inside tile was not as per drawing (see Figure 10). Gas head distance inside the burner tile should have been 51 mm (2 inch) which was not maintained during installation phase. Burner head was suspected to be touching outer edge of tile.

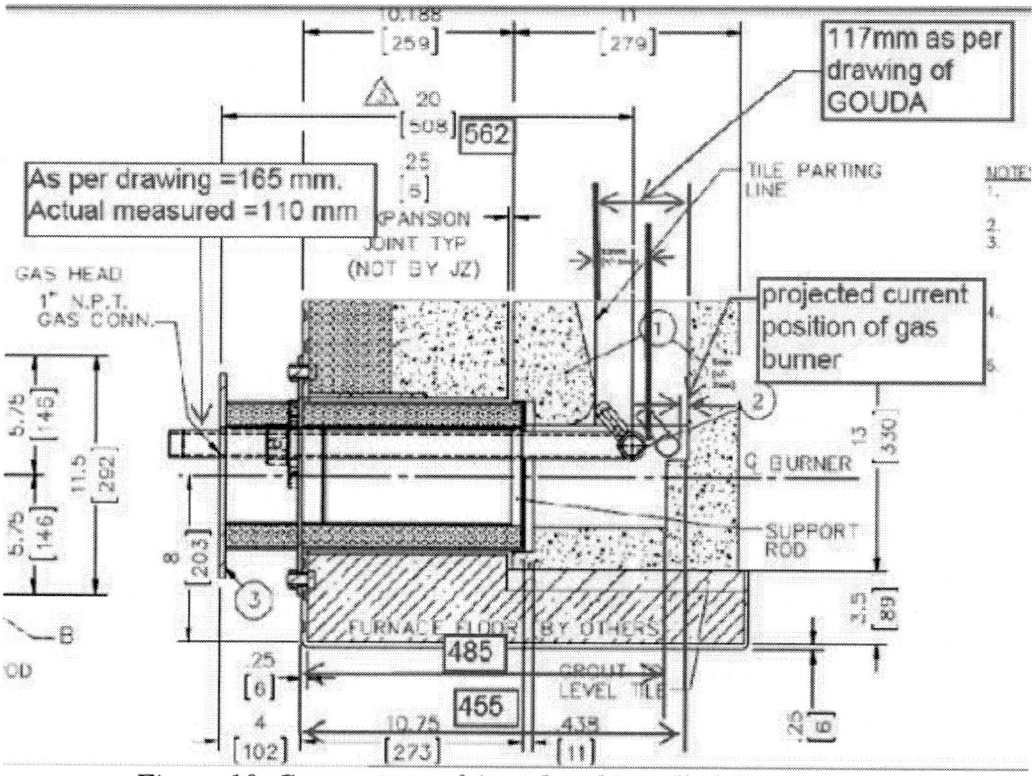

Figure 10. Comparison of Actual and Installed Gas head.

As part of the ammonia plant revamp, the burner design change was due to changes in fuel gas composition and increased heat load requirement. The burner revamp study, new design and performance testing of one burner was carried out in the manufacturer work shop. Based upon final design, fabrication of all gas heads was carried out locally.

Corrective actions in 2017

The burner inspection was carried out in March, 2017 and a detailed action plan was prepared and implemented in April, 2017. All activities were performed within the 2017 turnaround timeline:

- New burner heads were prepared in-house. Dimensional checks were ensured through a detailed quality check program. Every burner head was checked and details were logged.
- Deviation in angles of firing and ignition port holes across the gas head were corrected during drilling phase. A mechanical assembly was prepared to ensure all drillings are inline and angles do not deviate from drawings. Burner hole angles (ignition, firing, and air baffle) were adjusted for the 80 bottom terrace burners. The air baffle hole size was increased to 9.6 mm (0.37 inch) for 120 burners
- The welding point of the gas pipe to the burner head was checked according to the drawing. For most of the burners inspected, the angle deviation was due to this error. Pipe welding was not as per the required angle from centerline.
- Burner head distance inside the burner tiles were checked as per actual drawing and dimensions of each burner logged. Burner head distance inside the burner tiles were adjusted for 120 burners.
- The full tightening of the combustion air damper controlled by 1PIC010 was checked. No major leak was reported during the inspection.
- Gear boxes for the combustion air header valves were repaired, and the position of the valves were marked.
- Air registers for the individual burners were checked.

Results

- A considerable decrease in the energy index of the furnace fuel was observed. It decreased by 0.50 GJ/MT (0.52 MM BTU/ST) due to the improved combustion.
- The post combustion issue was resolved. Flame condition and length was also improved. Control of excess oxygen was improved with no major fluctuations. The FD fan (B-201) is no longer needed. Refer to Figure 11 at the end of this paper. Excess O2 in Furnace after burners adjustments attached to the end of this paper
- The average tube metal temperature (TMT) decreased by approximately 10°C (18°F) as a result of good burner performance and better flue gas distribution inside the furnace (see Figure 12 below). The overall impact will be an increase in catalyst tube life.

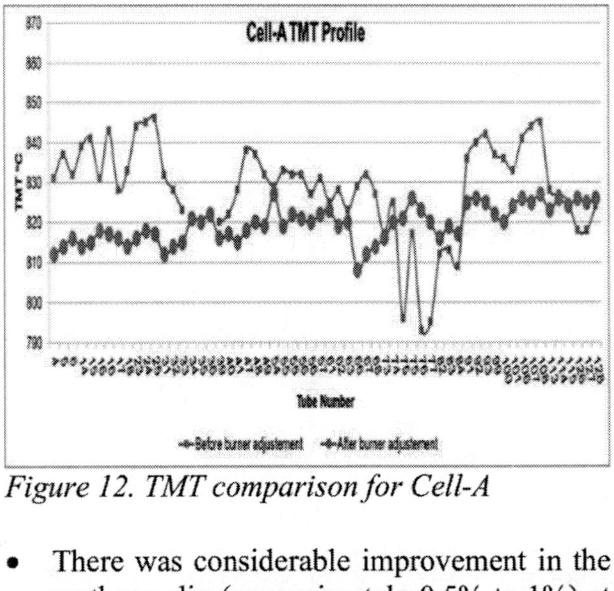

Figure 12. TMT comparison for Cell-A

- There was considerable improvement in the methane slip (approximately 0.5% to 1%) at approximately 6 to 8°C lower hot collector temperature (11 to 14°F) as a result of the combustion improvement in the Primary Reformer. Previously the hot collectors were maintained in range of 728 to 730°C (1342 to 1346°F) and are now operated 8°C (14.5°F) lower (see Figure 13 below).

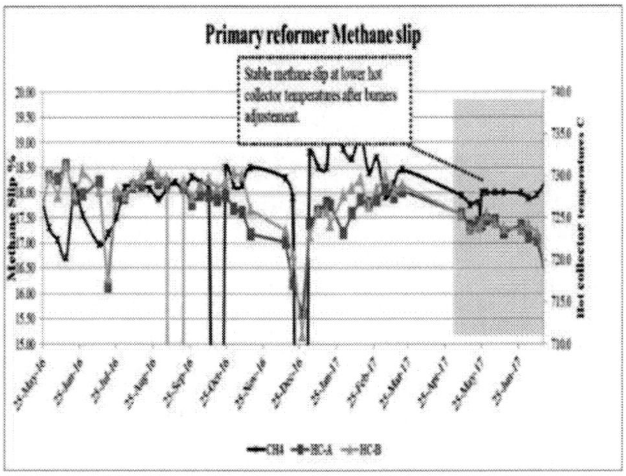

Figure 13. Trend of Methane slippage (%) vs hot collector temperatures (°C) before and after burner's adjustments

- In February, 2018, the maximum CO slip measured on level-1 burners was reduced to less than 5 ppm. Figure 14 shows that a few burners still showed higher CO (approximately 400 ppm max). Inspection of those burners revealed blockage of the air path with refractory pieces. These have since been cleaned.

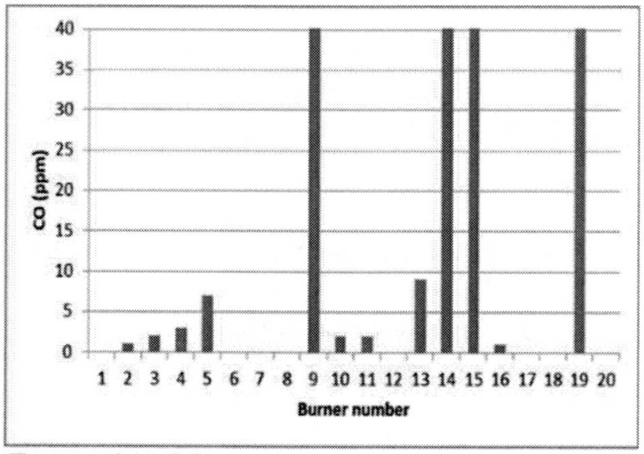

Figure 14. CO concentrations (ppm) at Cell-A level-1 after burners' adjustment Feb-2018.

Root cause

For the Fatima ammonia plant revamp, a new burner was properly designed, and performance testing of the burner was also carried out at a test bench to verify its operation under different fuel conditions and flame stability. The test results were normal; however, during manufacturing of the burners and their installation, deviations from the design resulted in overall poor combustion in furnace. The importance of the burner dimensions and installation were not considered critically enough. Quality control was ensured with respect to basic mechanical values (material, welding procedures, etc.), but quality hold points were missing for items like:

- Dimensional inspections during installation of gas heads.
- Tip orifice, its projections and orientations during fabrication of gas heads.

Conclusion

After fabrication and installation of new burners as per the original revamped design, post combustion issues were resolved and provided significant safe operating margins. The reliability and life of the reformer has been improved in terms of lower tube metal temperatures, and elimination of post combustion which could lead to severe failures. A significant gain was also observed in furnace thermal efficiency and methane conversion.

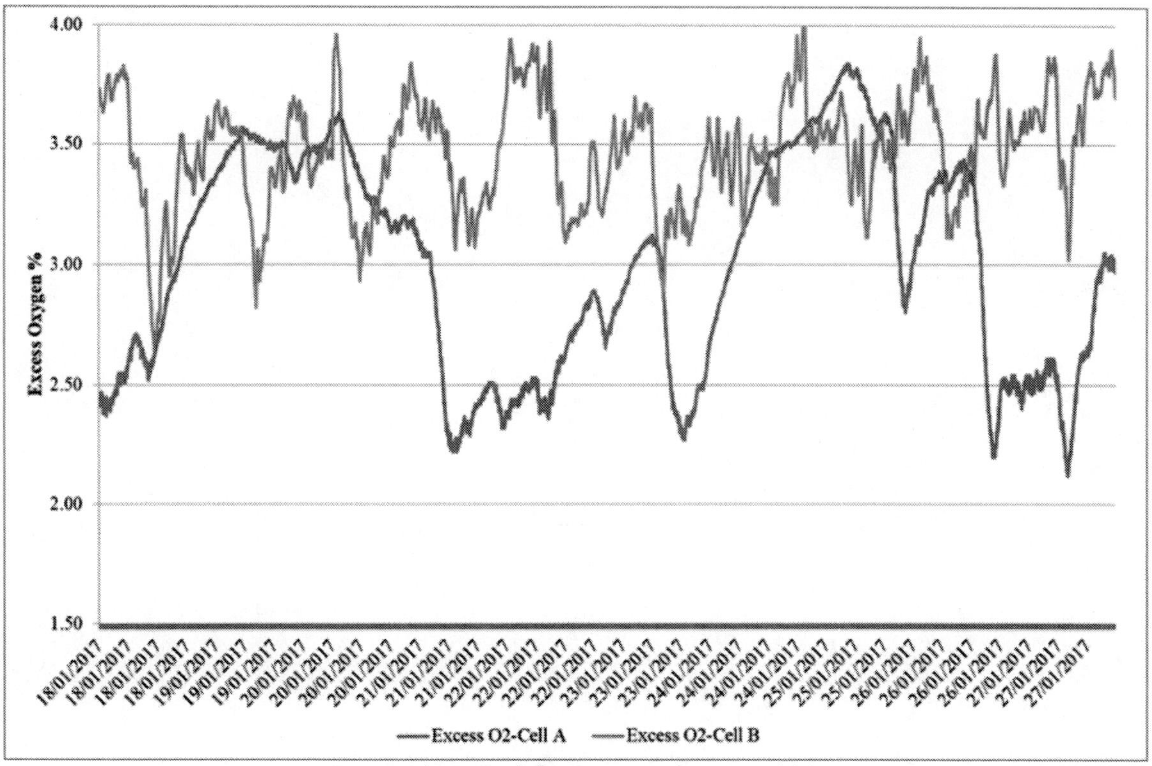

Figure 3. Trend for Excess Oxygen-2017

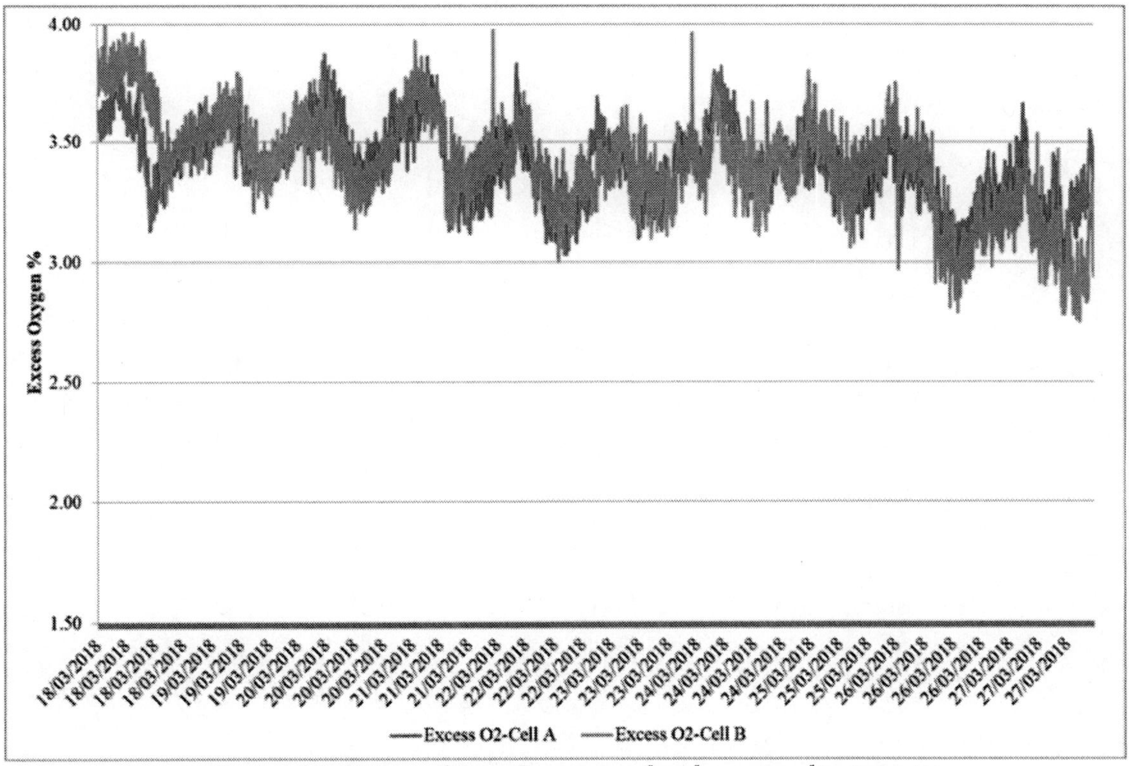

Figure 11. Excess O2 in Furnace after burner adjustments

Challenges in Commissioning and Operation of OASE® Solvent Systems

MDEA based solvents for CO_2 removal in ammonia plants came into existence in the early 1980s, and energy consumption of these systems were the lowest. Since then, more than 200 ammonia plants are operating with BASF's OASE® solvent system which was earlier known as aMDEA®. There have been many new ammonia projects that have come on stream in all parts of the globe, and hence it is important to recap the lessons learned from these recent projects. In this paper, the challenges in pre-commissioning, commissioning and operation of OASE® solvent systems are described.

Dr Torsten Katz and John Nichols
BASF

Venkat Pattabathula
Incitec Pivot

George Colman
Kellogg Brown & Root (KBR)

Introduction

The growth of ammonia industry has been languishing in the USA for about 30 years. However, a step change has occurred since the shale gas industry was developed and the gas prices became affordable. Hence, more ammonia capacity is being added to supplement existing operating capacities to meet the USA's ammonia demand.

BASF's OASE® white Process

The Acid Gas Removal Units (AGRU) in Ammonia plants are usually designed in line with the Basis of Design Information, provided by the main engineering company. For most plants, an operating expenditure optimized design using a two-stage absorber is the preferred setup. On the flipside, those designs lead to higher capital expenditure. Figure 1 shows the most common configuration in the industry. The synthesis gas, which shall be purified from CO_2, enters the absorber column at the bottom.

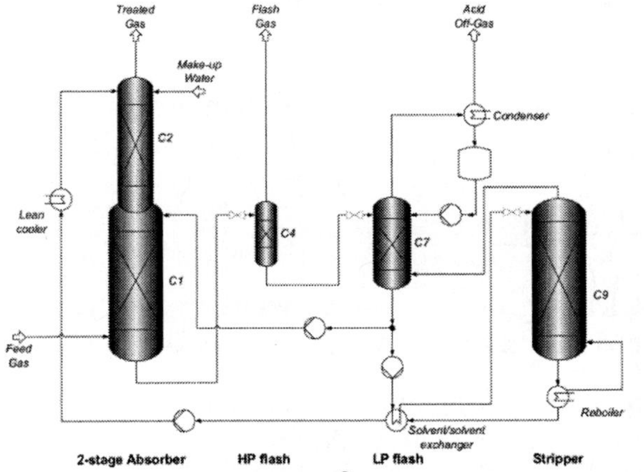

Figure 1. Typical OASE® white configuration

The majority of CO_2 is removed from the gas in the lower bulk absorber section (C1). Co-absorbed H_2 and other inert gases are released in the HP flash (C4), and the rich amine solution is partly regenerated to the semi-lean quality in the LP flash drum (C7). A minor part of the semi-lean amine solution is fully regenerated in the stripper (C9) to a lean solution quality. The lean amine solution enables a deep CO_2 removal in the lean absorber (C2), down to less than 500 ppmv in the Treated Gas.

As a heat integration measure, a solvent/solvent exchanger and the recycle of the stripper vapor into the LP flash sump is implemented. With the two solvent loops of lean amine and semi-lean amine, the process offers high operational flexibility and is very forgiving in case of upsets. The hot synthesis gas serves as the heating medium in the reboiler before it is further cooled down in several heat exchangers and before it enters the absorber as Feed Gas. Consequently, the Steam to Carbon ratio in the steam reformer defines the available energy input which will end up in the AGRU. Usually, the split between lean and semi-lean flow rate is well balanced for the available amount of energy in the hot synthesis gas.

Pre-commissioning and commissioning guidelines

Some Items to Be Considered When Planning for the Pre-Commissioning of an OASE® white system

Allow a minimum of 6 weeks in the overall plant pre-commissioning schedule for cleaning of the OASE® white system (note: this duration excludes pre-cleaning).

Cleaning of the OASE® white system can only proceed after the following items are in place:

a) All packing and internals have been installed in the towers and drums.

b) All machinery is in place and all pre-commissioning steps such as motor solo runs, alignment, lube oil system flushing, etc. have been completed.

c) All pipe work is in place and hydro tested

d) All instrumentation is in place and commissioned.

e) A suitable supply of clean, filtered, low chloride (< 50ppm$^{w}/_{w}$) water is available.

f) A suitable supply of demineralized water is available.

g) A suitable supply of nitrogen is available for pressurization of the CO_2 absorber and HP flash.

h) A suitable system is in place for the disposal of waste water and alkaline solutions generated by the cleaning activities.

i) A suitable supply of electricity and/or steam (if steam turbine driven pumps form part of the system) is available.

j) Pump related interlocks have been functionally checked and are in service.

k) The plant's Distributed Control System (DCS) is available for use.

l) Sufficient pipe fitting resources are available to support the cleaning of pump suction strainers

Pre-commissioning of an OASE® white system generates a significant quantity of:

a) Rust contaminated water during the pre-cleaning and cold / hot water recirculation steps.

b) Weak alkaline solution and high pH water during the alkaline degreasing and water rinsing steps.

Before starting to pre-commission, the OASE® white system, plans shall be in place to capture the waste streams and dispose of them in accordance with company and local authority regulations.

Main OASE® white Solution System Pre-Commissioning Activities

Cleanliness of the OASE® white system is the key in ensuring a trouble-free start-up. Contaminants such as oils, greases, particulate matter, and construction dirt/debris result in serious foaming issues during the initial start-up and operation of the system.

OASE® white system cleaning is split into 10 steps:

i. Pre-cleaning of towers and drums
ii. Pre-cleaning of large bore pipe work
iii. Loading tower and drum packing
iv. Cold and hot water recirculation
v. Alkaline degreasing
vi. Water rinsing
vii. Internal inspection of towers and drums
viii. Cleaning and loading of vortex packing
ix. Final water rinsing
x. OASE® white solution preparation and filling

Pre-cleaning of towers and drums comprises of power brushing the walls of all towers and drums to remove particulate matter, i.e. mill scale and surface rust, from the vessel walls. This activity is best completed when each item of equipment is in the horizontal position while vessel "dress out" activities are ongoing.

Pre-cleaning of large bore OASE® white pipe work is best affected by either air blowing or once through water flushing. The most effective form of cleaning large bore pipe work is to use the tower or drum as a volume tank and perform a rapid decompression air blow of the pipe work. Loading of tower and drum packing may proceed after pre-cleaning activities have been completed. However, the vortex breaker packing located in the bottom of both the CO_2 Absorber and HP Flash should not be loaded until after the final water rinse is completed.

Further removal of particulate matter from the OASE® white system is performed by using the installed pumps to recirculate demineralized or reverse osmosis water around the system and capture any remaining particulate matter on a fine mesh fitted to each pump suction strainer. Prior to the start of water recirculation, plan to bypass any plate and frame exchangers in the system and remove all erosion susceptible instrumentation. To be able to establish water recirculation it is necessary to pressurize both the CO_2 Absorber and HP Flash vessel with nitrogen. To achieve effective particle removal, one should aim to attain a water recirculation flow rate of greater than 90% of the OASE® white solution process flowsheet. Before starting this activity, it is essential that all interlocks provided to protect the recirculating pumps have been functionally tested and placed in service. Additionally, throughout water recirculation, all precautions must be taken to ensure that the recirculation pumps are not cavitated or that any pump suction strainer is not collapsed due to excessive pressure drop. To aid in cleaning of the system, every 36 hours, the entire contents of the system should be drained and replenished with clean water. Typical acceptance criteria for this step is, with a water recirculation flow of greater than 90% of the OASE® white process flowsheet flow, a recirculating water temperature of 50 - 70°C (120 - 160°F), and 80 mesh screens or larger fitted to all pump suction strainers, to maintain these conditions for 48 hours without it becoming necessary to clean any pump suction strainers.

All the wetted surfaces of the OASE® white system are to be degreased by using the installed recirculation pumps to circulate a weak alkaline solution at greater than 90% of the process flowsheet of OASE® white solution. The system temperature is maintained at between 50-70°C (120-160°F) for 12–24 hours. The most effective alkaline degreasing solutions are 3-5%$^{w}/_{w}$ K_2CO_3 solution, 3%$^{w}/_{w}$ KOH or 3-5%$^{w}/_{w}$ NaOH solution. To eliminate corrosion concerns, the concentrated KOH or NaOH used for the preparation of the dilute alkaline degreasing solution must be low in chlorides, less than 100 ppm, membrane grade material. The alkaline degreasing solution is prepared by diluting the concentrated alkali with demineralized water in a temporary vessel outside of the AGRU. If after the prescribed cleaning period the alkaline degreasing cleaning solution is very dirty, replace the dirty solution with clean solution and repeat the operation. When the appearance of the used alkaline degreasing solution is determined to be acceptable, fully drain the solution from the system.

Trisodium phosphate (TSP) can also be used for degreasing the system. If the supplied chemical was in TSP form, the required chemical strength in the solution should be 3-5%, whereas, if it was supplied in the form of dodeca-hydrate (Na3PO4.12H2O), its strength should be a minimum of 7%.

Following alkaline degreasing, it is necessary to rinse the entire system with demineralized water until the pH of the rinse water is < 9.0 and the foaming tendency of a 40%$^{w}/_{w}$ OASE® white solution prepared using a sample of the rinse water and a sample of the OASE white Premix, has a foam height less than 350 mls and a collapse time of less than 35 seconds. Experience shows for a large OASE® white system, 4 – 5 water rinses are necessary to achieve the acceptance criteria.

After water rinsing has been completed, an internal inspection of all towers and drums is to be performed and any damaged components must be repaired. In advance of reaching this step in the sequence, the tower vortex breaker packing must be degreased, and water rinsed external to the system in readiness for installing it in the CO_2 Absorber and HP Flash after the inspection.

After closure of the system, 2 or 3 water rinses should be performed to remove any dirt that may have entered the system during the inspection and vortex breaker packing loading activities.

Once the final rinse has been completed, 45%$^{w}/_{w}$ OASE® white solution which has been prepared by diluting OASE® white concentrate with degassed demineralized water is charged to the system.

Main OASE® white Solution System Commissioning Activities

Immediately following the charging of the OASE® white solution to the system, recirculation of the solution should be started, and the side stream filtration unit commissioned with 100-micron elements and placed in service.

To prevent contamination of the OASE® white solution from upstream equipment it is essential that:

a) Upon completion of the Low Temperature Shift (LTS) catalyst reduction that the LTS catalyst bed is effectively de-dusted to the front-end flare.

b) That equipment and pipe work upstream of the CO_2 Absorber is blown clean by venting process gas immediately up stream of the CO_2 Absorber at a flowrate equivalent to 40~50% of the flowsheet value.

Before the introduction of process gas to the CO_2 Absorber, the anti-foam injection system shall be made ready for service.

Prior to the introduction of process gas to the CO_2 Absorber it is necessary to passivate all the wetted surfaces within the OASE® white system. Before the start of the system passivation, sufficient anti-foam is charged to establish an anti-foam concentration of approximately 50ppm$^{w}/_{w}$ in the recirculating OASE® white solution The unit should be operated with the cooling medium for the lean cooler blocked in and the amine circulated at design conditions and at a temperature close to but less than 70°C (160°F) or as close to design conditions as can be achieved throughout the OASE® white system for 72 hours.

Passivation is completed after the amine solution has been circulated for 72 hours at the conditions specified above.

Challenges Experienced in Commissioning

In several new ammonia plants, ineffective cleaning of the OASE® white system has resulted in foaming events either in the CO_2 Stripper/LP Flash which have led to OASE solution loss of containment from the CO_2 vent into environment surrounding the plant, or in the Absorber which results in not meeting the CO2 treated gas specification. Some of these cases are discussed below.

Case Study 1

In one plant due to an ineffective alkaline degreasing after charging the OASE white solution to the system, instead of it being colorless it had a yellow discoloration. The discoloration can be an indication that some of the contaminates that should have been removed in the commissioning steps were not removed and remained in the system. This yellow discoloration was removed by passing the solution through a carbon bed for approximately 1 week.

Case Study 2

In one plant, it was not possible to operate at plant loads above 60% due to the CO_2 vent valve being fully opened and the CO_2 Stripper/LP Flash pressure increasing above the design value. An investigation determined that the CO_2 vent silencer was partially plugged with construction debris. (See Figure 2) The root cause was ineffective cleaning of the CO_2 product pipe work during pre-commissioning.

Figure 2. Partial plugging of CO_2 vent silencer.

Case Study 3

In several new plants, large spills of OASE® white solution have occurred when temporary gaskets, which remained undetected from plant pre-commissioning, failed during plant operation.

Case Studies of Failures in Operating Plants

Operational issues have occurred over the years due to design recommendations not being followed. Some of those instances are discussed in the case studies presented below.

Case Study No.1

Observations:

- Leakage in semi-lean line pump discharge (Original material of construction was carbon steel)
- Line thickness found to be paper thin
- 8-10 mm pitting in flow control valve in the same line
- Flow control valve became non-operational

Root cause

- Material of construction was not in line with BASF's recommendations
- Line velocities should not exceed 2 m/sec (6.5 ft/sec) (max for carbon steel) because at these velocities erosion corrosion have been reported in API 945

Corrective action

- Upgrade to stainless steel material for replacing the affected portion of the piping in line with BASF 's recommendation

Case Study #2

Thinning of a pipeline due to erosion corrosion was also reported in another plant in the semi-lean piping downstream of a flow orifice (see Figure 3). Also, the body of a block valve in the same pipe showed major corrosion attack, (see Figure 4).

Figure 3. Corrosion of carbon steel semi-lean pipe downstream of flow orifice

Figure 4. Corrosion of CS steel block valve body downstream of control valve in semi-lean pipe

Case Study #3

Corrosion occurred in another plant which resulted in BASF modifying its design recommendations for the metallurgy.

Observations:

- Corrosion of the carbon steel parts of the body of the semi-lean and rich control valves (See Figure 5)
- Corrosion of packing internals (See Figure 6)
- Solids blocking exchanger inlet (See Figure 7)
- Solids blocking plate and frame plate (See Figure 8)

Corrective action:

- Upgrade to stainless steel material - BASF's recommendation
- To ensure continuous operation and flexibility, an additional lean/semi-lean exchanger installed in parallel to existing one.

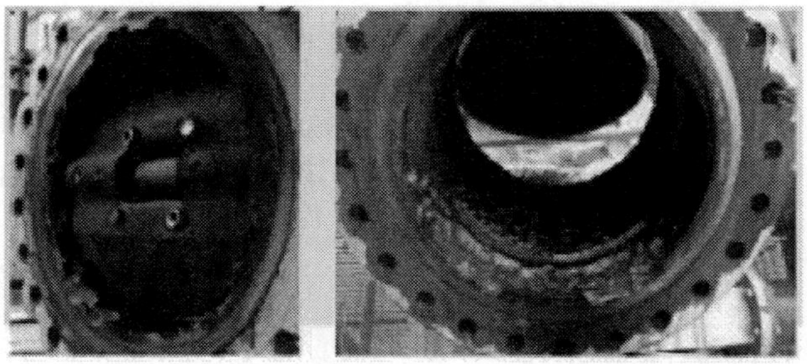

Figure 5: Corroded valve parts and inlet piping

Figure 6: Corroded packing

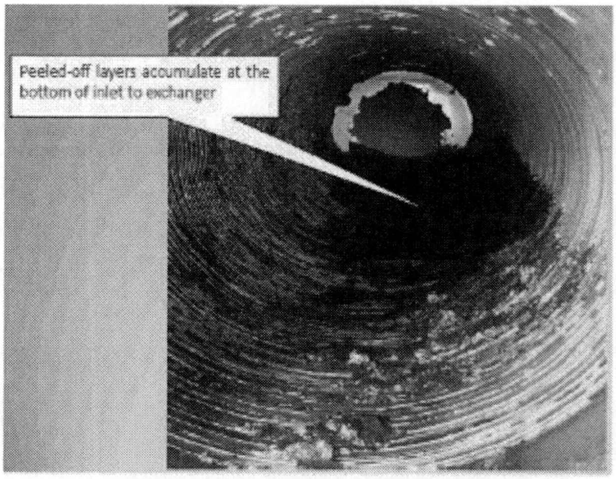

Figure 7: Solids blocking exchanger inlet.

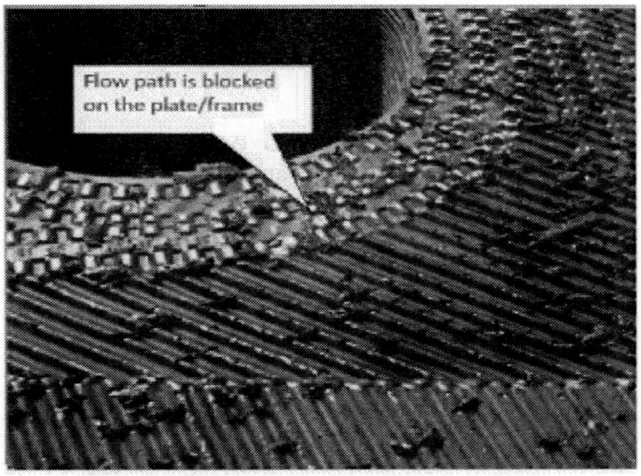

Figure 8: Solids blocking exchanger plate

Case Study #4

In this case study, an ammonia plant was having the following issues:

- High CO_2 slip
- Low purity CO_2
- Fouling of exchangers
- Leaking lean solvent FCV

After reviewing the operational data of the plant, a plan of attack was developed. The plant immediately

- Increased the anti-foam dosage
- Increased the solvent strength to 45% wt.

At the next shutdown, the following actions were taken:

- Upgraded the stripper packing to stainless steel material - BASF's recommendation
- Cleaned the lean / semi-lean exchanger to bring the lean amine temperature closer to design value
- Installed a strainer upstream of lean / semi-lean exchanger
- Changed the lean inlet flow control valve material of construction to stainless steel - BASF's recommendation

The plant has operated without any issues after these changes were made.

Case Study #5

Flooding in LP Flash Vessel

A two-stage amine system in an ammonia plant was experiencing flooding of the LP flash. The indications of the problem were:

- Abnormal behaviour of tower's high and low-level switches
- Wrong actuation due to foaming and accumulation of antifoam in dead legs of level switches
- Too much antifoam dosing – its deterioration products resulted in more foaming, and the spent antifoam accumulated in the system
- Strong amine smell in CO_2 stream due to trimethyl amine formation
- Amine spilled from CO_2 Vent

The following corrective actions were implemented:

- Adjustments in column levels were done very slowly and step wise
- Minimum concentration was to be maintained in the range of 38 to 42%
- Antifoam injection directly into LP flash vessel
- Dosing rates were optimized from 4 to 1 litre/day (0.3 gal/day)
- Steam / carbon ratio to be maintained as per design conditions

Case Study #6

OASE entrained vapor escaped through the CO2 vent silencer on two occasions.

The first occurrence was due to increased foaming when OASE solvent was pumped from the sump into the system. The level transmitters were invariably inaccurate and the level control valve from the HP flash to LP flash column opened wide. The result was about 10 m3 (2600 gal) of diluted OASE solvent flowed through the CO2 vent silencer and across the site.

The antifoam dosing procedures and makeup from the sump to the system were revised. Side stream filter change out frequency was increased.

The second event occurred soon after startup. Operators observed erratic level in the LP flash column and downstream knock out drum. The levels through the system was behaving erratic as well. Antifoam was added but this did not help. Eventually the level in the knock out drum went >110%, and OASE® white went out through the silencer.

Further investigation revealed that the cause was most likely the gas blow through. The HP flash vessel wet leg drained out when the plant tripped and caused the level to remain at 55%. The fix was to ensure that these transmitters were all checked whenever the system tripped before restart.

Troubleshooting in normal operation

Higher than design steam to carbon ratio

Many NH_3 plants are being operated with higher than design Steam to Carbon Ratio in the steam methane reformer because operations often want to minimize the risk of coking on the primary reforming catalyst. As a result, this can cause an energy oversupply to the AGRU: due to the higher water content, the hot synthesis gas contains more energy. Figure 9 shows that a slightly higher than design reboiler duty has only a marginal impact on the maximum achievable feed gas capacity of the AGRU. However, due to the increased hydraulic load in the stripper, flooding is more likely to happen. In the example of Figure 9, flooding in the stripper starts at reboiler duties larger than 110% of the design value.

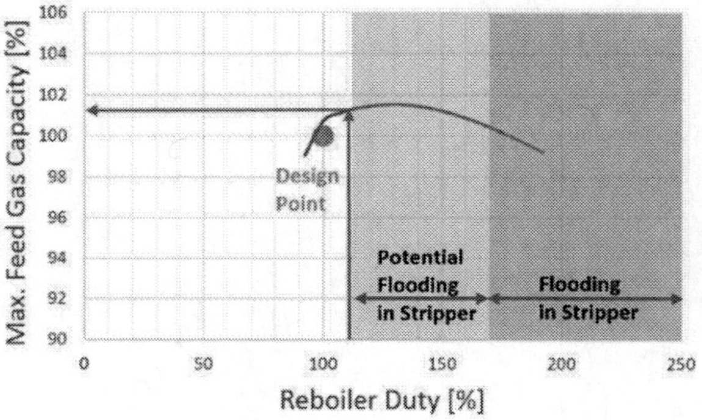

Figure 9. Maximum feed gas capacity versus reboiler duty

Since the nature of the curve in Figure 9 shows only a flat optimum, certain deviations of the design Steam to Carbon Ratio can be managed by the AGRU process, but the differential pressure across the stripper should be monitored to avoid malfunction.

Operation at capacity limit

Whenever the ammonia price is high, operations tend to push the production capacity to the limits. For those plants, where the AGRU is the bottleneck in the overall process, a mid- and long-term damage of the plant can be the consequence. Often, operations try to upgrade pumps to increase the lean or semi-lean amine solution flow rates. A velocity of 2 m/s (6.5 ft/s) in carbon steel liquid piping however shall not be exceeded, otherwise erosion corrosion is likely to occur. For plants with stainless steel piping, this criterion is not relevant, except if the flanges at the vessels are made of carbon steel and the velocity limit is exceeded there. Figure 10 shows corrosion in a temporary bypass line between the absorber and the HP flash due to too high velocity *(5.5 m/s, 18ft/s)*.

Figure 10. Corrosion in carbon steel amine solution piping due to too high velocity (5.5m/s, 18ft/s)

Another corrosion pattern, which has been observed in plants that are pushing the capacity to CO_2 breakthrough limits, goes back to operation at pinched conditions. In such a scenario, the upward flowing gas stream is in equilibrium condition with the down coming amine solution. Nearly constant temperature and CO_2 gas phase profiles, as shown for the red curves in Figure 11 in the lower part of the lean absorber, are the consequence.

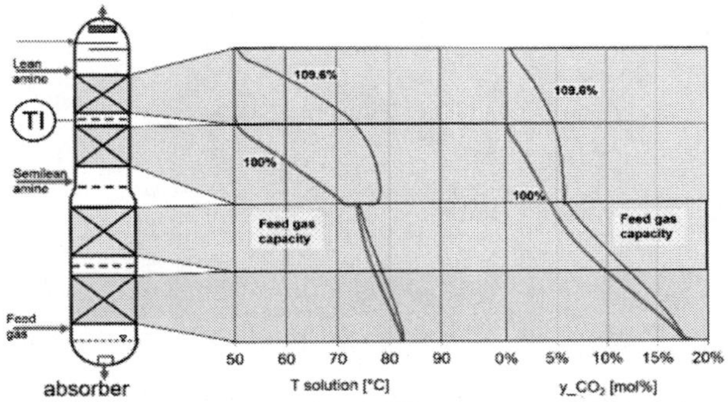

Figure 11. Corrosion in CS amine solution piping due to too high velocity

The increase in temperature goes hand in hand with breakthrough of the CO_2 profile (see left graph in Figure 11). Under the red curve conditions, the plant was still able to manage NH_3 production. However due to minor local maldistribution effects, or splashing of the loaded amine solution on the liquid re-distributors, local CO_2 degassing took place. This led to severe corrosion damage as shown in Figure 12. To avoid operating under such critical conditions, which affect the long-term integrity of the equipment, a temperature indicator (TI) with an appropriate warning to avoid operation under pinched conditions is a simple countermeasure.

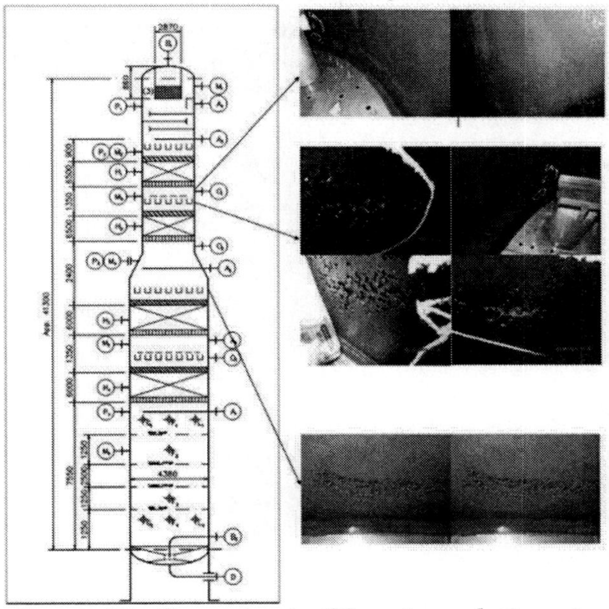

Figure 12. Corrosion in CS amine solution piping due to too high velocity

Conclusion

After over 40 years of experience in designing and operating OASE Acid Gas removal units, BASF continues to make changes to its procedures and design practices to improve the commissioning and operation of Ammonia plants. Because of the various lessons learned over the years, BASF has recommended and implemented various changes to their design best practices. Some of these include:

i. Install stainless steel packing in all the columns.
ii. Follow BASF guidelines in materials of construction for piping, elbows, bends etc.
iii. Follow BASF recommendations for amine velocity in CS piping
iv. Conduct routine analysis of amine strength and foam tests.

v. Send lean amine solvent samples and plant operating data to BASF every quarter for complete analysis.
vi. Ensure CO2 absorber operates far away from high temperature zone that could contribute to potential corrosion.
vii. Perform reclaiming operation if there were any buildup of degraded products.

Poor Ammonia Converter Performance Caused by Choked Ammonia Separator (U0802)

In QAFCO (Qatar Fertilizer Company) Ammonia-4, ammonia synthesis converter-1 developed a peculiar characteristic of one side cooling down at high load and winter conditions. One of the probable reasons was the gas inlet temperature being too close to the catalyst activation temperature. In the 2013 shutdown for inspection of internals, the catalyst was replaced. Since 2015, the cooling of the converter was caused by high inlet ammonia presence due to the choked ammonia separator. This paper will discuss the poor ammonia converter performance caused by choking of Ammonia Separator, U0802.

Neelakantan Rajasekaran
Qatar Fertilizer Company, Mesaieed Qatar

Introduction

Qatar Fertilizer Company (QAFCO) is the world's largest single site ammonia/urea fertilizer producer. The site is located in the Mesaieed Industrial Area of Qatar. Presently, the QAFCO complex comprises six completely integrated Ammonia/ Urea Plants. In the Ammonia-4 (A-4) unit, ammonia synthesis converter-1 (R0801) developed a peculiar characteristic of one side cooling down at high load and winter conditions. This paper will discuss the poor ammonia converter performance caused by choking of Ammonia Separator, U0802.

Description of the A-4 Process

Ammonia-4 (A-4) is a state of the art natural gas based UHDE plant with a design capacity of 2000 MTPD (2200 STPD), commissioned in 2004. Product CO_2 is delivered to either Urea-4 or the CO_2 network. Product ammonia is delivered to Urea-4 or to the refrigerated atmospheric storage. The synthesis section of A-4 is designed with two converters and a boiler in between. The 1st converter is a radial flow, two bed converter with inter bed heat exchanger, the 2nd converter a radial flow single bed converter. The A-4 plant has a proven capacity of 2400 MTPD (2650 STPD).

A drawing of the synthesis loop is shown in Figure 1.

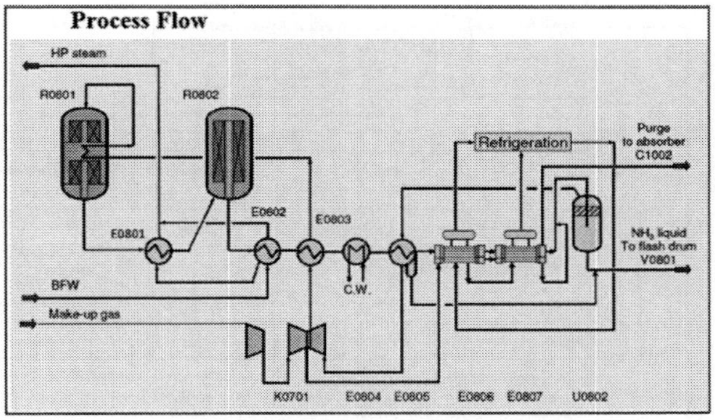

Figure 1. Schematic Process Flow

Concerns

In 2005, the ammonia synthesis converter-1 developed a peculiar characteristic of cooling down of one side during high load/high circulation rate condition. The cooling down generally happened for a few hours and then normalized back after reducing the plant load/circulation rate (across R0801) and adjusting the parameters to further heat up the converter. Over the period of 2005 till 2014, twenty five cooling cycles, mostly during winter, were experienced causing a loss of 36000 MT (39300 ST) of production. Since 2015, the cooling frequency and severity increased.

Since February 2018, it was observed that a phenomenon of very sudden and heavy carryover of liquid ammonia from the Ammonia Separator U0802 was occurring with U0802 level indications, including the trip level indications, reading reducing to zero. Indications became normal after a few minutes by itself. This phenomenon continued until May 2018 shutdown.

Actions from 2010 to 2013

1 - In April 2010, ammonia converter related data were collected for evaluation of catalyst life to determine if the catalyst was contributing to the cooling down. Based on a study, the catalyst activity was calculated to be 83% of the expected activity for a fresh Yara AS4F, 1.5 to 3 mm catalyst, or 92% of the expected activity for a catalyst being 6 years in operation (Figure 2).

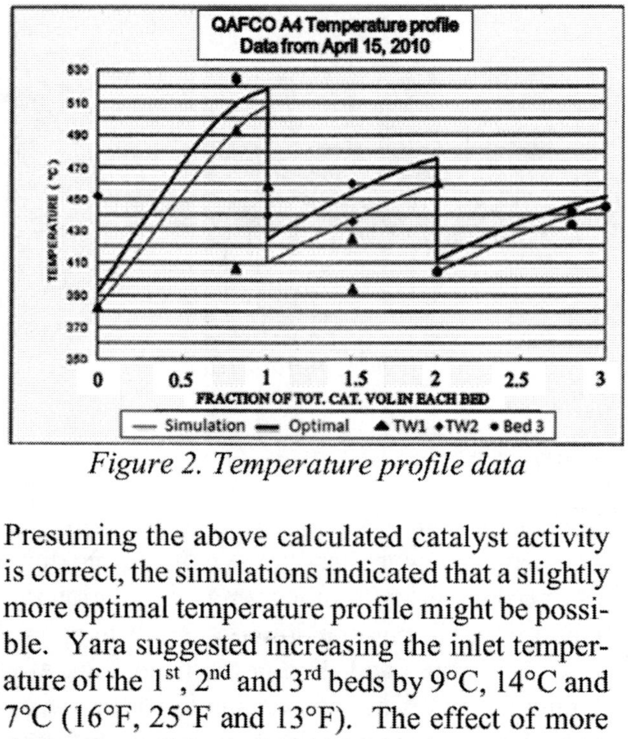

Figure 2. Temperature profile data

Presuming the above calculated catalyst activity is correct, the simulations indicated that a slightly more optimal temperature profile might be possible. Yara suggested increasing the inlet temperature of the 1st, 2nd and 3rd beds by 9°C, 14°C and 7°C (16°F, 25°F and 13°F). The effect of more optimal conditions should be in the order of 23 MTPD (25 STPD) production. The position of the thermocouple in the gas path is 81%, 50% and 80% from the bed inlet for bed# 1/2/3 respectively.

2 – A poor gas distribution to the 1st bed inlet and/or a inter bed heat exchanger leakage were considered as well. During planned shutdown in 2013, all three beds of catalyst in the converter were replaced by a Topsoe KMIR (1.5- 3.0 mm) type. Converter internals/inter bed heat exchanger were inspected and the following were the results:

- R0801 basket and inter bed heat exchanger was found normal. (Figure 3).
- A minor crack was observed in the expansion bellow of the start-up heater outlet line joining to R0801. As start up heater is offline during normal operation with inlet valve in closed condition and the location of the bellow is in the top dome above the catalyst basket (which is not in the main gas inlet flow path), the bellow damage could not have caused any cold gas bypass.
- The outer basket mesh was found slightly damaged.
- The bottom pipe sleeve was slightly buckled.

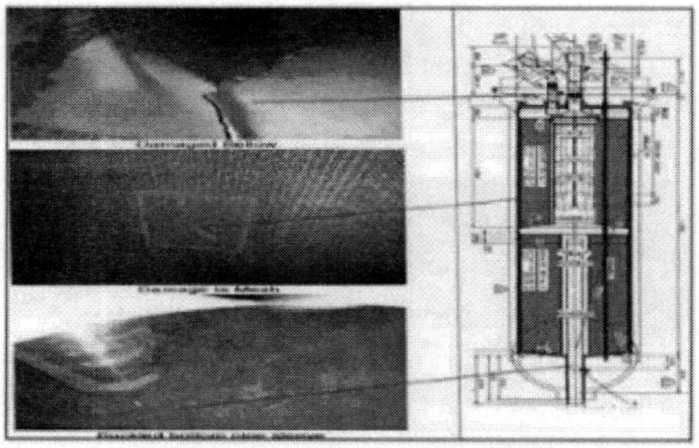
Figure 3. R0801 internals during shutdown 2017

None of the found damage could explain the peculiar cooling down of the converter.

The plant was re-started in November, 2013 and achieved maximum loads:

Dec 2013-	118 %
2014:	120 %
2015:	117 %

However the frequency of the cooling down phenomenon increased significantly after November, 2015 and restricted the load to 108% and below with time.

Topsoe site visit December-2015

Haldor Topsoe was contacted to look into the various parameters, and recommended adjustments were made. The cooling down phenomenon happened two times in their presence when a load increase was attempted, and both were visible as a temperature decrease around thermocouple nozzle N4A (see Figure 4).

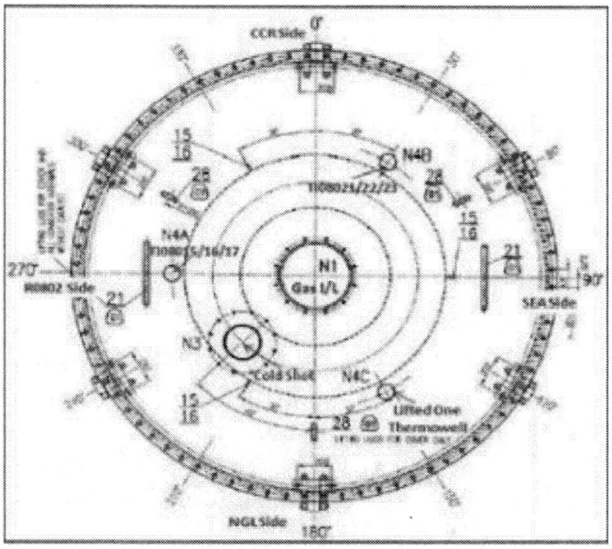

Figure 4. R0801 Top View

The converter is designed with three thermowells (A, B and C), placed at 120° from each other (Figure 4). In the thermowells various thermocouple elements of different lengths are inserted measuring temperatures inlet (one per well), inside 1st bed at 51 % of travel (two per well), inlet 2nd bed (one per well), and inside the 2nd bed at 81 % of travel (two per well). Of the three thermowells, only A and B are in permanent use and hooked up to the DCS. In well C, temporary thermocouple elements were inserted and hooked up to some of the shell temperature measurements in order to see them on DCS.

The 1st bed inlet temperatures appeared to indicate higher than expected. Several attempts were made to adjust the thermocouple height to establish the correct 1st bed inlet temperature but without success. Correct temperature monitoring for R0801 beds was a challenge.

Topsoe Conclusion

Topsoe indicated a flow distribution problem most likely due to misdistribution in the inter bed U-tube heat exchanger, caused by increased flow (above 110% of design). Such misdistribution could be created by a small asymmetrical defect in the inlet system or a tiny leakage in one compartment of the heat exchanger, evident by the fact that just one of the three sides always got colder than the other two. When increasing the load/flow, the colder side gets just cold enough to actually lose the reaction at that side.

UHDE and Topsoe Inputs

Following the visit of Haldor Topsoe, Uhde was contacted as well and recommended to stop the synthesis section and then perform a startup with the heater, which was taken on 13th Jan 2016. The problem however became worse with the plant load limited to 80%. A Root Cause Analysis (RCA) with involvement of QAFCO, Uhde and Haldor Topsoe was performed on 25th of January 2016, split into three categories as follows:

1. Equipment

The cooling down of converter-1 can happen due to passing of the cold shot and the startup heater valves. However, the passing of these valves was ruled out as the valves were in a closed position in the field and were checked for any abnormal sound.

Potential exchanger tube leakage, blockage by external material or catalyst dust, damage to baffles and misalignment of the exchanger were ruled out based on the inspection during the 2013 shutdown and other evidence.

The converter-1 inlet line expansion bellow leak was ruled out based on the inspection in 2013 shutdown as well as a boroscope in February, 2016, both indicating a good condition.

The converter-1 inlet temperature was running around 282°C (540°F), while the PFD value of 305°C (581°F) was never achieved. To increase this temperature, the Waste Heat Exchangers (E0801/2) heat recovery were reduced to operate the 3rd bed at slightly higher temperature resulting in higher temperature across gas/gas exchanger (E0803). By doing this, the R0801 inlet temperature only increased by 4 to 5°C (7 to 9°F). UHDE agreed to look into the general design problem.

2. Catalyst

Catalyst deactivation due to temporary or permanent poisoning was also ruled out due to the fact that it would affect both sides of the catalyst beds. In addition, a chance of poisoning is small as oxides from the Methanator were well below the limit. The catalyst was only 2.5 years old and has never been exposed to very high temperature. Hence, thermal sintering was ruled out as well.

3. Operation

The plant was being operated with a high H_2/N_2 ratio of 3.5 and above. An attempt was made to lower it down to a normal value below 3 but this was not successful as cooling down started to happen due to a different density causing a higher circulation rate/higher dP across the agglomerator U0802.

RCA Conclusion

A- It was observed that whenever the cooling down happened the concentration of ammonia in the syngas inlet to the converter also increases. The source of high ammonia in the inlet gas is either leakage from the cold exchanger (E0805) or the ammonia separator (U0802).

B- Level indications were rechecked and density corrections were confirmed. Therefore it could be concluded that the separator has either a problem in the agglomerator, vane pack or drain line.

Most Probable Root Causes

- Carryover of ammonia from U0802 due to malfunction of the internals.
- Design of converter-1 inter bed heat exchanger not allowing proper mixing of shell side gas causing one side to cool down.
- Excess heat recovery in the synthesis loop (E0802) causing too low gas temperature entering converter-1.

Shutdown February 2016

Based on the RCA recommendations an unplanned shutdown was used to inspect U0802 and the following observation/findings were made:

1. Severe deposits seen on the bottom bed limiting mesh up to total blockage (Figure 5). Rings were unloaded and a complete cleaning was done.

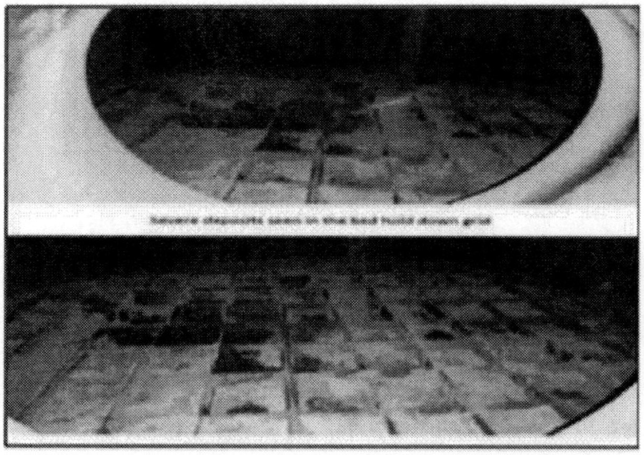

Figure 5. Deposits on bottom hold down grid

2. Visual inspection of the top perforated plate of agglomerator showed severe blockage (see Figure 6).

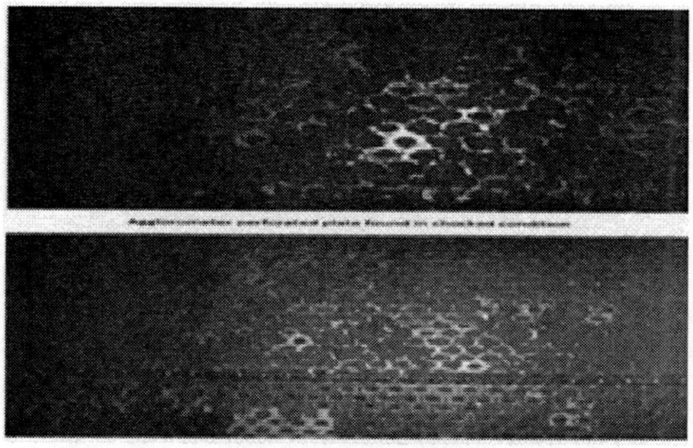

Figure 6. Agglomerator perforated plate choking

3. Inlet side of the agglomerator was found completely choked on the first few layers and the remaining part was found clean. The agglomerator was cleaned and re-used.

Figure 7. Agglomerator demister choking

4. Inlet line and diffuser were found with deposits and cleaned by flushing with water.

Figure 8. U0802 Inlet line deposits

Figure 9 provides a reference between where the pluggage locations were relative to the vessel drawing.

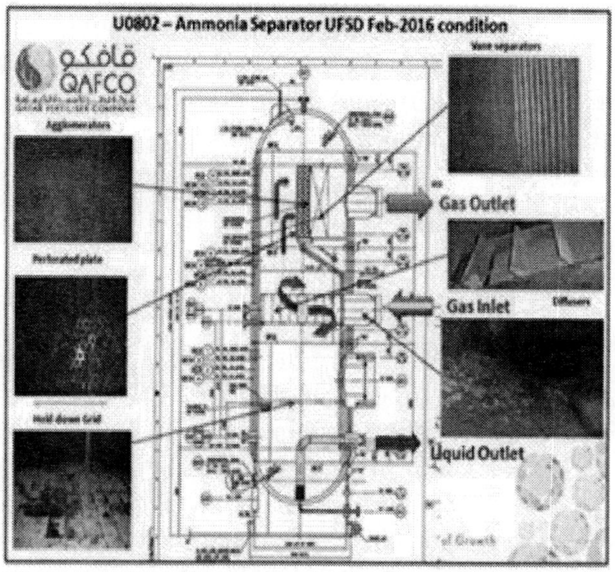

Figure 9. U0802 Internal condition during shutdown 2016

The following maintenance was carried out:

- U0802 inspection and maintenance.
- U0802 Pall rings unloaded, washed and loaded back.
- U0802 all 4 level transmitters calibrated.
- Many other locations in the loop were inspected/checked for deposits and condition but no deposits or catalyst dust were found.

These observations supported the RCA analysis of high differential pressure due to blockage, finally causing siphoning of ammonia liquid through the 3 inch down comer inside the separator.

U0802 Deposit Sample Analysis

Collected deposits were analyzed in laboratory for composition leading to the following general conclusions:

1. The presence of ammonium carbonate was proved by IC analysis, by ammonium analysis, and by behavior to heat of deposits when they lose significant weight after heated over 60°C (140°F).

2. The presence of aluminum and potassium elements, expected if there would be a migration of catalyst from ammonia converter, was not confirmed.
3. Major element found was iron, most likely from equipment body corrosion (rust).

CO/ CO_2 Analysis

Based on the finding of the choked agglomerator in U0802 due to carbonate formation, it was decided to perform a regular analysis of CO/CO_2 Prism separator feed inlet and outlet product H_2 as well as gas exit Methanator. Most of the time, less than 1 ppm (CO/CO_2) was found for Prism feed inlet and outlet product H_2. Methanator exit gas analysis showed some slip of CO_2 in the range of 1 to 10 ppm, while online analyzer indication was regularly much lower than the lab analysis. For supporting calculations, please see the appendix at the end of this paper. A potential minor leak in the Methanator exit gas/gas exchanger (E0601) was suspected.

Major shutdown November 2016

Following up on the severe blockage of the bottom hold down grid, agglomerator and internal piping during the shutdown February 2016, it was decided to again inspect the U0802 internals during major shutdown in November 2016. The inspection found much lower amounts of deposits. (Figure 10)

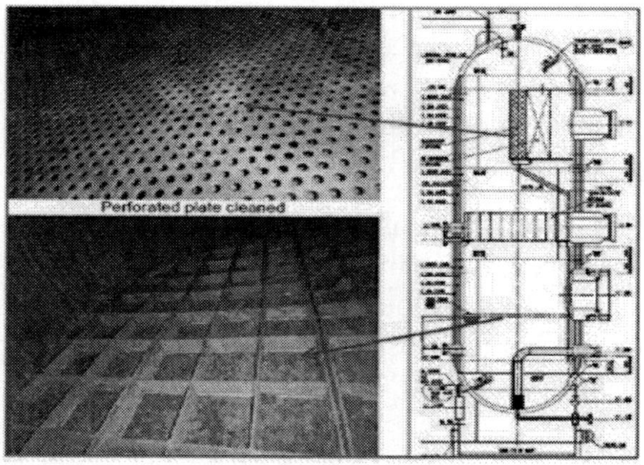

Figure 10. U0802 Internal condition during November 2016 shutdown

In addition, an inspection was performed on gas/gas exchanger (E0601 shown in Figure 11) including a helium test.

- Helium leak test indicated leak from the bundle gasket face, gasket was replaced.
- Shell and tube bundle high pressure cleaning executed.
- Many baffles were found bent, and some cracks were seen and repaired on the baffle welds.
- A new the leak test done at 1 barg pressure indicating no leak.

After restart the plant achieved a load of 110%.

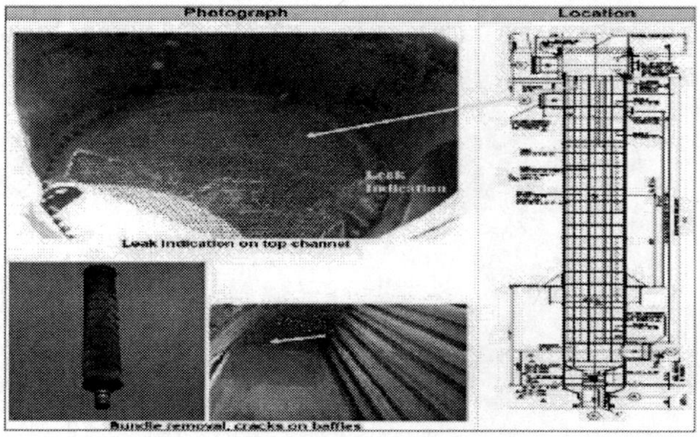

Figure 11. E0601 Internals and tube bundle

Status beginning 2017

Until January 2017 plant production load had to be reduced again to 92 % and was slowly coming down further due to converter-1 cooling down phenomenon due to liquid ammonia carry over from separator U0802.

Whenever the differential pressure across the agglomerator exceeded the calculated design value of 27 kPa (3.9 psi), the evidence of liquid carry over could be observed. Finally it was decided to shutdown and to attend the above problem on 31st of March 2017.

Removal of carbonate by hot gas flushing

By heating to around 60°C (140°F) or above (considering equipment design) solid Ammonium carbonate decomposes to its constituents $2NH_3$ and CO_2. Decomposition will be enhanced by raising the temperature while keeping the equipment design temperature in mind. Decomposition will also be favored by minimizing the NH_3 content in loop.

After executing of a HAZOP study, heating up of the synthesis loop cold part was done with warm synthesis gas through: E0602- K0701- bypass HV07005- E0806- E0807- U0802 and then vented through HV08014 (Figure 12). Maximum gas temperature achieved at U0802 outlet was 38.2°C (101°F).

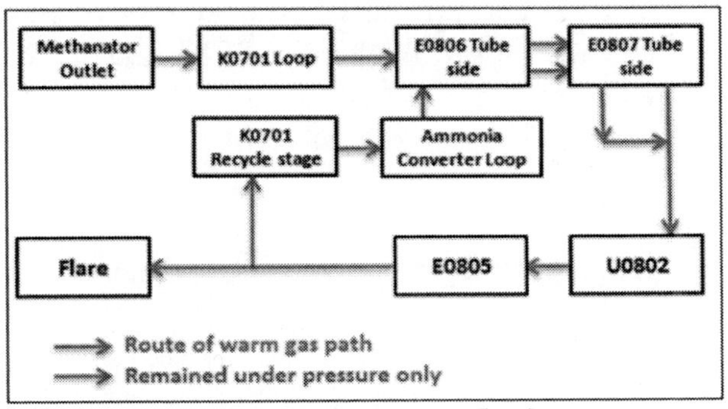

Figure 12. Syn loop hot gas flushing route

Activity during shutdown March 2017

U0802 inlet line, hold down grid, packing and demister were found blocked with relatively less deposits (compared to shutdown January 2016), presumably due to hot gas flushing. Lab analysis of the deposits showed that the majority of the samples were iron oxides, most likely due to the warm synthesis gas flushing performed before the stop.

Additional inspections were performed in U0802 and adjacent synthesis loop equipment/piping indicating:

1. Bottom bed hold down grid was found with deposits. (Figure 13).

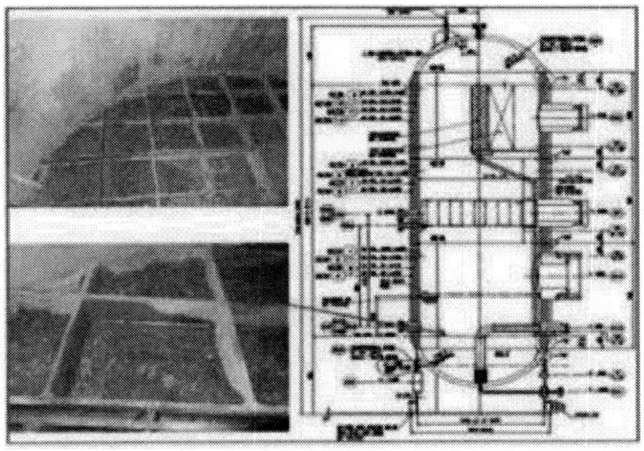

Figure 13. U0802 Internals condition March 2017

2. Gas inlet diffuser vanes were found with deposits, U0802 shell bottom had 3-4 mm thick deposits.
3. E0805 shell side and U0802 outlet lines were found clean. Pipe from E0806 to E0807 no deposits found
4. Boroscopic inspection of U0802 gas inlet pipe indicated up to 5 mm thick deposits (see Figure 14).

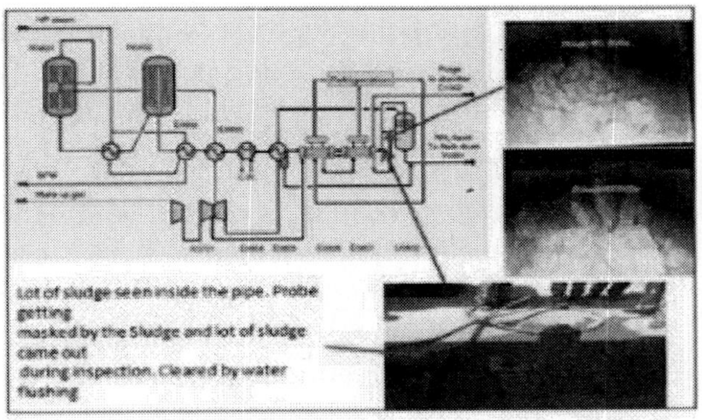

Figure 14. U0802 gas inlet pipe line

Figure 15 shows U0802 internal details where maintenance was performed. The vessel and associated lines were cleaned from deposits. The U0802 demister was replaced with a new agglomerator.

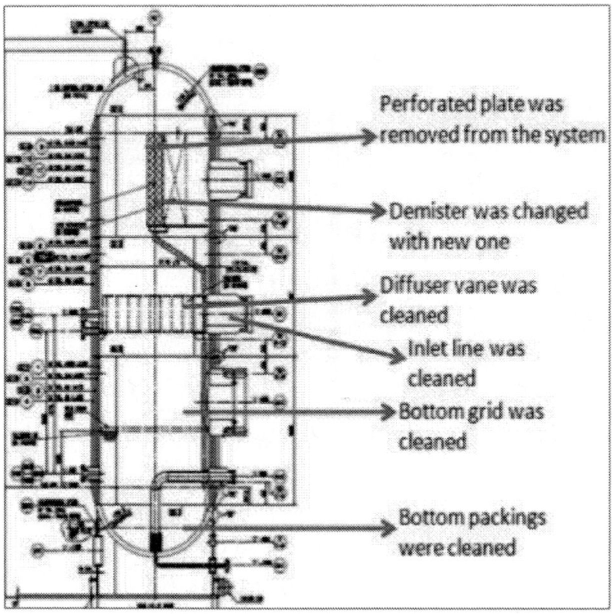

Figure 15. U0802 Internal details where maintenance was performed

After completion of maintenance activities, the plant was started and production commenced on 06/04/2017.

New RCA Conclusion

1. The majority of the deposits were found in U0802 inlet line after mixing point, inlet diffuser, packing and vessel wall till the demister, but no deposit visible in gas outlet line.
2. There were no signs of catalyst dust within the loop, neither promoter elements (K, Al, Ca) were detected in the deposits and there were no blockages of control valves / lines. Thus a damage of the converters (R0801/02) is not expected;
3. Carbon oxides can ingress to the loop if there is leak in the E0601 or during start up as excursion or by poor performance/slip of methanator catalyst.

Actual Status 2018

1. Converter-1 cooling phenomenon was still dominantly during winter at cooling water temperature around 22 – 24°C (72 - 75°F).
2. Since January 2018, the liquid carry over from U0802 was very frequent and severe due to expected choking of the agglomerator resulting in 1st converter bed#1 cooling.
3. As a consequence the plant load had to be reduced to 100% and U0802 level was maintained low to minimize the likelihood of carryover. In addition an ammonia carryover from E0805 to U0802 through tube side exit case was fostered in order to enable the dissolution of ammonium carbonate in the produced ammonia.
4. The poor performance forced a new shut down May- 2018. Similar findings of a choked U0802 were found and rectified.
5. Finally the methanator catalyst was replaced. Figures 16 and 17 show Methanator exit gas analysis data (lab analysis / online analyzer) from November 2017 to June 1, 2018.

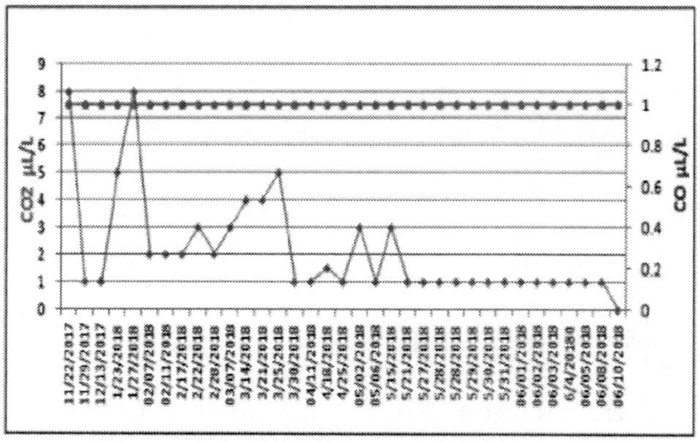

Figure 16. CO/CO_2 exit Methanator (Lab analysis)

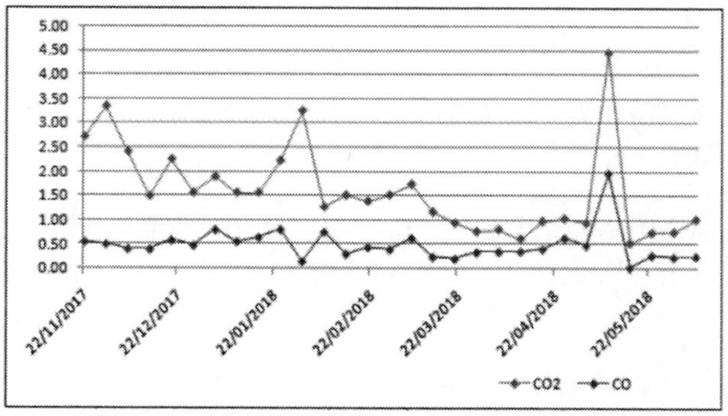

Figure 17. CO/CO_2 exit Methanator (Online Analyzer)

Discussion and Conclusion

Based on shutdown observations and deposit analysis, there is sufficient evidence that ammonium carbonate formation had occurred. Even though lab/online analyzers indicated constantly a very low methanator slip for CO at <1 ppm and for CO_2 at <5 ppm, as indicated in Figure 16 and 17, still the formation is experienced. Contamination from other sources at the inlet to the synthesis gas compressor was verified and ruled out by positive isolation. This leads to the actions and conclusions as follows:

1. Replacement of Methanator catalyst as done during shutdown May 2018.
2. Lab/online analysis accuracy shall be improved for low level of CO/CO_2.
3. R0801 inlet gas temperature enhancement study shall be continued.

Appendix

Formation of Ammonium Carbamate in Synthesis Loop of Ammonia Plant

The make-up purified synthesis gas containing H_2 and N_2 in the ratio of 3:1 and maximum of 10 ppm ($CO+CO_2$) is compressed to syn loop pressure. This gas is cooled in the loop by 1st chiller (E0806) and 2nd chiller (E0807), and is then mixed with recycle gas, which contains a considerable quantity of condensed ammonia, and is cooled to -1°C (30°F).

Theoretical Analysis of carbamate Formation:

$2\ NH_3 + CO_2 = NH_4O\text{-}CO\text{-}NH_2$
$K_P = P^2_{NH3} \times P_{CO2}$
Where
K_P = Equilibrium constant
P_{NH3} = Partial pressure of ammonia
P_{CO2} = Partial pressure of CO_2

Temp °C	0	10	20
K_P (atm)	6.9×10^{-7}	8.4×10^{-4}	7.9×10^{-5}
Temp °C	40	60	80
K_P (atm)	4.91×10^{-3}	1.86×10^{-1}	4.64
Temp °C	100	120	140
K_P (atm)	8.23×10	1.09×10^{3}	1.12×10^{4}

The above values might not be accurate at high pressure and/or low temperature as ammonia and CO_2 will not then behave ideally. Again below 10°C (50°F), the reaction rate approaching equilibrium from either side of reaction is too slow to permit an accurate vapor pressure to be measured. However these Kp values can be taken as guide.

At the synthesis loop condition: 210 kg/cm^2 (2990 psi), 4.0% NH_3, and 0°C (32°F), the calculated equilibrium CO_2 concentration is 5.0×10^{-5} ppm. Thus carbamate can be formed and will be stable at concentrations of 5.0×10^{-5} ppm CO_2 or above.

The solubility of carbamate in liquid NH_3 is rather small. Even at 60°C (140°F), the solubility is not more than 0.1%. It is expected to be much lower at 0 - 6°C (32 – 42.8°F). This means that less than 9.5 m3/hr (340 ft3/hr) corresponding to <100 ppm CO_2 can be dissolved as carbamate. Higher content would circulate as solid carbamate and can deposit anywhere in loop.

Assuming that CO_2 concentration at the Methanator exit is 10 ppm, then CO_2 carryover (as carbamate) will be as low as 0.0002 ppm at the inlet to U0802. This will be favorable for carbamate formation where make up gas mixes with circulating gas inlet to U0802. Carbamate solidification can occur in separator or its outlet when its concentration is 0.1% or more.

Recommendation from this analysis is to minimize the CO_2 slip from methanator, as CO_2 is not easily methanated as easily CO, and avoid carryover and minimize the NH_3 content in circulating gas exit the separator. Also, hot gas circulation in syn loop is helpful in removal of deposited carbamate. The temperature is limited by the design of equipment.

References

1. Article published in Technical Journal as: UREA: Its properties and Manufacture by George Tsemi Yuchao (Pg 315)

Experience with Second Ammonia Converter (R-0502) Bypass at Ammonia Plant of Kaltim-1A

The R-0502 of Kaltim-1A plant experienced 2 major problems in the last 18 years. The first problem was an internal failure of the catalyst basket in 2003. The second problem was a crack in the bottom head in 2017. During both cases, a bypass line was installed to keep plant in operation at 90% production rate. This paper provides a description of each of the problems encountered. This paper also provides a description of the operational differences that were experienced when the plant operated without the R-0502.

Achmad Rois
PT Pupuk Kalimantan Timur

Jaka Kirwanto
PT Pupuk Kalimantan Timur

Introduction

Kaltim-1A started first commercial production in February, 2000 and has been in operation for over 17 years. During this period, the Second Ammonia Converter (R-0502) experienced two different failures. First occurrence was an internal catalyst basket failure in 2003, and the second one was a bottom head failure in 2017. During both failures, the unit was restarted with the R-0502 bypassed. This paper describes each of the failures experienced and the operating conditions that occur when operating the R-0502 bypassed.

Background

PT Pupuk Kalimantan Timur (PT Pupuk Kaltim) operates 5 ammonia plants which are designated Kaltim-1A, Kaltim-2, Kaltim-3, Kaltim-4 and Kaltim-5. The total capacity of ammonia is 2,765,000 MTPY (3,047,000 STPY) and urea is 3,430,000 MTPY (3,780,000 STPY). PT Pupuk Kaltim also operates 2 NPK plants and other supporting facilities such as coal boilers, steam turbine generators and gas turbine generators.

Kaltim-1A plant is located in the Kaltim Industrial area and is integrated with an adjacent plant. Kaltim-1A is licensed by HaldorTopsoe. The plant capacity is 2000 MTPD (2200 STPD) and was commissioned in 2000. The plant initially represented the Kaltim Pacific Ammonia (KPA) plant with the Build Operate Transfer scheme. Since the transfer of assets in 2014, the KPA ammonia plant was operated by PT Pupuk Kaltim. The plant has an S-200 for the 1st ammonia converter and an S-50 for the R-0502.

Figure 1 demonstrates the original flow path through R-0502. The flow enters nozzle A traveling up from the bottom along the shell, through the catalyst bed, down the center collector, and exits at nozzle B.

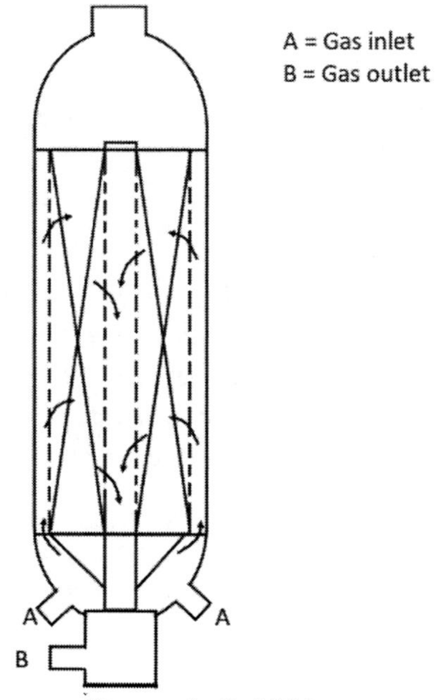

Figure 1. R-0502

R-0502 Technical Data Sheet

Diameter	: 3000 mm (9.8 ft)
Operating pressure	: 135 bar (1957 psi)
Operating Temp.	: 376–420°C (708–788°F)
Design Pressure	: 152 bar (2204psi)
Design Temperature	: 450°C (842°F)
Catalyst Volume	: 137.6 m^3(4900 ft^3)
Catalyst type	: KM1
Material of vessel	: SA-542 Type B Cl 4
Shell thickness	: 119 mm (4.685 inch)

Bottom head thickness: 85 mm (3.346 inch)
Material of center screen: A 240 TP 321
Material of outer screen: A240 TP 321

First Problem: Internal failure

From initial start up to the end of 2001, the synthesis loop performance was normal with the synthesis pressure at a normal 1977 psi. Starting early 2002, the synthesis loop pressure had to be increased to maintain maximum production. The pressure was increased gradually. In July 2003, the synthesis loop pressure reached 2062 psi. Finally in September 2003, the plant could not be operated due to a sudden increase of the pressure and had to be shut down for repair (see Figure 2).

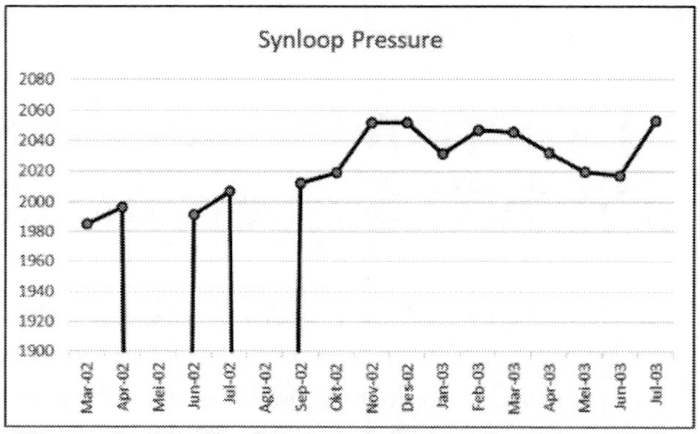

Figure2. Gradual increase of synloop pressure.

The increasing synthesis loop pressure was suspected to be derived from a problem with the internals of the R-0502. Due to the long duration of repair, it was decided to bypass the R-0502. There are three ways to bypass the R-0502:

1. Install a new line outside of R-0502.
2. Open the cap of the top center screen so the catalyst is bypassed.
3. Empty all of the catalyst inside the R-0502.

Installing a new bypass line outside of the R-0502 was chosen as the best option. The line was a connection from the inlet to the outlet of the R-0502 (see Figure 3). Due to limitation of piping material, the new bypass line used existing inlet piping material.

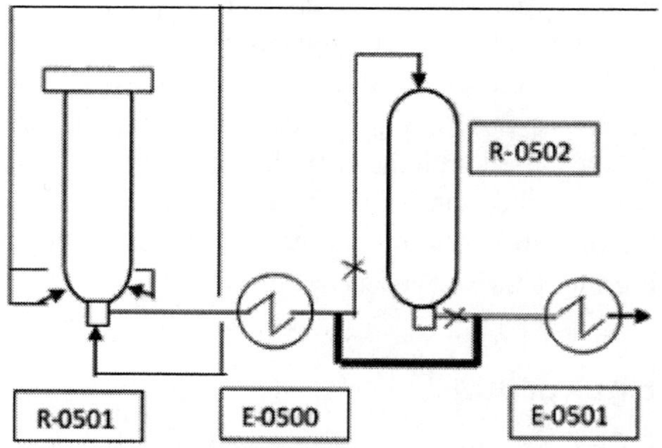

Figure 3. Bypass line arrangement

The advantage of the bypass method was the R-0502 could be repaired without interfering with operation of the plant. To continue with repairs, cooling for the R-0502was done with a portable system to make it safe for worker (see Figure 4).

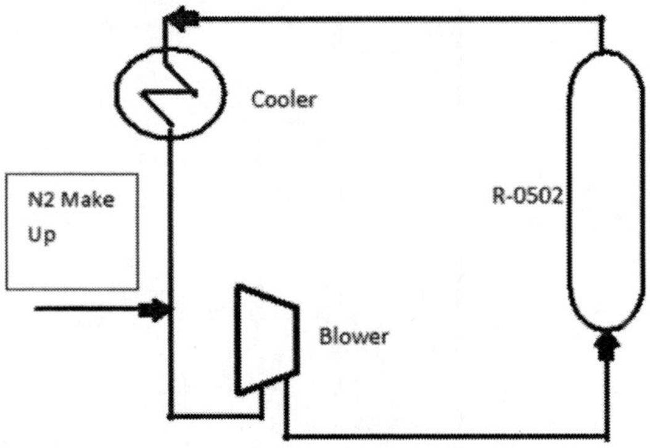

Figure 4. Schematic diagram of Cooling down of the R-0502

Results from R-0502 Inspection

Results of inspection of the R-0502 were as follows:

1. Vessel and bottom head - There were no defects found in the vessel and bottom head using visual inspection and photographs.
2. Center Screen - Center screen had a minor bulge and twist in 6 joints. Major deformation was found at 4500 mm (14.7 ft) from bottom (see Figures 5 and 6).

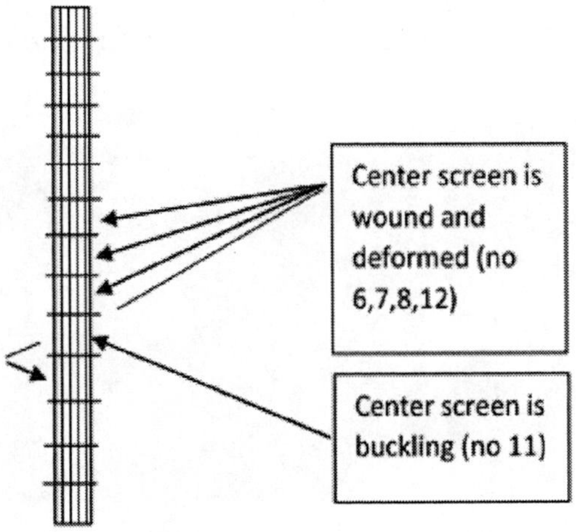

Figure 5. Center screen damage location

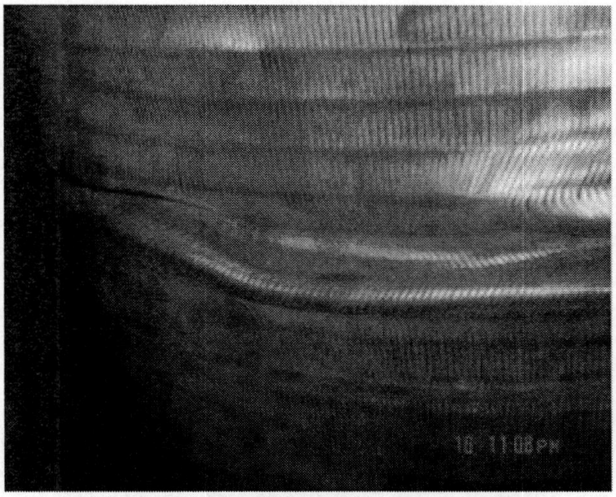

Figure 6. Buckling of center screen

3. Outer Screen Panel–The R-0502 is equipped with 36 outer screen panels. 12 screen panels were filled with catalyst and some of them were damaged (see Figure 7).

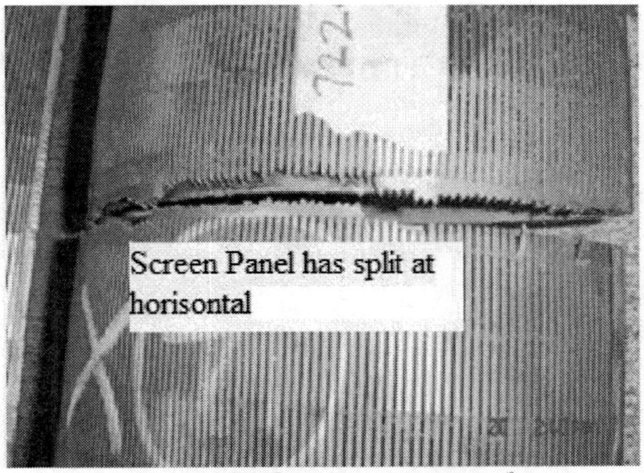

Figure 7. Outer screen panel

4. Catalyst - Catalyst settled approximately 2500 mm (8.2 ft) from the top of the scallop. This catalyst volume is approximately 16 m^3(565 ft^3), and was found at the bottom portion of the outer screen panels, inlet line and cone.

The catalyst level dropped down due to the damage of the outer screen panels. The damage allowed the catalyst to pass to the bottom of the R-0502 (see Figure 8). The damaged outer screen panels were repaired and the center screen was replaced with a new one.

Figure 8. Catalyst found in inlet of the R-0502

Discussion from first problem

The root cause of the catalyst basket failure was leakage across the tubesheet of E-0500 (see Figure 3 for orientation). Steam and liquid water entering the R-0502 caused high pressure drop across the outer screen panels. Since the outer screen panels are fixed, the mechanical stress from the high pressure drop caused a horizontal split of the outer basket.

The catalyst was damaged both from the liquid and high flow across the outer basket failure location. Catalyst dust and particulate accumulating against the center collector resulted in a high pressure drop and subsequent failure of the center collector.

Revamping of The R-0502

During the repair of the R-0502 in September 2003, the flow direction was changed from a bottom stream inlet to a top stream inlet. The purpose of changing the direction of flow was to reduce the effects of increasing pressure when the outer screen panel was damaged, like what happened in the September 2003 incident. Figures 9, 10 and 11 show the modifications to the flow direction.

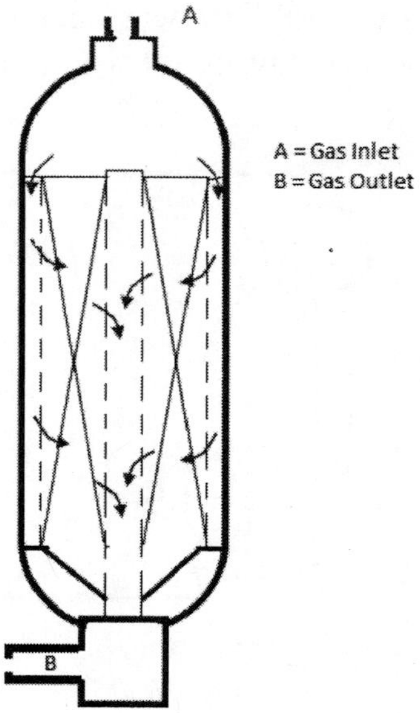

Figure 9. Revamping of the R-0502

Figure 10. Original piping arrangement

Figure 11. Revamping of the R-0502

Completing the repair, the R-0502 was put back online in June 2004 and continued to operate normally. Thirteen years later, the R-0502 operated without any constraints until finally in November 2017 a crack occurred in the bottom head.

Second Problem: Bottom Head Failure

After operating for 17 years since the first commercial production, or 13 years since the internal failure, there was leakage in the bottom head of the R-0502. The leak was found at 02:00, November 27th, 2017 and it was identified from a strange noise below the converter. Since the leak was suspected to be very explosive synthesis gas, the area was checked using an infrared (FLIR) camera and also an explosimeter. Figure 12 shows the infrared camera reading. The temperature at the location of leakage was 260^{0}C (500^{0}F) and the explosive was 100%.

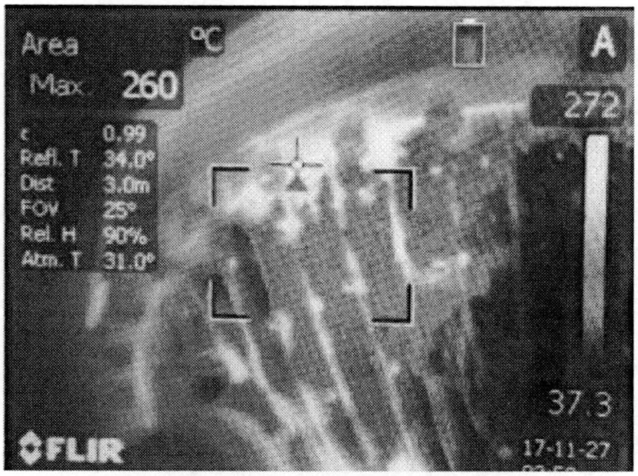

Figure 12. Infrared camera reading

It was too dangerous to keep the plant running, so the plant was shut down in a controlled manner to allow cooling of the synthesis loop. Cooling down of the synthesis loop was done by circulation of syngas for 16 hours, and the lowest temperature achieved was 90^{0}C (190^{0}F). Continued circulation of syngas for a longer time did not result in a lower temperature.

Inspection of the bottom head using die penetrant test at low temperature found a 700 mm crack (see Figure 13). Based on the drawing, the crack was at the weld support of the outer screen panel. Due to the expected long duration of repair, the R-0502 was bypassed for a second time. While doing the bypass installation, an ultrasonic test (UT) was performed and found the crack to be 2000 mm (6.56 ft) around the head of the vessel (see Figures 13, 14, 15, 16).

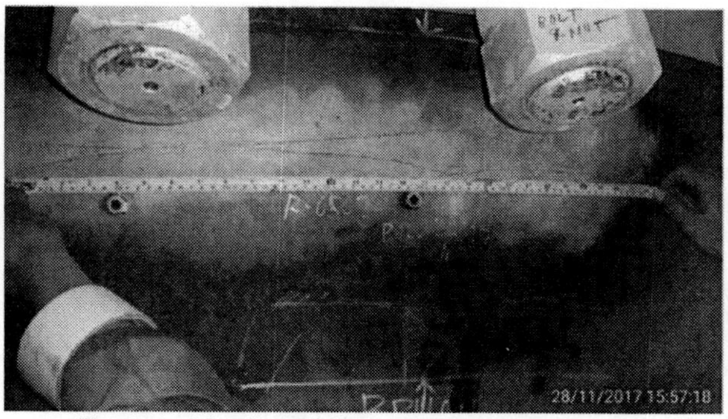

Figure 13. Crack length at bottom head

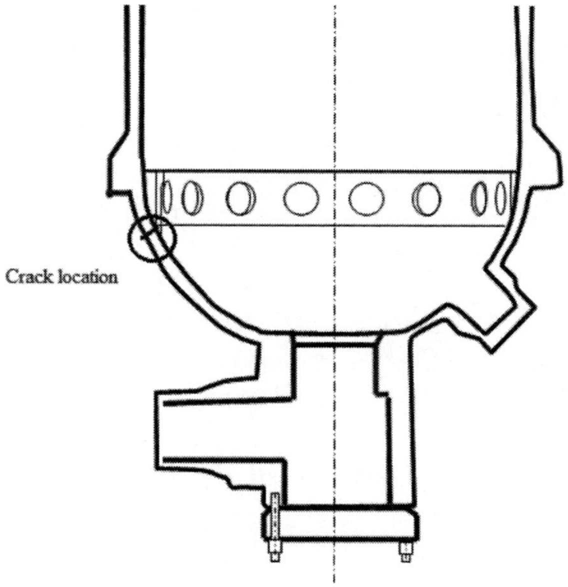

Figure 14. Crack Location

Figure 15. Crack location and area of repair.

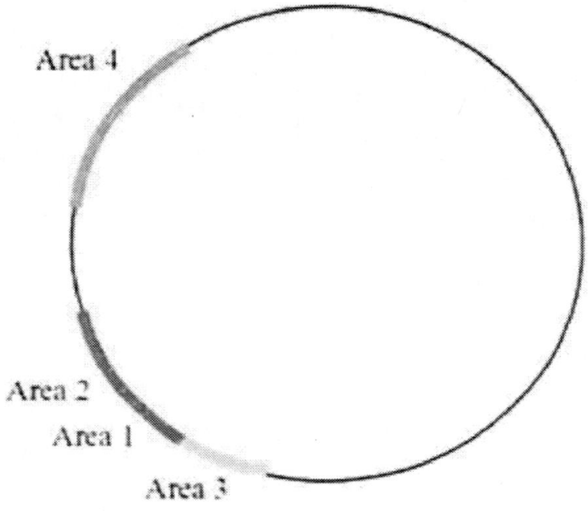

Legend :

Area 1: Crack from inside to outside = 600 mm (1.96 ft)

Area 2: Crack from inside but not through outside = 600 mm (1.96 ft)

Area 3: Crack from inside but not through outside = 500 mm (1.64 ft)

Area 4: Crack from inside but not through outside = 500 mm (1.64 ft)

Figure 16. Area of repair

Repair of the R-0502

Based on the weldability test, repair of the crack could be done from the outside because the test found that the crack didn't spread in an irregular direction. Some other considerations for outside repair: it was a temporary solution while waiting for replacement of the equipment; shortening duration of repair and saving of catalyst cost.

Discussion from second problem

The possible causes of the crack are:

1. Since the material of converter is SA-542 Type B Cl 4 (2 ¼ Cr-1 Mo enhanced), it has high strength but is easier to crack. The crack can also be driven from the vessel design itself. The bottom head holds the weight of catalyst.

2. Modification of inlet gas from bottom to the top of vessel has made the bottom head experience stress from the piping of gas outlet.

For the next plan, the entire vessel will be replaced with a new one (not only the head). The new vessel material will be changed to SA-542-Type B (2 ¼ Cr-1 Mo) that has lower strength but less prone to crack. The vessel design will also change in order to avoid excessive stress at the bottom head.

Operation Without R-0502

Operating conditions change with the R-0502 in service versus when bypassed. Table 1 below shows the operating conditions before the ammonia converter was offline compared to after it was offline and bypassed. An explanation of the main differences follows.

Parameter	Before offline	After offline	Units
Front End rate	100	90	%
Production	2028 2235	1821 2007	MTPD STPD
Primary Reformer outlet temperature	1427	1423	^{0}F
Methane slipPrimary Reformer	12.85	14.72	%
Methane slipSecondary reformer	0.64	0.43	%
Secondary Reformer outlet temperature	1749	1767	^{0}F
Steam production	339 373	314 346	MTPH STPH
E-0500 outlet gas side temperature	707	757	^{0}F
Syngas Compressor Speed	10,187	9,878	rpm
Circulated syngas flow	834,000 31.1	812,000 30.3	Nm^3/h mmscfh
Ammonia concentration outlet R-0502	20.39	17.07	%
Purge Gas Flow	26,195 0.979	28,300 1.06	Nm3/h mmscfh
Synloop Pressure	1927.26	1993.83	psi
Hydrogen production from HRU	15,082 0.564	18,656 0.697	Nm3/h mmscfh

Table 1. Conditions before and after the R-0502 was offline

Front end rate : Maximum front end process gas rate was only 90% due to limitation of high pressure steam production in ammonia plant. There was not enough high pressure steam for the syngas and refrigeration compressor turbines. Supplemental high pressure steam is not available from adjacent units.

Secondary Reformer : Secondary Reformer was operated at a slightly higher outlet temperature compared to before the R-0502 was offline because of more hydrogen from Hydrogen Recovery Unit (HRU). Higher temperature increased front end steam production, which helped offset decreasing steam production from the back end due to decreasing heat of reaction in the R-0502.

High Pressure Steam Production: High pressure steam production was about 346 ST/hr, only 92% of normal production rate.

Syngas Circulation: Synthesis loop circulation was 97% of normal circulation rate even though the front end process gas rate was 90%. The higher circulation rate compared to front end process gas was chosen because less hydrogen and nitrogen is converted to form ammonia per pass. Before the R-0502 was offline, ammonia product from the converter was 20.39% and after it was offline it decreased to 17.07%.

Synloop Pressure: Synthesis loop pressure had to be increased from 1927 psi to 1994 psi to maintain conversion and ammonia production. The increased pressure compensated for lower conversion of hydrogen and nitrogen.

Purge Gas Flow: Purge gas flow was increased to 108% to reduce the synthesis loop inert level. Even though this was much higher than normal, the Ammonia Recovery Unit and the HRU could still handle the purge gas flow without any obstacles.

Hydrogen Production from HRU: Hydrogen production from the HRU was 123% of normal production rate due to the increase of purge gas feed.

Conclusions

1. In more than 17 years in operation, there have been two problems with the R-0502. The first problem was the internal failure in 2003. The second problem was the bottom head failure in 2017.
2. The R-0502 has material SA 542 type B Class 4, (2 ¼ Cr - 1 Mo enhanced). The material has high a strength advantage so that the vessel weight can be lower, but this material has a deficiency in that it is easier to be crack.
3. During repair of both problems, the bypass line was installed to keep the plant running with 90% rate of process gas.
4. There were some differences observed in operating conditions with and without R-0502, i.e.:Secondary reformer outlet temperature, high pressure steam production, synthesis gas circulation in the synthesis loop, pressure of

synthesis loop, purge gas flow and ARU/HRU rate.

5. Repair of the bottom head from outside was a temporary solution to shorten repair duration while waiting for the new equipment to replace.
6. The new R-0502 will use material SA-542 type B (2 ¼ Cr-1 Mo) for construction. The strength of material will be lower and the weight will be higher than before.

Acknowledgement:

We would like to express our gratitude to all those who gave us the possibility to complete this paper. We want to thank:

1. PT. Pupuk Kaltim Board of Directors, who give us the opportunity to attend the conference.
2. AICHE, who invited us to attend the "63rd Annual Safety in Ammonia Plants and Related Facilities Symposium"

CFCL-2 Ammonia Plant Operating Experience of a KBR True Cold Wall Add-on Converter

Chambal Fertilizers and Chemicals Limited (CFCL) currently operate two ammonia plants in Kota; Rajasthan, India and a third plant is in construction. The new plant is based on KBR purifier technology. To increase the plant capacity and improve energy efficiency, the existing plants have considered installation of add-on ammonia converters for increasing per pass conversion. The ammonia concentration from the add-on converter is in the range of 20 – 23 mole %. At CFCL-2 the new reactor was successfully commissioned during a short turn around in April 2017 and has achieved all capacity and energy targets. This paper presents the analytical design, installation and operating experience at CFCL-2.

Inderjeet Singh
CFCL, Kota, Rajasthan, India

Mahesh Gandhi
KBR Technology, Houston, Texas, USA

Introduction

Chambal Fertilizer & Chemicals Limited (CFCL) is located in Kota, Rajasthan, India. This fertilizer complex has two ammonia plants and a third plant is in execution. This new plant is based on KBR Purifier® Technology with expected start in 1st quarter of 2019.

The first ammonia plant has a design capacity of 1,500 MTPD (1650 STPD) and is based on Haldor Topsoe technology. KBR revamped this plant in April 2009 and increased the capacity to 1,900 MTPD (2,090 STPD). Major features of the revamp were installation of KBR Reforming Exchanger System (KRES), a new secondary reformer and a new add-on converter.

The second ammonia plant, CFCL-2, was originally commissioned in mid-1999 with a design capacity of 1,500 MTPD (1,650 STPD). This plant is based on KBR's conventional technology. CFCL-2 was also revamped by KBR in April, 2009 with a revamped capacity of 1,780 MTPD (1,960 STPD).

CFCL decided to enhance the capacity further by carrying out a retrofit in 2017. To limit shutdown time for capacity expansion, CFCL opted to add a new cold wall add-on converter downstream of the existing converter. KBR's true cold wall converter was commissioned at CFCL-2 in April – May 2017.

Add-on converters are more often considered for revamping the existing ammonia plants to boost per pass ammonia conversion to increase plant capacity and also to enhance process energy efficiency.

When feed gas to the add-on converter is hot, technology providers face challenges in using it with cold wall pressure vessels. Typically, hot wall add-on converters have been used in the industry for such service which is known to be unreliable due to cracks experienced in the weld seams, particularly on the hotter end of the converter due to hydrogen-induced cracking and nitriding in the presence of hydrogen and ammonia. All components in the synthesis loop are in some way affected by the combination of nitriding and high temperature hydrogen attack

(HTHA). [2] Both mono wall and multi wall vessel designs have experienced such failures. There are numerous examples of premature failures of the hot wall converters worldwide. [1] [2], [3]

This is unlike the cold wall main ammonia converter shells, which face no issues even after decades of being in service. Keeping in view this fact and utilizing KBR's classic "Slim-Jim" (a tall and skinny vessel, which was the early version of a standard 1,000 STPD ammonia plant converter) features, true cold wall design based add-on converter has been incorporated in CFCL ammonia plant-1 (2016) and plant-2 (2017). These plants were successfully commissioned after installing the add-on converter during a short turnaround.

The effluent gas from the main ammonia converter is passed through a loop boiler used for generating HP steam. Then the gas is sent to a new single-bed add-on ammonia converter, where its ammonia concentration is increased further by about 3 mol%. Refer to Table 2 for add-on converter in out composition. The bed inlet temperature is controlled using the loop boiler bypass at a temperature in the range of 350°C to 380°C (662°F to 716°F). There is no feed/effluent exchanger in these converters. To keep the converter shell cool using the cold-wall concept, a sweep gas arrangement is provided. Cold syngas from the synthesis loop is passed through the converter shell annulus space to keep the shell cool. In this case, the cold syngas is taken downstream of feed/effluent exchangers in both the CFCL plants. The sweep gas temperature is in range of 65°C to 75°C (149°F to 167°F).

CFCL-2 Add-on Converter

The CFCL-2 add-on converter was commissioned in May 2017. The synthesis loop of CFCL-2 is designed for 1,940 MTPD (2,140 STPD), but all the capacity upgrade features have not been implemented. So, the current capacity is close to 1,800 MTPD (1,980 STPD). Table 1 below shows the operating parameters of CFCL-2 Add-on Converter, 1105-D.

Parameter (July 8th, 2017)	1105-D (CFCL-2)
Inlet Pressure (Kg/cm^2 G) / psig	160.4 / 2282
Outlet Pressure (Kg/cm^2 G) / psig	160.3 / 2280
Pressure Drop (Kg/cm^2) / psi	0.14 / 2
Inlet Temperature (°C) / (°F)	350 / 662
Outlet Temperature (°C) / (°F)	398 / 748
Sweep Gas Inlet Flow Rate (T/Hr) / ST/Hr	92.6 / 102.0
Sweep Gas Inlet Temp (°C) / (°F)	75 / 167
Sweep Gas Outlet Temp (°C) / (°F)	81 / 178
Gas-Gas Exchanger Outlet Temp (°C) / (°F)	75 / 167

Table 1 - 1105-D Operating parameters

Increase in ammonia concentration is 3 mole % as indicated in the Table 2.

Component	H_2	N_2	CH_4	Ar	NH_3
1105-D Inlet	53.47	18.93	8.28	2.58	16.74
1105-D Outlet	50.64	18.22	8.71	2.71	19.72

Table 2 - 1105-D In / Out Molar Composition

Sweep Gas

The temperature rise in the sweep gas is approximately 6.0°C (10.8°F). Nearly 30 % of the total circulation is the sweep gas flow. This flow keeps the converter shell cool.

Converter Shell Temperature

The converter shell is at a uniform temperature of 75°C (167°F). This shows that sweep gas flow is sufficient and the arrangement is efficient for keeping the shell cool. The pressure drop in the annulus space for sweep gas flow is 0.1 kg/cm^2 (1.4 psi).

KBR's True Cold Wall Add-on Ammonia Converter Concept

The synthesis gas entering the add-on converter is at a high temperature 350°C to 430°C (662°F to 806°F). However, it gives significant advantage from an equipment design perspective if the wall of the add-on converter is not exposed to this high temperature gas and the pressure shell temperatures are maintained low. It is well known that the cold walled main ammonia converters operate reliably without any issues even after decades of being in service.

Cold wall design for the add-on ammonia converter is achieved by passing a relatively cold "sweep gas" through the annulus between the shell and the basket. A slip-stream of cool converter effluent gas taken from downstream of the existing feed-effluent exchanger is used for the sweep gas, similar to KBR's classic "Slim-Jim" quench converter circuit design. The sweep gas, having a temperature of approximately 70°C (158°F), remains isolated from the main process gas and is returned to the existing circuit upstream of the cooling water exchanger.

A sketch depicting the cold wall design of the add-on converter is shown in Figure 1. The process gas enters the equipment through a nozzle at the top and directly enters the catalyst filled basket without making contact with the pressure shell of the converter. The basket in turn is insulated on the outside thus preventing any heat transfer between the high temperature process gas inside the basket and the sweep gas that flows through the annular space between the basket and the pressure shell - thus keeping the pressure shell at a lower temperature.

Although the sweep gas temperature which is in contact with the pressure shell is low, KBR has selected a low alloy steel material for the pressure shell. Low alloy steel provides a conservative design considering the exposure to high pressure synthesis gas containing hydrogen.

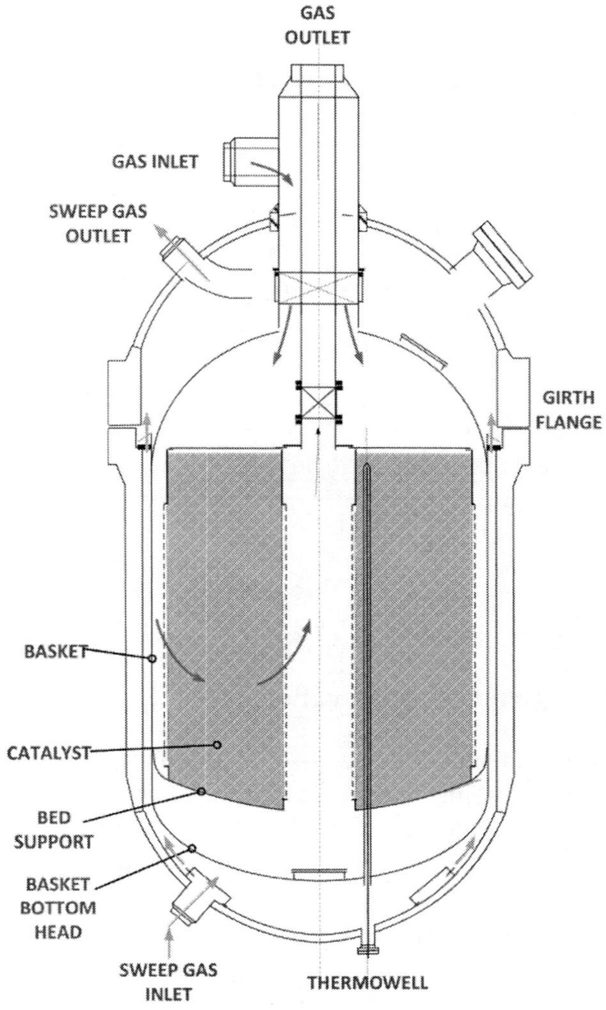

Figure 1 - KBR's Proprietary True Cold Wall Configuration for Add-on Ammonia Converter

The Flow Scheme

Figure 2 shows the process flow scheme. The compressor discharge is pre-heated in the Feed / Effluent Exchanger, 121-C before the synthesis gas is sent to the Horizontal Converter, 105-D. The main converter (105-D) effluent is cooled in the new HP Steam Generator 1123-C. The gas exit the steam generator is sent to the Add-on Converter, 1105-D.

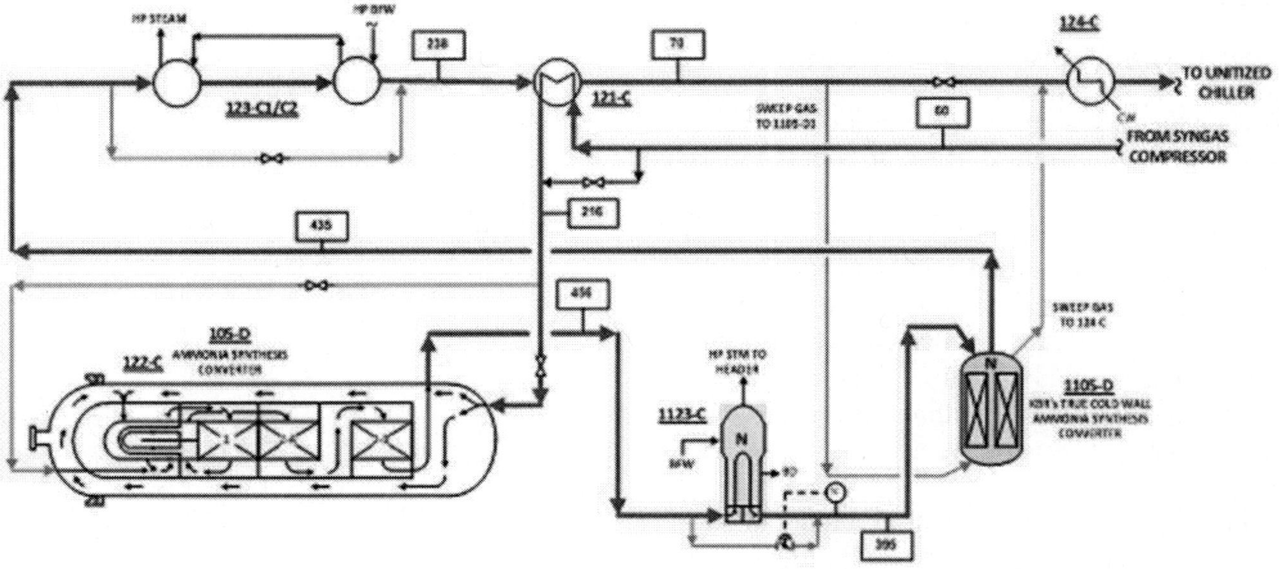

Figure 2: Process Flow Diagram of CFCL-2 Synthesis Loop

Equipment Configuration

The add-on converter in CFCL-2 is configured vertically and is supported on a skirt resting on a concrete foundation. It is installed in the vicinity of the existing ammonia converter.

Pressure Shell: Add-on Ammonia Converter

The external pressure shell is provided with a full diameter girth flange which enables a removable top head assembly that provides full access for lowering an insulated catalyst basket into the pressure shell. The basket rests securely on a ledge which is integral with the shell girth flange thus enabling free expansion of the basket shell in the downward direction. This top hung basket arrangement is a unique feature of KBR's designed add-on ammonia converters.

The girth flange of the pressure shell is sealed by a highly reliable double conical gasket which is a special gasket for high pressure equipment such as the ammonia converter.

Being high pressure equipment, KBR has chosen ASME Section VIII Division 2 code for achieving optimum wall thickness for the pressure components.

Figure 3: Fabricated Pressure Shell after Hydrotest in the Fabrication Shop [4]

The gas inlet and outlet connections between the basket and the converter are located on the top dished head, and these pipes are provided with expansion bellows. This is to take care of the differential thermal expansion between the basket and the pressure shell.

A manway located on the top head allows access to the top side of the basket for catalyst loading/unloading and other maintenance activities. The pressure shell is made of 1.25 Cr – 0.5 Mo low alloy steel material.

Internal Catalyst Basket

The basket is made of stainless steel grade 304 with 3" thick insulation on the outside. The catalyst is contained by outer and inner screens made from state of the art Profile V wire (see Figure 4) construction which is robust and reliable even when subjected to the start-up and shutdown cycles. These screens also have less friction with the catalyst particles thus allowing longer life for the catalyst and lower mechanical loads on the screens.

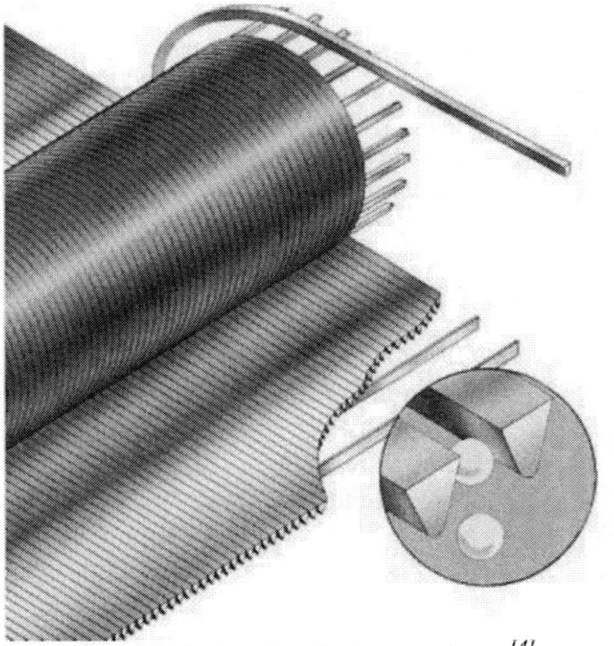

Figure 4: Profile V wire products [4]

Figure 5: Stainless Steel Basket with External Insulation Encased in SS Sheath Being Made Ready for Dispatch from Fabrication Shop

As shown in Figure 5, the basket is supplied in a specially designed and fabricated transportation container to protect it during dispatch to the site and during erection prior to insertion in the pressure shell in the field.

Field Installation and Catalyst Loading

In the field, the pressure shell is first erected on the foundation, and the top head assembly is disassembled. Thereafter, the transportation container with the basket inside is made vertical next to the pressure shell. The basket is then extracted vertically from the container and inserted into the pressure shell (Figure 6).

Figure 6: Converter Basket Insertion [5]

Once the basket is lowered, the top head is reassembled, and the internal piping connections, including the expansion bellows, between the basket and the process nozzles are welded. The basket is now ready for catalyst loading.

Catalyst loading is performed through the manhole in the top head assembly and through another manhole in the basket head right below the outer manhole. After completion of catalyst loading, the basket manhole is seal welded and dye penetrant examined prior to boxing up of the outer manhole on the top dished head. This ensures that the process gas circuit is isolated from the sweep gas circuit. At CFCL-2, this activity was completed in April 2017 and the plant was commissioned between April and May 2017.

Once the plant is commissioned, after installation and catalyst loading, the sweep gas, at about 70°C (158°F), flows through the annulus between basket and shell. The sweep gas keeps the pressure shell at a very low temperature. The catalyst basket which is insulated on the outside retains the heat generated through the exothermic catalytic reaction. The sweep gas stream and the process gas stream are isolated from one another thus achieving true cold wall design.

True pressure vessel cold wall technology provides unmatched flexibility in operations. The catalyst bed can be operated at optimum hotter temperatures to maximize conversion, as temperature of the pressure vessel remains totally independent of the synthesis gas being processed through catalyst bed. Feed and product synthesis gas can have hotter temperatures as required over life cycle of catalyst to maximize conversion.

Conclusions

1) Hot wall ammonia converters have had a history of failures. Many articles have been published which records these failures.[1] [2], [3] All components in the synthesis loop are affected by the combination of nitriding and HTHA.[2] The most critical common denominator is the presence of hydrogen-induced micro-cracks in the weld.[3]

2) The cold walled ammonia converter shell, which has shell cooling flow, has had no issues even after decades of being in service. Prime examples of this are KBR's classic "Slim-Jim" converters which are still in operation after over forty years of service.

3) KBR's true cold wall converters implemented at CFCL 2 plant has shown successful process performance. The sweep gas is about 30 % of the total circulation. The converter shell remains at around 75°C (167°F).

Acknowledgement

The author would like to thank CFCL process team Inderjeet Singh and Ajay Tayal for their help with this article.

References

(1) S. G. Choudhary; M. S. Prasad and A. K. Singh: "Window Repair of site joint circumferential weld seam crack of Ammonia Converter-II (S-50)", AIChE Ammonia Safety Symposium, Henderson, US, 2007

(2) Bernhard Geis and Hans-Jűrgen Bassler: "Experience with 2.25 Cr-1Mo Steel Equipment in the Ammonia Synthesis Loop after 30 Years in Service", AIChE Ammonia Safety Symposium, Vancouver, Canada, 2014

(3) J. D. Campbell and G. R. Prescott: "Weld Cracks in an Ammonia Converter", AIChE Ammonia Safety Symposium, Charleston, US, 1998

(4) Johnson Screens® (A Weatherford Company) catalog

(5) R. Subramanian: "KBR's True Cols Wall Add-on Ammonia Converter", Nitrogen + Syngas 2018, Sweden

Promoting Safe Operation of Steam Reformers

Safe and reliable operation of the ammonia plant is key to any plant owner. A well-functioning steam reformer can greatly impact the on-stream performance because it is one of the most complex unit operations in the plant. Carbon laydown, which is often experienced in top-fired furnace designs, will lead to higher tube wall temperatures and, in the extreme, to premature tube ruptures.

To avoid these issues, a potassium-promoted catalyst is typically installed in the top part of the steam reformer; however, potassium also reduces the activity of the catalyst. Topsoe's new alkali-promoted steam reforming catalyst, RK-400, utilizes a unique potassium promotion technology to obtain the perfect balance between carbon resistance and activity.

Mads Feddersen, Todd Allward, Nicolai Petersen, Charlotte Vinding Ovesen
Haldor Topsoe A/S, Denmark

Introduction

The steam reformer is one of the most complex units in a synthesis gas producing plant, which is evident not only from an equipment design perspective but most certainly also from an operating perspective. Ensuring a safe and reliable operation is one of the main challenges that producers face during their daily work.

The catalyst used in the reformer is key in this aspect, as it has a huge impact on the integrity and lifetime of the catalyst tubes. If excessive carbon formation occurs, operation of tubes beyond design temperatures will result. Long term operation in these conditions creates the risk of premature tube ruptures and uncontrolled plant shutdowns.

Unlike many of the adiabatic reactors in the plant, it is difficult to have a complete picture of the state of the catalyst in the different zones of the catalyst tubes. Visual inspection of reformer tubes can sometimes allow detection of large temperature differences (see Figure 1); however, temperature differences between tubes and along tube length are unfortunately not always so easily detected.

Figure 1. Hot zones on reformer tubes in a top-fired furnace.

The old-fashioned way of monitoring the performance of the reforming catalyst is by measuring tube wall temperatures (TWT) manually with a pyrometer. The pyrometer method is not reliable because it is encumbered with some uncertainties, and the results depend very much on the people taking the readings and the frequency of the measurements. Online digital monitoring solutions, such as Topsoe Furnace Manager, are a much more efficient and effective way of monitoring the TWT.

When the catalyst has reached end-of-run, the replacement of the catalyst is a time-consuming activity. Even though loading methods have improved, it still normally requires 10–14 days of interrupted production to do a reload. Any unplanned reloads or skims will significantly influence plant economics.

It is thus in the interest of operators to use a catalyst system that is reliable and robust to be able to handle the operating conditions in the reformer for many years. Topsoe has developed a new way of designing such catalysts, which will be presented below together with industrial data from a couple of installations.

Causes of carbon formation in top-fired reformers

When looking at the reasons for unplanned shutdowns in ammonia plants, comparative surveys of ammonia plant operation taking place in the time period from 2002–2014 show that failure related to the primary reformer is one of the top three contributors to lost production days [1].

In top-fired furnaces, carbon laydown in the catalyst tubes continues to be an operational challenge, and ways to prevent and/or monitor this phenomenon are high on most operators' agendas. Why then is carbon laydown of special concern in top-fired reformers, and what exactly is carbon laydown/formation?

Carbon formation or carbon laydown is the result of higher hydrocarbons being thermally cracked inside the top part of the reformer tubes. Top-fired reformers are more susceptible to carbon formation where the temperatures are higher in the top of the tubes compared to other reformer designs.

At ideal conditions carbon formation will not occur, as the catalyst used in the top of the tubes will have sufficient activity to reform the higher hydrocarbons before the temperature in the tube reaches the critical level where carbon is formed. The majority of all plants with top-fired reformers use potassium-promoted catalysts which help suppress the carbon formation. Unfortunately, in the real world things are not ideal, and carbon formation does occur.

Aging catalyst

As the reforming catalyst ages, it will lose activity due to sintering of the active nickel crystals on the catalyst surface. The sintering phenomenon occurs over time, and the degree of sintering depends on the operating conditions as well as the composition and microstructure of the catalyst.

As the catalyst activity decreases, the conversion of higher hydrocarbons (as well as methane) slows, which allows the higher hydrocarbons to penetrate further down into the tubes. Furthermore, the temperature in the tubes also increases, as less heat is removed by the endothermic reaction (see Figure 2). Combined, these conditions result in higher risk of carbon formation, and the risk will continue increasing as the catalyst ages.

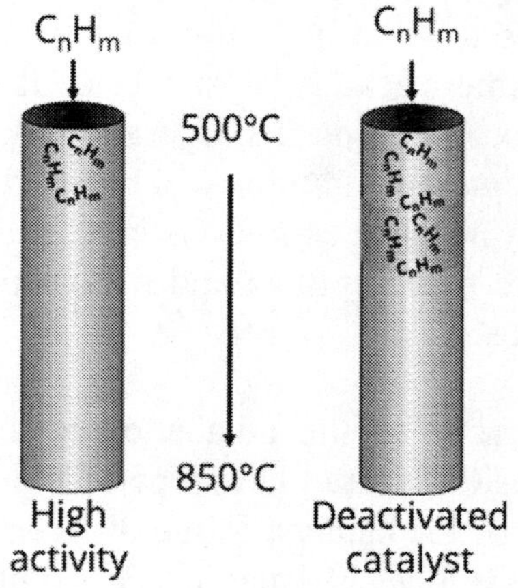

Figure 2. Conversion of higher hydrocarbons in the top of the reformer tubes.

In addition to losing activity, some catalyst types are also prone to lose the carbon-inhibiting potassium promoter as they age. This phenomenon will be described in detail later in this paper.

Sulfur poisoning

Low levels of sulfur (in the order of parts per billion) exiting the desulfurization section will slowly be adsorbed by the nickel-based reforming catalyst, reducing the catalyst activity slowly. A small and steady ingress is normally not that big of an issue and high activity catalysts may operate for years under such conditions. Sudden breakthrough of sulfur, however, will have a more dramatic effect and can result in significant catalyst deactivation, and ultimately lead to carbon laydown.

Uneven firing

Uneven firing can lead to significant temperature differences between the individual tubes, which can also be a contributing factor when it comes to carbon formation. Even a small difference in the tube temperature can be the "final straw", leading to the onset of carbon laydown. Uneven firing is often part of the reason that neighboring tubes do not always share the same fate, i.e. one tube experiences carbon formation while the neighboring tubes may not.

Uneven catalyst loading

The loading of all types of tubular reformers is a critical operation that will have an impact on the operation of the reformer and on the lifetime of the catalyst charge. Differences in the achieved loading density will affect the pressure drop across the individual tubes, and any difference in the pressure drop will affect the flow through the individual tubes. If the firing is not adjusted, a given tube receiving less flow will become hotter, thereby increasing the risk of carbon formation. As also described in the section about uneven firing, this may also explain why carbon formation is sometimes only observed in some and not all tubes.

Fluctuating feedstock composition

As carbon formation in top-fired reformers is typically caused by thermal cracking of higher hydrocarbons, it comes as no surprise that the amount and types of higher hydrocarbons in feedstock will influence the risk of carbon formation.

At start-of-run for a reforming catalyst charge, there will be a certain safety margin allowing some feedstock fluctuations with no negative consequences. However, as the catalyst ages, higher hydrocarbons will penetrate further down the tube into increasingly high temperatures. For a given feed composition and catalyst temperature there exists a critical steam/higher hydrocarbon ratio (S/HHC) (see Figure 3), and operation at a S/HHC ratio lower than the critical ratio can result in carbon laydown. Therefore, if the composition of the feedstock fluctuates, especially with regards to the content and/or types of higher hydrocarbons, an otherwise well-functioning reformer might suddenly experience problems with carbon laydown.

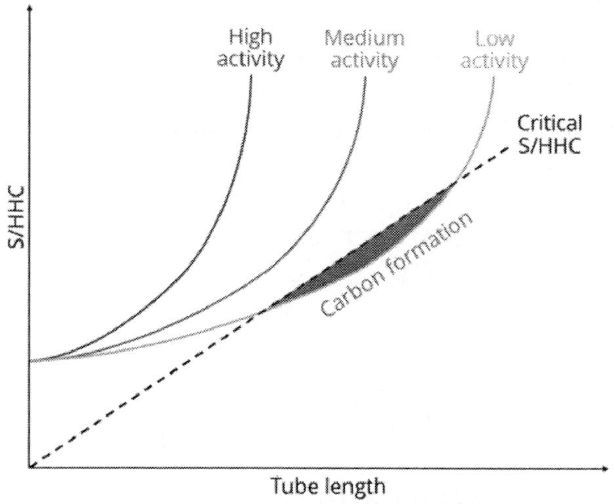

Figure 3. Carbon formation as a function of S/HHC and catalyst activity.

Consequences of carbon formation

Carbon formation in a reformer can have severe operational, safety, and economic consequences. Depending on the root cause (as discussed in the previous section), carbon formation can occur slowly over the course of many weeks or even months, or it can happen in a matter of minutes. Obviously, the latter is more difficult to handle, and the severity of the consequences is generally proportional to the rate of carbon formation.

Decreased tube lifetime and increased risk of tube rupture

From a safety point of view, the occurrence of carbon formation is highly undesirable, as the risk of operating outside the design limits of the tubes will be significantly increased. Tube wall temperatures above design will lead to decreased lifetime of the affected tubes, and if the proper precautions are not taken it may, in the worst case, lead to tube rupture.

Increased inspection frequency

When the first signs of carbon formation have been observed, most plants would increase the frequency of the visual inspection of the reformer. While increasing inspection frequency is highly recommended (and will serve to reduce the overall risks and consequences), it also increases the risk level to plant personnel, as these workers spend more time in the physical vicinity of the reformer, opening inspection windows, shooting temperatures, etc. These risks can of course be reduced if Topsoe Furnace Manager is installed in the furnace chamber [2].

Shutdown of the plant

In severe cases, or in less severe cases that are allowed to progress over a period of time, it will be necessary to shut down the plant both for safety reasons as described above and to minimize the necessary repairs described below.

Mitigation

When the first signs of carbon formation are observed (e.g. increased tube wall temperatures, hot spots, giraffe necking or hot-banding), it will often be necessary to modify the operating conditions (increase S/C ratio and reduce outlet temperature) or to decrease the plant load to avoid acceleration of the carbon formation and to protect the tubes.

Regeneration, catalyst replacement and tube cleaning

In less severe cases, a so-called "regeneration" procedure (involving steaming and controlled oxidation/combustion of the carbon in the tubes) may help to remove the carbon deposits and any sulfur if sulfur poisoning has taken place. However, in some cases such a procedure will not be sufficient to fully restore the reformer performance to the level before the carbon formation occurred, and replacement of the catalyst may be necessary.

Prior to the loading of the new catalyst charge, the affected reformer tubes should be thoroughly inspected to determine if the integrity of any tubes have been compromised from operating at a higher temperature. Furthermore, the inside of the tubes should be cleaned for any remaining carbon deposits, otherwise the tube temperatures will still be elevated after start-up of the new catalyst charge. Depending on the severity of the carbon deposits and the number of affected tubes, the cleaning of the tubes may delay the loading of the new catalyst charge, thereby increasing the days of lost production.

Solutions for avoiding carbon formation

As discussed previously, carbon formation is of special concern in top-fired reformers, but it can happen in any design. It is therefore advantageous to avoid carbon formation and ensure early detection if it does happen.

Increased focus on the desulfurization section

One of the main causes of severe carbon laydown is sulfur. One of the best ways to minimize sulfur poisoning is to ensure a well-functioning desulfurization section at all times. In some plants, it may be advisable to increase the frequency of the regular sulfur measurements carried out both as a final check of the desulfurization section and to determine the remaining lifetime of the installed catalyst. In other plants, it may be advisable to reconsider the replacement strategy for lead-lag configurations of sulfur absorbents, as keeping the catalyst in the lag reactor for too many replacements of the lead reactor may result in increased risk of sulfur break-through.

Installation of a prereformer

A revamp of the plant including the installation of a prereformer can significantly reduce the risk of carbon formation in the reformer, and the use of a top layer of alkali-promoted catalyst would no longer be required. A prereformer effectively pre-converts the higher hydrocarbons (and some of the methane) in the feed and acts as a final sulfur guard, thereby eliminating some of the most serious contributing factors for carbon formation. Sulfur poisoning is easily visible in the temperature profile of a prereformer, which will enable early detection of any issues in the desulfurization section; furthermore, prereformer catalyst replacement activities are faster and simpler than the methods that are used for catalyst replacement in tubular reformers.

Increased monitoring of the reforming section

Increased monitoring of the reforming section alone will not prevent carbon formation. However, it will allow for earlier detection of the problem, which potentially could reduce the severity of the carbon formation. Early detection results in earlier action to identify and address the root cause. Furthermore, increased monitoring after detection of carbon formation will reduce the risk of exceeding the design temperature of the tubes, and will help to identify the tubes that have been subjected to the most severe conditions for close monitoring and inspection.

Traditional monitoring would consist of tube wall temperature (TWT) measurements taken with the use of a pyrometer. More advanced continuous online monitoring of tubes, refractory and burners, such as the Topsoe Furnace Manager (TFM), is now available. The TFM allows 24/7 measurements using proprietary technology based on heat sensitive cameras [2].

If traditional monitoring with pyrometers is used, it is very important to choose the optimal type of pyrometer and to ensure proper calibration of the pyrometer. In addition, it is important that the measured TWTs are corrected based on the actual conditions in the reformer, as reflected heat, back-scatter, and other sources of interference could lead to incorrect measurements [3].

Catalyst formulation

Potassium-promoted catalysts have long been used to reduce carbon formation, especially in top-fired reformers. Today, the vast majority (>85%) of top-fired reformers in ammonia, methanol and hydrogen industries employ this type of catalyst in the top part of the tubes. The rest of the plants operate either on very lean natural gases, or have a prereformer upstream, so that unpromoted catalyst can be used in the entire tube.

The catalyst carrier system will also influence the tendency of carbon formation, which increases with the acidity of the carrier system (see Figure 4).

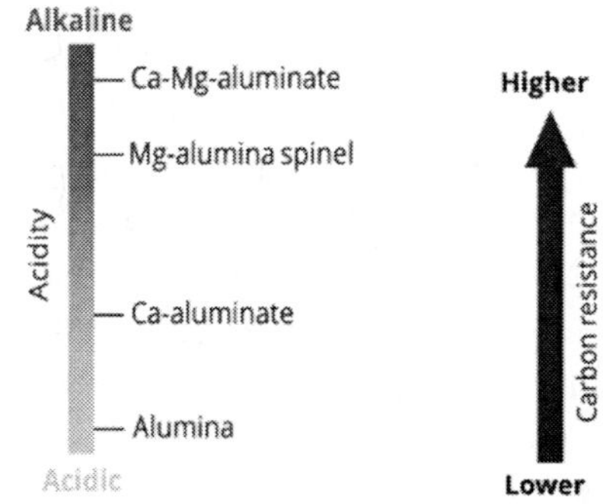

Figure 4. Different reformer catalyst carriers and their resistance to form carbon.

New generation of reforming catalysts

While the old generation of alkali-promoted reforming catalysts has been successfully used for many decades, the use of these traditional alkali-promoted catalysts is not without problems. Over time, some alkali will slowly be stripped from these catalysts, resulting in lower carbon resistance and potentially causing alkali fouling of downstream equipment. However, in recent years, Topsoe has introduced a new generation of alkali-promoted catalysts, which minimizes these issues, promising more reliable and efficient protection from carbon formation.

Theory behind potassium promotion

The development of the Topsoe RK-400 catalyst is based on a detailed understanding of the mechanism of carbon formation and the role potassium has in mitigating it. Addition of alkali metal prevents carbon formation in the following two ways: it promotes gasification reactions, and by adsorption on step sites it inhibits nucleation of carbon islands, which leads to carbon formation.

Mechanism of carbon formation and role of potassium

The formation of three types of carbon is possible under steam reforming: whisker carbon, encapsulating carbon (gum), and pyrolytic carbon (see Figure 5). Conditions in tubular steam reformers favor whisker carbon and pyrolytic carbon formations.

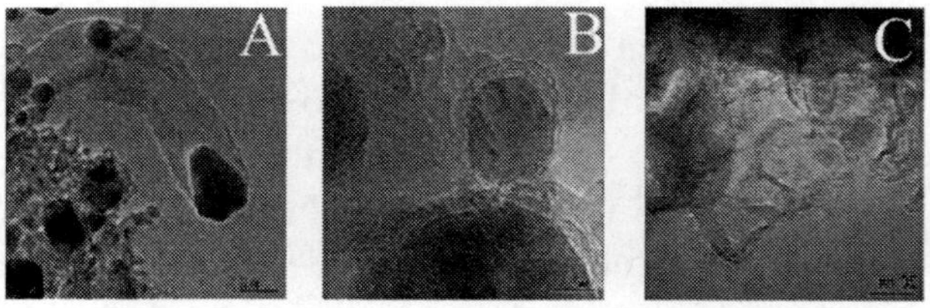

Figure 5. Electron microscopy images of whisker carbon (A), encapsulating carbon (B), and pyrolytic carbon on an $MgAl_2O_4$ carrier (C) on $Ni/MgAl_2O_4$ reforming catalysts [4].

The formation of whisker carbon (see Figure 5A) is a risk in feeds with a low steam-to-carbon (S/C) ratio, in feeds with a high carbon/hydrogen (C/H) ratio, and in CO-rich feeds. The insight into the mechanism for formation of whisker carbon has emerged from a combination of in-situ electron microscopy studies and density functional calculations [4,5,6]. In the formation of the whisker, the step sites on the Ni-particle play a crucial role. Adsorbed atomic carbon is much more stable at steps than at terrace sites. The step sites act as nucleation sites for the formation of graphene islands. The carbon atoms are bound more strongly in a graphene layer, creating a driving force for graphene growth from the step sites. The whisker grows from the step site and by diffusion of carbon from the Ni-particle surface to the graphene layer as illustrated in Figure 6. A certain size of graphene island is needed for whisker growth and a small island will disappear and not lead to whisker carbon. As the whisker grows, it will eventually fill out the pores of the catalyst tablet and the tablet will break.

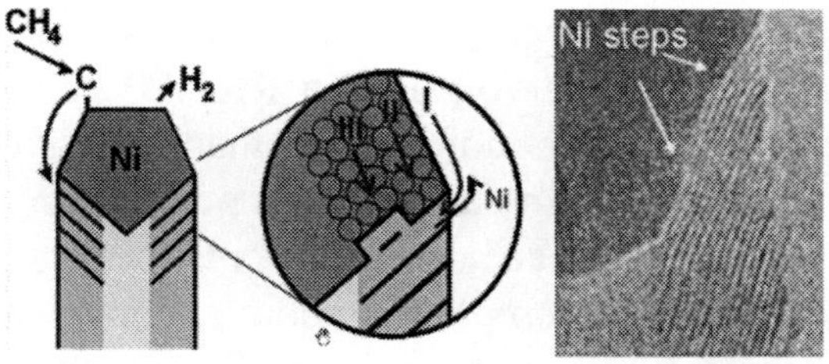

Figure 6. Illustration of the whisker growth mechanism based on graphene on step site and HREM image revealing step edge assisting with the growth of the whisker [5].

The nucleation of graphene islands can be avoided by blocking the step site. Potassium will preferentially adsorb on a step site, as the energy for adsorption is stabilized on a step site and is further stabilized if covered with oxygen [6]. Thereby potassium prevents nucleation of the carbon and prevents whisker formation. The side effect of promotion of the catalyst with potassium is a reduction in the rate of the steam reforming reaction. To avoid the nucleation of the graphene island, only a partial blocking of the step sites is needed. The optimal catalyst has just enough potassium to prevent graphene nucleation and yet has sufficient free sites to maintain a high activity. This optimal amount of potassium is the underlying design behind Topsoe's tubular reforming catalyst RK-400.

Pyrolytic carbon is formed from thermal cracking of hydrocarbons, leading to an amorphous type of carbon (see Figure 5C). The hydrocarbons are converted into radicals and unsaturated hydrocarbons. These compounds undergo polymerization and dehydrogenation to form pyrolytic coke. The formed coke will be deposited on the walls of the reformer tube and the catalyst outer surface. Alkali metal can slow the formation of pyrolytic carbon by enhancing the reverse carbon forming reaction: the gasification reaction. Potassium reacts with water forming potassium hydroxide, thereby stabilizing the concentration of hydroxide on the surface and increasing the rate of gasification.

Pyrolytic carbon: $CnHm \rightarrow nC(s) + m/_2\, H_2$
Gasification: $C(s) + H_2O \rightarrow CO + H_2$

The hydrocarbons will still crack at high temperatures, but as potassium promotes the gasification reaction and increases the rate of formation of gasification products, the net formation of pyrolytic carbon is prevented.

Potassium stability

Potassium reacts with the steam in the feed gas, forming potassium hydroxide. Potassium hydroxide is volatile and can leach from the catalyst. To minimize the leaching of potassium from the catalyst, the potassium can be bound in a reservoir in a more stable phase in the catalyst. By carefully designing this reservoir, the release of potassium to the Ni-particle can be controlled, ensuring only enough potassium is released to the surface of the catalyst to prevent carbon formation, thereby ensuring maximum catalytic activity. RK-400 has been designed to meet these requirements [7]. Potassium is an integral part of the carrier and is bound in potassium beta alumina phases and in particular magnesium-substituted potassium beta alumina phases (see Figure 7).

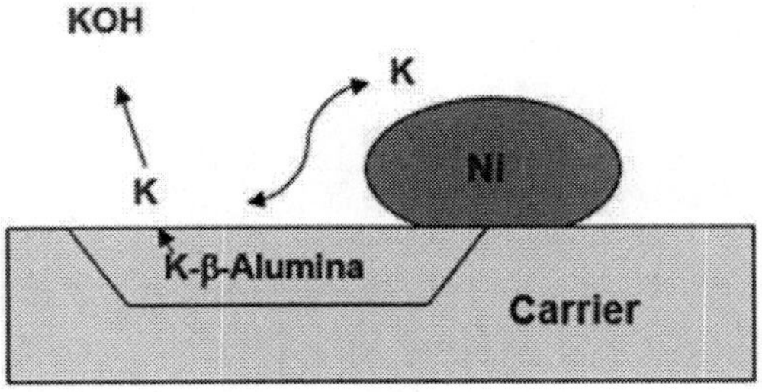

Figure 7. The design of RK-400 with the unique reservoir

RK-400 was designed to ensure a slow and stable release of potassium to the Ni-particle to partially adsorb on the step sites, preventing nucleation of graphene islands and preventing formation of whisker carbon. This design minimizes the leaching of potassium from the catalyst, which can be a problem to downstream exchangers. As mentioned, Figures 8 and 9 illustrate the superiority of Topsoe RK-400 with respect to potassium stability and activity compared to conventional catalysts.

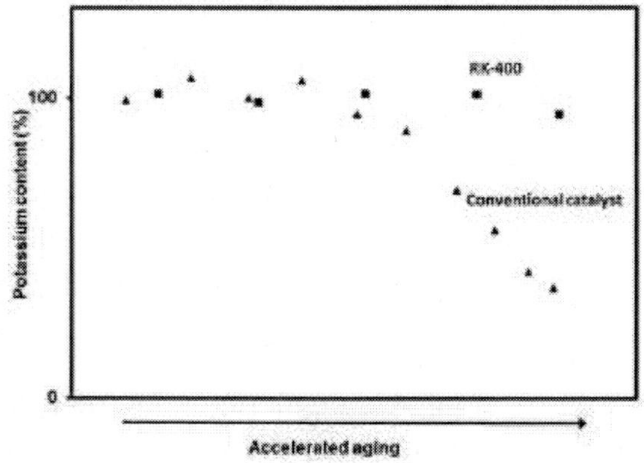

Figure 8. Potassium content in spent catalyst after exposure to gas mixture of $H_2O/H_2/CO_2$ at different aging temperatures.

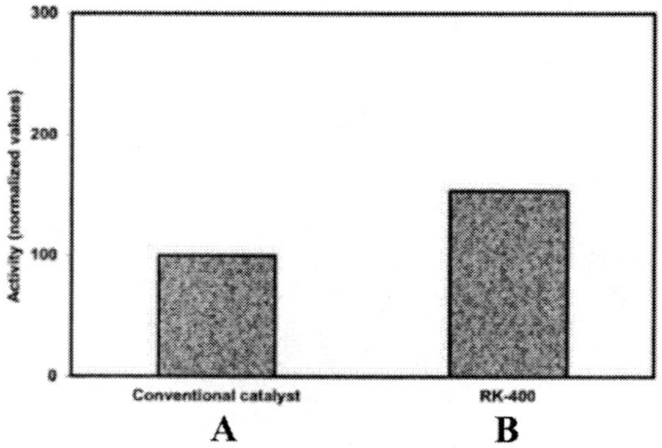

Figure 9. Steam reforming activity of conventional catalyst (A) and RK-400 (B).

The RK-400 catalyst was studied in a bench-scale reactor in an accelerated potassium loss experiment. By exposing the catalyst to a gas mixture of $H_2O/H_2/CO_2$ in a reactor at different temperatures, the potassium stability and catalyst activity could be probed. The potential potassium-releasing reactions are as follows:

1. $K_2CO_3(s) + H_2O(g) \rightleftarrows 2\ KOH(g) + CO_2(g)$
2. $KOH(s) \rightleftarrows KOH(g)$
3. $K^+(s) + H_2O(g) \leftrightarrows H^+(s) + KOH(g)$

In this scheme, Reaction 3 represents potassium bound in the reservoir in the RK-400 carrier. This design is far more stable than potassium bound as potassium carbonate or potassium hydroxide at the surface and represents the superiority of RK-400 compared to conventional catalysts that only contain potassium in the form represented by Reactions 1 and 2.

The design of RK-400 assures that only a small amount of the surface is covered by potassium, ensuring sufficient carbon resistance, maximizing the activity, and minimizing potassium migration.

Industrial experience

A poorly performing tubular reforming catalyst can dramatically impact the overall performance of a plant. This impact can be especially true with alkali-promoted reforming catalysts, as these catalysts are often employed in plants where carbon formation is likely. Figure 10 contains reformer tube wall temperatures measured at a world-scale Latin American methanol plant operating at an average plant load of 97% in the period.

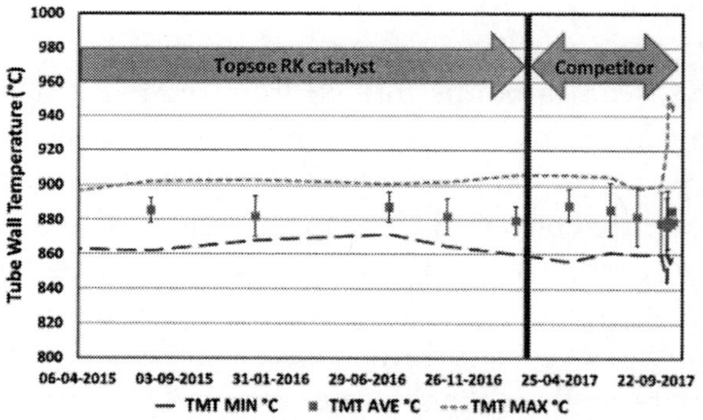

Figure 10. Comparison of TWT measurements for two alkali catalyst charges at a world-scale Latin American methanol plant.

The first catalyst charge (2010–2017) is Topsoe's RK-201 series catalyst; the second charge (2017–present) is reforming catalyst from a leading competitor. The RK-201 catalyst was installed for 82 months without any TWT issues, whereas the competitor product had excessive TWTs after only 8 months of operation. This plant purchased a new charge of Topsoe's RK-400 series catalyst (pictured in Figure 11) soon after experiencing these high TWT excursions with the competing product.

Figure 11. Photo of Topsoe's RK-400 catalyst.

Reformer operators can be assured of long-term stable reformer catalyst performance by selecting Topsoe's high-quality and well-referenced catalyst. The performance of an industrial charge of RK-400 catalyst is compared to that of a standard alkali catalyst in Figure 12 and Table 1. The RK-400 catalyst charge has a longer lifetime, a lower methane slip, and a lower pressure drop than those of the previous charge.

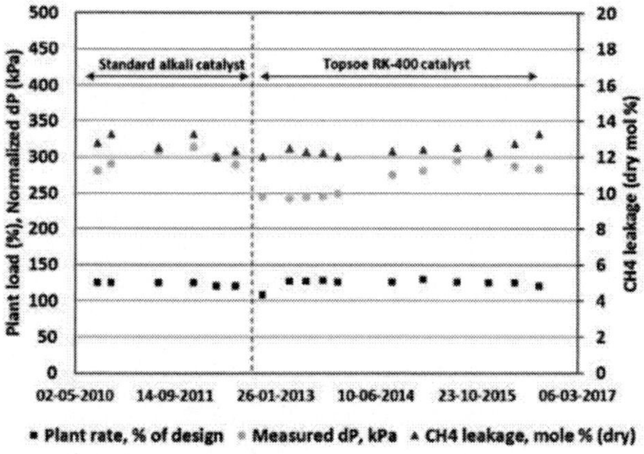

Figure 12. Comparison of performance of standard alkali reforming catalyst and Topsoe's RK-400R and RK-400 catalysts at a world-scale North American ammonia plant.

	Average plant rate, % of design	Average CH_4 leakage, mol % (dry)	Average measured dP, kPa (psi)
Standard alkali catalyst	123.7	12.7	297 (43)
RK-400	124.5	12.4	268 (39)

Table 1 Comparison of performance of standard alkali reforming catalyst and Topsoe's RK-400R and RK-400 catalysts at a world-scale North American ammonia plant.

Due to its excellent properties and high performance, Topsoe's RK-400 tubular reforming catalyst has quickly taken off in the reforming market, with 35 references to date (see Figure 13).

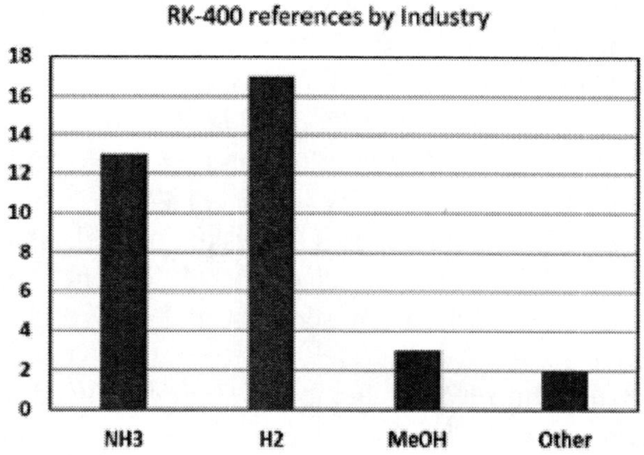

Figure 13. RK-400 references in the different syngas industries.

Conclusion

Prevention of and mitigation of consequences due to carbon formation in tubular reformers is a key aspect of ammonia plant management. The likelihood of carbon formation increases as the catalyst ages and is gradually deactivated by incoming catalyst poisons. Ensuring proper burner management and thereby an even TWT distribution is of critical importance, and can be accomplished either by regular visual furnace inspections or by online digital monitoring tools, such as Topsoe Furnace Manager. However, catalyst selection is also vitally important; fundamental studies on the effect of alkali promotion carried out by Topsoe have resulted in the development of a new and more effective alkali-promoted catalyst for top-fired reformers: RK-400. As of today, 35 references have been obtained in the ammonia, hydrogen, and methanol industries.

References

[1] Causes for Loss Time in Ammonia-Urea Plants, M. Goswami, A. Pandey and S. Nand, The Fertiliser Association of India, Indian J., Fert., Vol. 11(2), pp.16–22

[2] S. Sexton, Paper on Topsoe Furnace Manager Technology Benefits For Ammonia Producers, Presented at AIChE symposium 2016, Denver, USA

[3] K. H. Andersen & N. K. Nielsen, Paper on Pyrometry in Primary Reformers, Presented at AIChE symposium 2005, Toronto, Canada

[4] J. Sehested, Catal. Today 111, (2006) 103.

[5] S. Helveg, J. Sehested, and J. R. Rostrup-Nielsen, Catal.,Today, 178 (2011) 42.

[6] H. S. Bengaard, J. K. Nørskov, J. Sehested, B. S. Clausen, L. P. Nielsen, A. M. Molenbroek, and J. R. Rostrup-Nielsen, J. Catal. 209 (2002) 365.

[7] Joachim Harteg Jacobsen, Charlotte Vinding Ovesen and Christian Daugaard, Steam reforming catalyst and method of making thereof, Patent US9789470B2, Sept. 11, 2013.

Increasing Reliability of High Temperature Steam Super-Heater in Ammonia Production

The heat recovery equipment in Ammonia processes downstream of the secondary reformer typically consists of a process gas cooler and a steam super-heater. Those heat exchangers are major contributors to the overall process efficiency and reliability. The steam super-heater is especially important as the work horse in this heat recovery package. For a typical application the so called U-type steam super-heater was developed 15 years ago. However, the U-type is close to the mechanical threshold when gas and steam temperatures in the upper range are applied. Since the steam super-heater is a valuable but critical part, SCHMIDTSCHE SCHACK has designed and manufactured a steam super-heater, which allows higher steam parameters up to 540°C / 130bar (1004°F / 1885 psi). This design is a combination of proven features and is currently used in a Methanation process, but it can also be installed in Ammonia. This new steam super-heater provides the potential for improving the overall process efficiency and reliability.

Dr. Jörg Weidenfeller
Wolfgang Klecker
ARVOS GmbH | SCHMIDTSCHE SCHACK

Introduction

The steam system in Ammonia plants is often called the heart of the production plant, since steam at different pressure levels is used for the process itself, as well as for the driving power for the process compressors. The steam is produced by waste heat recovery from the process and additional natural gas fired auxiliary boilers. Depending on the plant situation, modern steam systems are designed either for zero steam export or using export steam to produce electricity or – in case of an Ammonia / Urea complex - to supply steam or electricity to the urea plant since this plant is a consumer for both. To improve the overall process efficiency it is obvious that the steam system is a big lever.

In ammonia plants, a major part of the steam generation by waste heat recovery happens downstream of the secondary reformer. The waste heat boiler package often consists of a horizontal process gas cooler (PGC) with a steam drum on top followed by a steam super-heater (SSH) to improve the overall process efficiency. The waste heat boiler package operates with natural circulation of the water / steam. A typical arrangement for two parallel lines is shown in Figure 1.

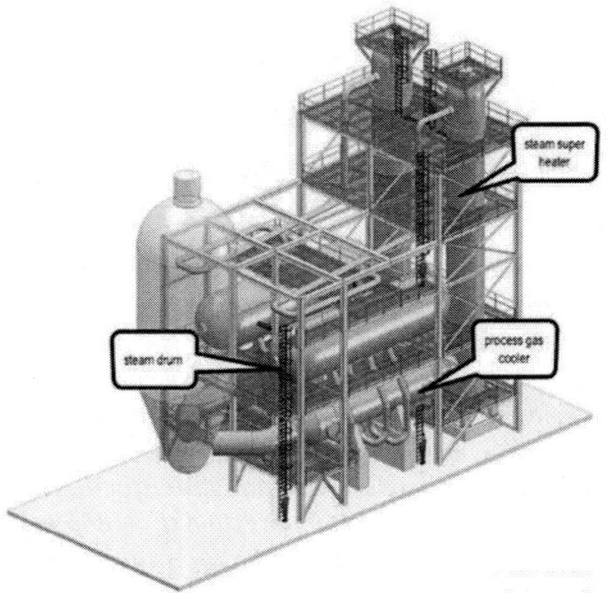

Figure 1. Typical waste heat boiler package

The process gas enters the waste heat boiler package typically at 900°C to 1000°C (1650°F to 1830°F) and is cooled down to 350°C (660°F) depending on further downstream process and requirements like the high temperature shift. By recovering the heat from the process gas, high pressure steam is generated and super-heated.

SCHMIDTSCHE SCHACK | ARVOS is the solution provider for this kind of waste heat boiler package and designs and manufactures the complete package as well as individual components. Based on executed and requested projects for all major licensors, the steam pressure and super-heated temperature is in a range of 100 bar (1450 psi) to 130 bar (1880 psi) and 330°C (625°F) to 470°C (875°F), whereas the majority of those designs have high pressure steam at 125bar (1810 psi) and 400°C (750°F), respectively.

Motivation to increase the super-heated steam temperature

An increase in super-heated steam temperature has a beneficial impact on the overall process efficiency and cost effectiveness. Since SCHMIDTSCHE SCHACK | ARVOS is not in the role of a process designer, the further assessment only focusses on the system of the waste heat boiler package by keeping constant the process conditions into the system. This is shown for the baseline or conventional scenario as well as for the improved scenario in the Heat-Temperature (Q-T) diagram in Figure 2. It can be seen clearly, that the process gas inlet and outlet temperatures (as well as pressure and gas mass flow) are not changed for both scenarios, whereas the super-heated steam temperature is increased for the improved scenario. Since the total available heat of the process gas remains constant, and the required heat for the steam super-heating increases, the remaining heat for steam generation in the PGC decreases.

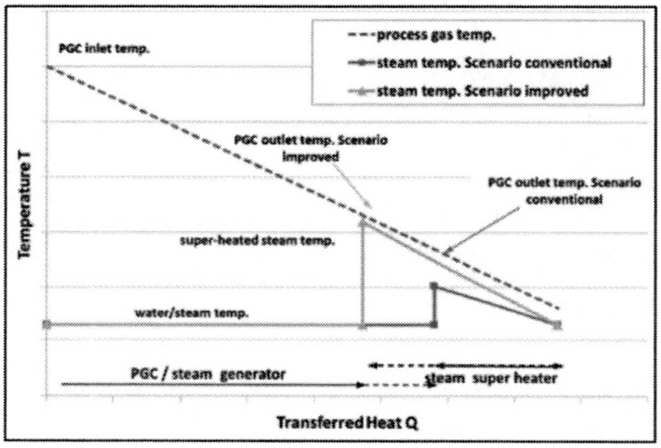

Figure 2. Q-T diagram for waste heat boiler package, conventional and improved scenario

To quantify this effect, a baseline of a super-heated steam temperature of 400°C (750°F) at 125bar (1810 psi) is chosen. Increasing the steam temperature by 15% means an increase of the specific steam energy by 21% and consequently, as described above, a decrease in steam mass flow by 12% occurs. But, overall the increase of SSH duty is 38% compared to the baseline scenario. By increasing the super-heated steam temperature by 30%, the overall SSH duty increases by 58%. Consequently this additional duty can be used for further electricity production or reducing the duty of natural gas fired auxiliary boilers.

A rough financial assessment for both improved scenarios is shown in Table 1, and clearly indicates the financial benefit of increasing the super-heated steam temperature, either in natural gas savings or additional revenue from electricity production.

Ammonia process, 2200 MTPD (2425 STPD) plant assessment for one year operation			
Super-heated steam temperature	**Duty increase**	**Natural gas savings**	**Revenue for electricity production**
400°C (750°F)	baseline	0	0
470°C (875°F)	7 MW	0.530 Million US$	0.590 Million US$
520°C (968°F)	11 MW	0.900 Million US$	0.810 Million US$
Assumptions: • Natural gas price 10,2 €/MWh (3,2 $/MMBTU) • Reimbursement for electricity 0.05 €/kWh • Plant operation: 7800 h/y • Efficiency steam to electricity 40% Steam turbine expansion to 70°C (158°F)			

Table 1. Exemplary financial assessment of increasing super-heated steam temperature in SSH

But a vital requirement for those scenarios is a mechanical robust and reliable steam super-heater with high availability and long life time!

Steam Super-Heater, Conventional Design

The steam super-heater is the critical component of the waste heat boiler package. Therefore special care was taken for a robust and reliable design. As a result, the so called U-type steam super-heater was developed and established first in Ammonia plants by SCHMIDTSCHE SCHACK | ARVOS approximately 15 years ago and after that also in Methanol plants. Also, a limited number of other heat transfer equipment suppliers have applied a comparable principle.

Figure 3 shows the design principle of the SCHMIDTSCHE SCHACK | ARVOS U-type SSH. The process gas enters into the gas inlet channel at the bottom and is guided upwards inside a central tube. At the central tube outlet near the tube sheet the gas is turned, enters the inner hot pass and flows downwards through the heating surface. At the bottom, the gas is turned again, enters the second cold pass of the U-type heating surface and flows to the top leaving the heating surface. After that the cooled gas flows downwards between the inner liner of the second pass and the outer pressure shell into the gas exit chamber. This means, that the hot process gas is at no point in direct contact with the pressure vessel wall, which is beneficial for the mechanical design of the outer shell.

The internal bypass valve controls the process gas outlet temperature by bypassing the hot gas directly into the gas exit chamber, where it mixes with the cooled gas to achieve the required gas outlet temperature.

The super-heater operates in counter-current flow, which means that the saturated steam flows into the U-tube from the second to the first pass. The upper head is designed as a combined steam inlet and outlet chamber, where the super-heated steam is collected in an internal steam chamber.

Refractory is always a risk for cracks and material losses, and damages require unexpected shut down due to immediate repair action. Therefore, this design is focused on a minimum use of refractory at the exit chamber only, which increases the availability of the steam super-heater and the Ammonia production line, respectively.

The effectiveness of the heat transfer design is based on the counter current flow of the media. This has a beneficial impact on the required size of the heating surface and therefore also on the size, weight and cost of the steam super-heater.

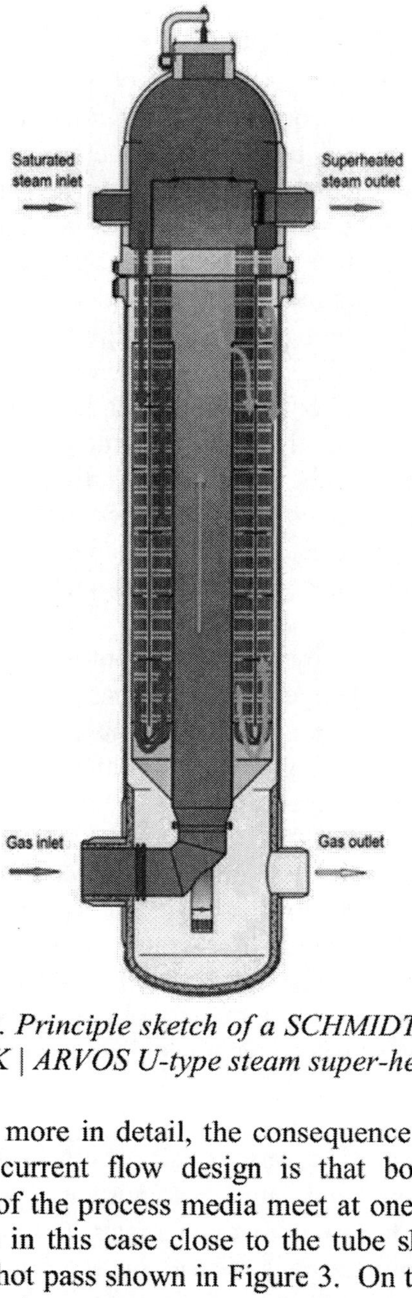

Figure 3. Principle sketch of a SCHMIDTSCHE SCHACK | ARVOS U-type steam super-heater

Looking more in detail, the consequence of the counter current flow design is that both hot streams of the process media meet at one point, which is in this case close to the tube sheet at the first hot pass shown in Figure 3. On the one hand, the hot process gas at shell side and the super-heated steam at tube side lead to high material temperatures for the U-tube, on the other hand, the high temperature at tube side is transferred into the tube sheet because of the hot U-tube leg. Only few possibilities are available to reduce this temperature impact by inserting ferrules, e.g., but this would increase also the pressure drop at steam side. Finally, the high temperature transfer into the tube sheet cannot be avoided.

This means that the temperature gradient caused by the media at shell and tube side is not only across the tube sheet, but a temperature gradient in the radial direction along the tube sheet is superimposed because of the U-tube principle with a cold and a hot leg.

The tube sheet, which is the heart of the steam super-heater, is loaded by the pressure of the steam and gas side as well as by the temperature gradients described above. A thick tube sheet is required to resist the primary stresses caused by the high steam pressure. But a thick tube sheet is also critical to thermal gradients, all the more to the thermal superposition mentioned above, which results in unfavorable secondary stresses.

For the current ammonia applications and the latest process temperatures the conventional SCHMIDTSCHE SCHACK | ARVOS U-tube design is still feasible and reliable based on the long term experience for the thermal and mechanical design. But in case of improving the process efficiency by increasing the superheated steam temperature beyond 470°C (875°F), the conventional U-type design is no longer robust. The non-uniform temperature distribution mentioned above gets worse and consequently the tube sheet expands thermally non-uniform with additional thermal, secondary stresses. This drives crack propagation at the tube sheet, tube to tube sheet welds and connecting welds. The situation will get worse for cyclic operation, which is characterized by high temperature and load gradients.

High Performance Steam Super-Heater

In general there are two steam super-heater design possibilities beyond 470°C (875°F), for which a U-type design is no longer reliable:

1. Cascade the steam super-heater in two exchangers, one hot and one cold apparatus as shown in Figure 4

2. Apply and combine proven design features to achieve a reliable solution, which fulfils reliability and life time criteria

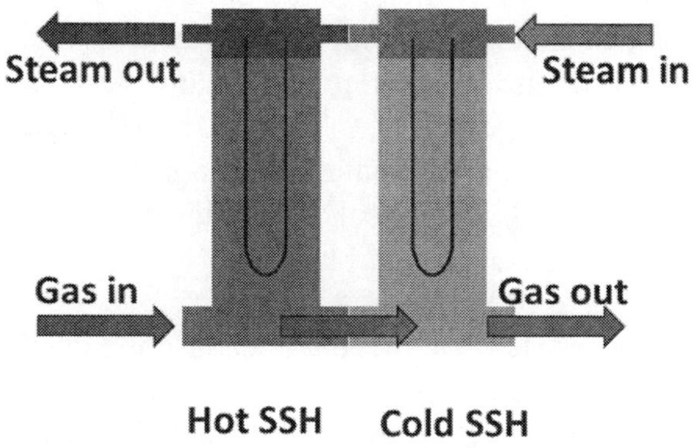

Figure 4. Sketch of hot and cold stage, option 1

Option 1 has the major disadvantage that an additional heat exchanger for the process is necessary and therefore an additional failure risk is applied, which reduces the overall availability and reliability of the production line. In addition, the hot stage is the critical SSH because

- High temperatures at the tube sheet. This has an impact on the mechanical integrity and life time.
- Shell is completely refractory lined. This reduces availability because damaged refractory has to be repaired immediately due to metal dusting atmosphere.

Therefore SCHMIDTSCHE SCHACK | ARVOS decided to go for option 2.

The design for the pressure vessel of the above described U-type steam super-heater was used, since this design uses internal cooling features and a minimum of refractory. For the heating surface the U-tubes were not used but the so called bayonet design. This design was successfully applied by SCHMIDTSCHE SCHACK | ARVOS more than 20 years ago for steam super-heating at comparable process temperatures as shown in Table 2.

Process / Year	**Steam Side** [bar / °C]	**Gas Side** [bar / °C]	**Used Design Feature**
Ammonia / 2000 until 2014	125 / 393 (1810 psi / 740°F)	40 / 560 (595 psi / 1040°F)	Pressure vessel of U-type
Styrol / 1994	0.9 / 840 (13 psi / 1545°F)	3.1 / 642 (45 psi /1187°F)	Bayonet tube
Methanation / 1981	123 / 550 (1780 psi /1020°F)	80 / 550 (1160 psi / 1020°F)	Bayonet tube
Methanation / 2014	130 / 520 (1885 psi / 968°F)	30 / 620 (435 psi / 1148°F)	Combination of Design features to next generation SSH

Table 2. Combination of proven design features for the high performance steam super-heater

It is worth mentioning that the bayonet tube applied at a process gas cooler for steam generation has several inherent reliability issues. For example, deposits in the boiler feed water can agglomerate in the end cap of the bayonet tube and reduce or block the cooling capability. This was reported in various conference papers [1, 2]. However, the application of the bayonet tube to super-heat steam does not have the risk of agglomeration of deposits because there are no deposits in the saturated and super-heated steam.

The bayonet tube is a tube in tube concept and is comparable with the SCHMIDTSCHE® double tube design, which is established for over 50 years in the ethylene industry at more than 7000 transfer line exchangers. The bayonet tube originally was not invented by SCHMIDTSCHE SCHACK | ARVOS, but special thermal and mechanical design expertise and manufacturing experience is required to achieve a reliable solution.

The design was performed with three dimensional numerical flow and heat transfer simulations accompanied by mechanical structure analysis. In addition, experimental tests were performed to verify the boundary conditions for the simulation. A principal sketch of the bayonet type steam super-heater is shown in Figure 5.

Figure 5. Principle sketch of the SCHMIDTSCHE SCHACK | ARVOS bayonet type steam super-heater

The steam super-heater operates in counter current flow like the U-tube design. Saturated steam flows downwards in the annular gap between the outer and inner tube. At the bottom of the outer tube, the steam is already super-heated and flows again upwards in the inner tube. The process gas enters the heating surface from the bottom and flows upwards turned by disc and doughnut baffles, which optimize the heat transfer but has less pressure drop compared to standard baffle elements. At the end of the pass, the cooled process gas flows downwards between the inner liner and the outer pressure shell, which is comparable to the U-type design. An internal by-pass is also applied to control the gas outlet and steam outlet temperature, respectively. This SCHMIDTSCHE SCHACK | ARVOS patented design has several thermal and mechanical advantages, which are described in more detail below.

The outer tube of the bayonet is connected to the tube sheet, which separates the steam and process gas side. This means that the tube sheet at steam side is only in contact with saturated steam, which is a major difference compared to the U-tube design. The super-heated steam flows in the inner tube of the bayonet, which is connected to the bottom of the steam outlet chamber. This design ensures that the outer tube as well as the inner tube can thermally expand independently without any constraint. Consequently no stresses due to thermal expansion occur at the outer and inner tube.

The major advantage of this design is demonstrated by the temperature distribution across the tube sheet in Figure 6 top.

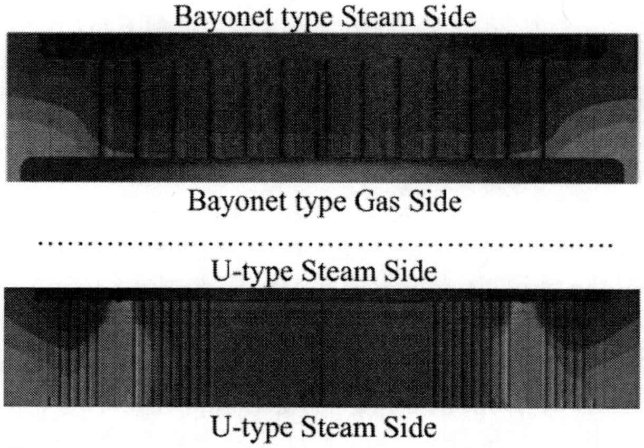

Figure 6. Temperature distribution across the tube sheet of the SCHMIDTSCHE SCHACK | ARVOS bayonet tube steam super-heater (top) and U-type (bottom). (Same temperature scale blue to red corresponds to cold to hot)

It can be seen clearly that the tube sheet is cooled by saturated steam, whereas at the gas side the tube sheet is in contact with the already cooled gas. This leads to the very homogenous temperature distribution at low level. Consequently this enables a design for a small tube sheet thickness which has a positive effect on the thermal stresses in the tube sheet and the tube to tube sheet welds. In addition the smooth temperature distribution at low level has a beneficial contribution to the thermal stresses. For comparison only, the temperature distribution across the tube sheet of a U-tube design for the same process and steam temperatures is shown in Figure 6 bottom.

Both figures have the same temperature scale. It has to be emphasized, that this U-tube tube sheet design is not robust and reliable but is only shown for comparison.

In contrast to the top figure, this tube sheet is in contact with saturated as well as super-heated steam due to the U-tube concept. This leads, as described above, to a superimposed temperature gradient in the radial direction, which has negative impact on the thermal stresses in the tube sheet and tube to tube sheet welds.

Design Comparison

As mentioned above, cascading the steam superheater by a hot and cold stage as shown in Figure 4 might also be an option to realize high temperatures. Besides the fact already discussed above that an additional heat exchanger introduces an additional failure risk, which reduces the overall availability, the bayonet type has further advantages. For this comparison a hot and cold stage super-heater as shown in Figure 4 were designed which have the same overall process parameters as the bayonet design.

Both concept are compared with respect to

- Weight of the shell: The bayonet design has a shell diameter which is 80% of the staged design. Due to high pressure at gas and steam side, this impacts the wall thickness of the shell. In addition two gas inlet chambers and two steam chambers are required. In addition the shell of the bayonet design has 48% less weight than the SSH cascade.
- Weight of the refractory: the hot stage of the SSH cascade is completely refractory lined whereas for the cold stage only the gas inlet chamber is refractory lined. Consequently, the weight of the refractory for the bayonet has 55% less weight.
- Weight of the heating surface: the bayonet is a tube in tube design which leads to 40% higher weight of the heating surface compared to the SSH cascade.
- Overall weight (see Figure 7): The overall weight of the bayonet design is 15% less than the cascade design. This effect will also become noticeable in the product cost.

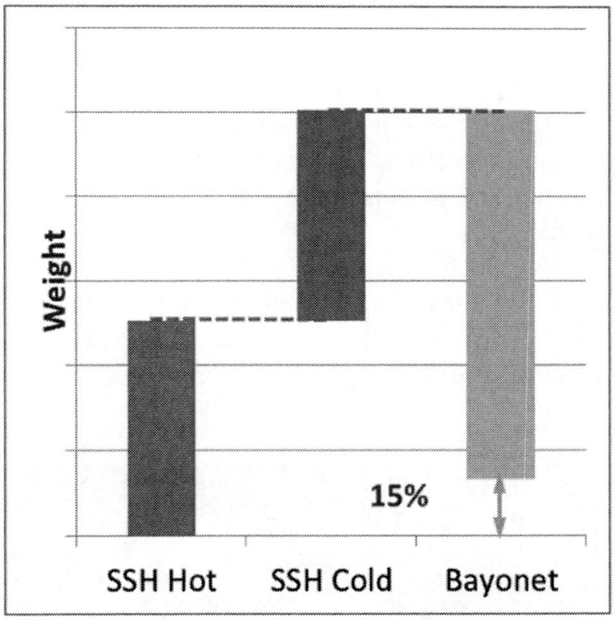

Figure 7. Weight comparison between bayonet SSH and SSH cascade

Conclusions

For the current process gas and steam conditions in ammonia and methanol applications, the steam super-heater with a U-tube design is still robust and reliable. However, based on the process flow scheme, the U-type is close to its mechanical threshold. For that reason, SCHMIDTSCHE SCHACK | ARVOS has developed steam super-heaters for the next generation of process temperatures, which are currently applied in a Methanation processes (see Figure 8). The new apparatus combines proven design features to ensure an evolutionary and reliable step in steam super-heater development. Special care was taken in the process flow concept, the design of the heating surface and the tube sheet to achieve a mechanically robust design, which supports a long life time and reliability of the super-heater. The application of internal cooling features significantly reduces the use of refractory, which positively supports high availability of the apparatus.

Therefore this design fulfils the critical criteria: life time, reliability, and availability, which are important factors to achieve the planned return on investment during the operating time of the plant.

This new steam super-heater therefore provides the licensor and the customer a more reliable potential for improving the overall process efficiency.

Figure 8. High performance steam super-heater ready for shipment

References

[1] A. Din, S. Hassan and A. Ahmad: A Waste Heat Boiler Replacement – From Bayonet to Fire Tube Design. Ammonia Technical Manual, 2008

[2] S. Singh, M. Srivastava and M. Masood Ahmad: Failure of Waste Heat Boiler in a Kellogg Designed Ammonia Plant. Ammonia Technical Manual, 2012

AUTHOR INDEX

AIChE
120 Wall Street, FL 23
New York, NY 10005-4020

ISBN 978-1-5108-8185-3

CURRAN ASSOCIATES INC.
proceedings.com